FOUNDATION

ORGANISATION AND THE HUMAN RESOURCE

BPP Publishing

First edition November 1993

ISBN 0 7517 5003 4

British Library Cataloguing-in-Publication Data

A catalogue record for this book
is available from the British Library

Printed in Great Britain by
Ashford Colour Press, Gosport, Hampshire

Published by

BPP Publishing Limited
Aldine House, Aldine Place
London W12 8AW

We are grateful to the Institute of Chartered Secretaries and Administrators for permission to include past examination questions in this text. The suggested solutions have been prepared by BPP Publishing Limited

PREFACE

The syllabus for the Foundation and Pre-professional Programme examinations of the Institute of Chartered Secretaries and Administrators is changing radically with effect from the June 1994 sittings. The ICSA's New Qualifying Scheme will be a demanding test of each student's knowledge and skills.

Thorough, up-to-date and effective learning and practice material is crucial for busy professionals preparing for professional exams. In our many years of publishing for ICSA exams, we have learnt this lesson well. All our Study Texts (and the accompanying Practice & Revision Kits) are written by qualified professionals who have themselves experienced the pressures of having to study as well as to work. Our material is therefore *comprehensive* - covering the *whole* syllabus - *on-target* - covering *only* the syllabus - and *up-to-date* at the month of publication.

This Study Text has been written specifically for the new Foundation Programme paper *Organisation and the Human Resource*. The syllabus on pages (viii) to (ix) has been cross-referenced to the text, so you can be assured that coverage is complete. There is also a study checklist so you can plan and monitor your progress through the syllabus.

The main body of this Study Text takes you through the syllabus in easily managed stages, with plenty of opportunities for skill - and exam question - practice. For a brief guide to the structure of the text, and how it may most effectively be used, see pages (vi) to (vii).

BPP Publishing
November 1993

HOW TO USE THIS STUDY TEXT

This Study Text has been designed to help students and lecturers to get to grips as effectively as possible with the content and scope of the Foundation Programme paper *Organisation and the Human Resource.*

The structure and topic order of the *syllabus* have been used as the framework for this Study Text. However we have also aimed to help those who choose to take a different path by indicating (in the Signpost section beginning each chapter) those areas which naturally precede the current chapter, and those chapters in which topics introduced can be further explored.

Syllabus coverage in the text is indicated on pages (viii) to (ix) by chapter references set against each syllabus topic. It is thus easy to trace your path through the syllabus.

As a further guide - and a convenient means of monitoring your progress - we have included a study checklist on page (xii) on which to chart your completion of chapters and their related illustrative questions.

Each chapter contains:

- a 'signpost' to indicate how the subject area relates to others in the syllabus
- clear, concise topic-by-topic coverage
- frequent exercises to reinforce learning, confirm understanding and stimulate thought
- a 'roundup' of the key points in the chapter
- a test your knowledge quiz
- a recommendation on illustrative questions to try for practice. These are provided in a bank at the end of the text, with suggested solutions.

Each chapter has at least one relevant illustrative question.

Exercises

All chapters contain a number of exercises to aid your studies and to enable you to check your progress as you work through the text. These come in a large variety of forms: some test your ability to do a task just described; others see whether you have taken in the full significance of a piece of information. Some are meant to be discussed with colleagues, friends or fellow students; others ask you to apply what you have learnt to your own experience.

If appropriate, a suggested solution is given, but often in an abbreviated form to help you avoid the temptation of merely reading the exercise rather than actively engaging your brain. We think it is preferable on the whole to give the solution immediately after the exercise rather than making you hunt for it at the end of the chapter, losing your place and your concentration. Cover up the solution with a piece of paper if you find the temptation to cheat too great!

Chapter roundup and Test your knowledge quiz

At the end of each chapter you will find two boxes. The first is the *Chapter roundup* which summarises key points and arguments, sometimes in diagrammatic form, and indicates what you should know having studied the chapter. The second box is a quiz that serves a number of purposes.

(a) It is an essential part of the chapter roundup and can be glanced over quickly to remind yourself of key issues covered by the chapter.
(b) It is a quiz pure and simple. Try doing it on the train in the morning to revise what you read the night before.
(c) It is a revision tool. Shortly before your examination sit down with pen and paper and try to answer all the questions fully.

Illustrative questions

Initially you might attempt such questions with reference to the chapter you have just covered. Later in your studies, it would be helpful to attempt some without support from the text. Only when you have attempted each question as fully as possible should you refer to the suggested solution to check and correct your performance.

Index

Finally, we have included a comprehensive index to help you locate key topics.

Further reading

Students are strongly advised to keep themselves up to date with organisational and human resource issues by reading:

(a) a good quality newspaper;
(b) relevant articles in *The Administrator*;
(c) if possible, the magazine *Personnel Management* and its fortnightly supplement *PM Plus*.

This will enable you to enhance your examination answers with fresh examples drawn from current best practice in real life. Do not be afraid to collect and use examples from your own country if you are living outside the UK.

A note on pronouns

On occasions in this text, 'he' is used for 'he or she', 'him' for 'him or her' and so forth. Whilst we try to avoid this practice it is sometimes necessary for reasons of style. No prejudice or stereotyping according to sex is intended or assumed.

SYLLABUS

General guidance

Note that the following general stipulations are made in the ICSA's 1993/94 syllabus booklet.

'All syllabi are based on English and EC derived law and practice, except where specifically stated.'

'Students are expected to keep abreast of changes in the law affecting the modules which they are studying. Generally, however, a detailed knowledge of new legislation will not be expected in examinations held within six months of the passing of the relevant Act. Syllabus changes will be notified to teaching establishments and will be published for the information of students in the Institute's journal *Administrator.*'

ORGANISATION AND THE HUMAN RESOURCE

Objective:

To provide an understanding of organisational structure, work environment and people at work.

A: ORGANISATIONS AND STRUCTURING THE WORK ENVIRONMENT

		Covered in Chapter
(i)	Organisations as bureaucracies, rules and norms, the problems of bureaucracy.	1, 2
(ii)	The nature of organisational roles; the roles of manager, supervisor and worker.	1, 6
(iii)	Sources of authority, power and control at work.	1
(iv)	Matrix organisations.	3
(v)	Choices of design of structure and jobs; choices of design of the workplace and allocation of tasks; ergonomic and man-machine systems.	3, 4
(vi)	The influence of size, market, environment and technology on organisational structure and functioning.	1, 3, 4

B: THE WORK GROUP AND LEADERSHIP

(i)	The development, operation and influence of work groups.	5
(ii)	Formal and informal work groups.	1
(iii)	Group pressures and conformity; influences on behaviour.	5
(iv)	Individual versus group performance.	5
(v)	Group cohesion and influences on group effectiveness.	5
(vi)	Theories of leadership.	6
(vii)	Leadership styles and effectiveness.	6
(viii)	The manager, supervisor and workplace representatives as leaders.	6

C: PEOPLE AT WORK

(i)	Recruitment and selection, job descriptions, personal specification, induction, job evaluation, motivation and appraisal, counselling, training and development.	7, 9-14
(ii)	Flexibility of labour, promotion, retirement, resignation and termination.	8, 15
(iii)	Formal and informal communications, internal and external communications: oral, visual and written.	16

THE EXAMINATION PAPER

Details of the format of the new examination paper had not been released at the time of the preparation of this Study Text. The ICSA is planning to produce pilot papers, and these will be reproduced, together with BPP's full suggested solutions, in BPP's Practice and Revision Kits which will be published in March 1994.

Analysis of past papers

There was no direct equivalent to the *Organisation and the Human Resource* paper under the old syllabus. The new syllabus draws upon several of the old papers.

An analysis of topics examined in the last 2 sittings under the old syllabuses for the Foundation Paper *Introduction to Organisational Behaviour* and the Pre-professional level paper *Personnel Administration* is set out below. Both papers contained 10 questions, of which candidates were required to answer 5.

Students should not assume that old syllabus examinations are representative of examinations which will be set under the New Qualifying Scheme.

Introduction to Organisational Behaviour

June 1993

1 Motivation, leadership and communication problems of flexible staffing
2 Organisational culture
3 Rewards and behaviour in conditions of change
4 Team building
5 Communication problems: poor listening
6 Stress at work
7 'Social class' and the behaviour of managers, clerical and manual workers
8 Styles of leadership
9 Age discrimination; organisations and older workers
10 Behavioural arguments in favour of empowerment, devolution, decentralisation, delegation and flatter hierarchies

Personnel Administration

June 1993

1 Buying/replacing a (computerised) personnel information system
2 Industrial relations: trends in the 1980s; possible future direction
3 Job evaluation and its validity
4 'Human resource management' v 'personnel' management
5 Reasons for high labour turnover
6 Dealing with redundancies from the point of view of costs, people and the law
7 Usefulness of courses for management training and development
8 Human resource specialists' role in relation to other line managers
9 Evaluation of recruitment and selection procedures
10 Active promotion of equality for women and minority groups

Introduction to Organisational Behaviour

December 1992

1 Are organisations really 'formal, precise structures of co-ordinated human activity', and should they be?
2 'Classic' studies of group behaviour: effective/ineffective teams; decision making in teams; roles in teams
3 Job design: fitting jobs to people or people to jobs
4 Usefulness of contingency theory; better alternatives
5 Impact of new technology on one of: manual work; white collar work; the employment of women
6 Leadership and followership
7 Similarities between behaviour in organisations and acting in the theatre, especially in communication and managing conflict
8 Stress at work: discuss five propositions
9 The learning process
10 Vocational choice: non-work structural factors; individual factors and work sphere structural factors

Personnel Administration

December 1992

1 Usefulness of psychometric tests
2 Counselling interviews (select one of three case-examples)
3 HR vision
4 Manpower planning issues for a large organisation in the 1990s. (Pick and name a real example)
5 Sources of power for people in the personnel function
6 One-to-one interviews v panel interviews
7 Performance appraisal: likely developments
8 Information technology and personnel information
9 Relevance or otherwise of the 'Industrial Relations' section of the ICSA syllabus for Personnel Administration
10 Is it worthwhile to provide fringe benefits?

STUDY CHECKLIST

		Text chapters *Chpt no/Date comp*		Illustrative questions *Ques no/Date comp*	
PART A: ORGANISATIONS AND STRUCTURING THE WORK ENVIRONMENT					
1	The nature of work organisations				
2	Introduction to organisation theory				
3	The structure and cultures of organisations				
4	The work environment and job design				
PART B: THE WORK GROUP AND LEADERSHIP					
5	The nature of work groups				
6	Leadership at work				
PART C: PEOPLE AT WORK					
7	Human resource management				
8	Manpower planning				
9	Recruitment				
10	Selection				
11	Training and development				
12	Performance assessment				
13	Motivation and rewards				
14	Job evaluation				
15	Termination of contract				
16	Communication in organisations				

PART A

ORGANISATIONS AND STRUCTURING THE WORK ENVIRONMENT

Chapter 1

THE NATURE OF WORK ORGANISATIONS

This chapter covers the following topics.

1. The nature of work organisations
2. Bureaucracy
3. Large and small organisations
4. Authority, responsibility and power in organisations

Signpost

- This chapter describes the nature of bureaucracies and the sources of authority and power at work, as well as giving a general introduction to organisations.
- In doing so this chapter touches upon a great many of the themes and topics that will be developed more fully later in this Study Text. For example, the conflict between corporate and personal objectives is discussed more fully in Chapter 13; communication in organisations is dealt with in Chapter 16; leadership at work is the subject of Chapter 6; organisational culture is explained in Chapter 3, and so on.

1. THE NATURE OF WORK ORGANISATIONS

1.1 Writings on organisations over the years have taken very many different perspectives and angles on the subject. Organisational metaphors have varied from the primitive tribe with its constituent hunting bands, to the ship with its crew, to the biological organism, according to the emphasis which the writer wished to put on particular aspects of organisational life.

1.2 Buchanan and Huczynski put forward the following definition of organisation:

> 'social arrangements for the controlled performance of collective goals'.

They point out that the difference between organisations (and particularly work organisations) and other social groupings with collective goals (the family, the bus stop queue) is:

(a) the preoccupation with performance; and
(b) the need for controls.

1.3 (a) Organisations are collections of interacting individuals, occupying different roles but experiencing common membership. However, this embraces a wide variety of behaviours; the relationship between them may be co-operative or coercive; their rules may be ill-defined or clearly-defined, overlapping, conflicting etc.

(b) Organisations are created because individuals need each other in order to fulfil goals which they consider worthwhile. However - as we have seen - one individual's goals may be very different from another's and from those of the organisation as a whole.

(c) Performance must be controlled in order to make best use of human, financial and material resources, for which individuals, groups and organisations compete. The need for controlled performance leads to a deliberate, ordered environment, allocation of tasks (division of labour), specialisation, the setting of standards and measurement of results against them etc. This implies a whole structure of 'power' or 'responsibility' relationships, whereby some individuals control others.

Reasons for organisations

1.4 In general terms, organisations exist because they can achieve results which individuals cannot achieve alone. By grouping together, individuals overcome limitations imposed by both the physical environment and also their own biological limitations. Chester Barnard (1956) described the situation of a man trying to move a stone which was too large for him:

(a) the stone was too big for the man (environmental limitation); and
(b) the man was too small for the stone (biological limitation).

By forming an organisation with another man, it was possible to move the stone with the combined efforts of the two men together.

1.5 Barnard further suggested that the limitations on man's accomplishments are determined by the effectiveness of his organisation.

1.6 In greater detail, the reasons for organisations may be described as follows:

(a) *social reasons*: to meet an individual's need for companionship;

(b) to *enlarge abilities*: organisations increase productive ability because they make possible both:

(i) specialisation; and
(ii) exchange.

The potential benefits of specialisation were recognised by Adam Smith in his famous book *The wealth of nations* (1776). Specialisation permeates our modern industrial and commercial society;

(c) to *accumulate knowledge* (for subsequent re-use and further learning);

(d) to *save time*: organisations make it possible for objectives to be reached in a shorter time.

Personal objectives and organisations

1.7 Barnard described an organisation as a 'system of co-operative human activities' and it is important to be aware that:

(a) an organisation consists of people who inter-react with each other;

(b) the way in which people inter-react is designed and ordered by the organisation structure so as to achieve joint (organisational) objectives. Each individual has his own view of what these organisational objectives are;

(c) each person in the organisation has his own personal objectives;

(d) the organisational objectives as gauged by an individual need to be compatible with personal objectives of the individual if he is to be a well-integrated member of the organisation.

1.8 The task of management is:

(a) to recognise the personal objectives of individuals and to integrate these with organisational objectives;

(b) to reconcile and integrate the differing views of organisational objectives that are held by different individuals so that a total company objective may be recognised (Argyris);

(c) to make the optimal use of resources (materials, human abilities and efforts) in achieving these integrated personal and organisational objectives.

1.9 An organisation will be effective only so far as it helps individual members to achieve their own personal objectives, and a large part of the task of management is therefore concerned with this problem.

1.10 It is important that an individual should consider that the objectives of the organisation (and consequently his own efforts and actions within the organisation) are compatible with his various needs. Some personal needs will be more important than others and an individual will be prepared to sacrifice some of his objectives for his own greater benefit (an individual may be prepared to sacrifice some of his time in order to do a job which earns extra money, say).

The objects and objectives of an enterprise

1.11 Every business organisation has what we shall call its *objectives* - its main purpose for being in existence. These may be to make as large a profit as possible for the organisation's owners, in the case of private enterprise, or to provide a service in the case of non profit-making organisations or government departments and institutions.

1.12 Every business organisation also has its *objects* - in other words every business exists to carry out certain activities such as manufacturing and selling motor cars, providing retail services, providing telecommunications services. In the case of companies, their objects are described in their constitution (in the objects clause of the memorandum of association). Non profit-making organisations, public corporations and other government institutions might also have formally-stated objects in their constitution, whereas other organisations such as sole traderships do not have any formally-stated objects (even though they will still have objects of some description). As an example, a partnership of solicitors might be formed with the *objective* of making profits but with the *object* of providing legal services.

1.13 Business organisations differ from each other partly because they have different objects, despite similar objectives. A hairdressing business and a manufacturer of armoured tanks have the same profit objective, but radically different objects. The objects of a business also help to dictate what sort of organisation structure the business requires.

1.14 To integrate individual and organisational objectives it is necessary to formulate organisational objectives which are understood by all members as well as compatible with their personal goals. 'Organisational objectives should give the organisation meaning to man... and man meaning to the organisation.' (*Davis*)

1.15 From a different perspective, incidentally, some would argue that there are no such things as *organisational* objectives: these are really just the *personal objectives* of the people who happen to be running the organisation at the time the objectives are put together. This would explain why organisations sometimes abruptly change direction when a new chief executive takes over.

1.16 Later in your studies you will learn more formally about 'objectives' as well as 'mission statements', 'strategy' and so forth. There is a lack of general agreement about the meaning of such terms, and you should keep this in mind. We use the term ' objective' above in an everyday sense - 'aim', 'target' or whatever synonym you wish.

Formal and informal organisations

1.17 A formal organisation may be defined (Etzioni) as a social unit deliberately constructed to seek specific goals. It is characterised by:

(a) planned divisions of responsibility;
(b) power centres which control its efforts;
(c) substitution of personnel;
(d) the ability to combine its personnel in different ways.

1.18 Formal organisations have an explicit hierarchy in a well-defined structure; job specifications and communication channels are also well-defined.

1.19 An informal organisation, in contrast, is loosely structured, flexible and spontaneous. Membership is gained consciously or unconsciously and it is often difficult to determine the time when a person becomes a member. Examples of an informal organisation are managers who regularly go together for lunch in a local cafeteria or a clique of workmates.

1.20 Within every formal organisation there exists, to a greater or lesser extent, a complex informal organisation. The formal organisation is a structure of relationships and ideas; informal organisation, in practice, modifies this formal structure (Blau and Scott 1962).

1.21 The informal organisation of a company is so important that a newcomer has to 'learn the ropes' before he can settle effectively into his job, and he must also become 'accepted' by his fellow workers.

1.22 When people work together, they establish social relationships and customary ways of doing things.

(a) They form social groups, or cliques (sometimes acting against one another).

(b) They develop informal ways of getting things done - norms and rules which are different in character from the 'organisational manual' rules which are *imposed* by the formal organisation.

Social groups, or cliques, may act collectively for or against the interests of their company; the like-mindedness which arises in all members of the group strengthens their collective attitudes or actions.

Whether these groups work for or against the interests of the company depends to some extent on the type of supervision they get. If superiors involve them in decision-making, they are more likely to be management-minded.

1.23 The informal organisation of a company, given an acceptable social atmosphere:

(a) improves *communications* by means of a 'bush telegraph' system;

(b) facilitates the co-ordination of various individuals or departments and establishes 'unwritten' methods for getting a job done. These may by pass communication problems - for example between a particular manager and subordinate - or lengthy procedures; they may be more flexible and adaptable to required changes than the formal ways of doing things.

1.24 Certain individuals can have an important informal influence in an organisation. To take an illustrative example, the managing director of XY Company is a very remote individual and the production and sales directors have difficulty in communicating and working with him effectively. The financial director, however, has a remarkably good personal understanding with the managing director and can approach him readily in all matters at work. The sales and production directors therefore often ask the financial director to put their views informally to the managing director and to sound out his opinion before they approach him formally. This is the very essence of what is often called 'office politics'.

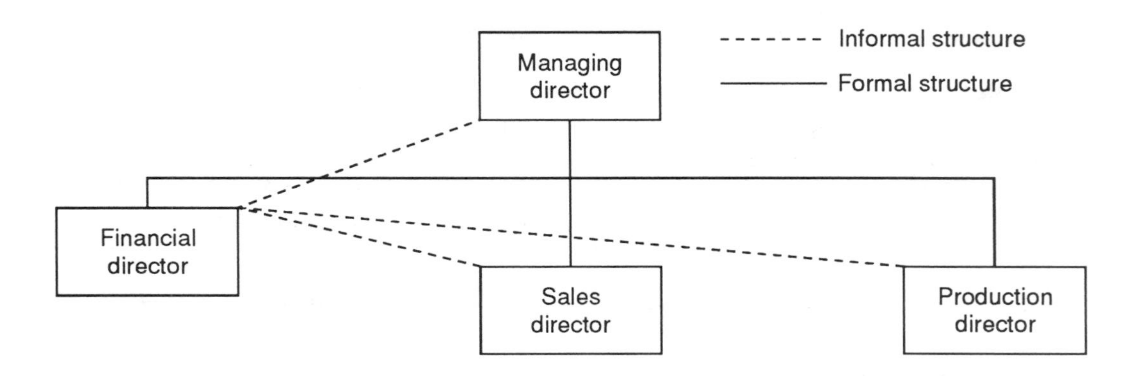

The formal organisation structure is therefore supplemented by an informal structure which improves the way in which top management sets about its job.

1.23 The informal structure of a company may 'take over' from the formal organisation when the formal structure is slow to adapt to change.

1.24 When employees are dissatisfied with aspects of formal organisation (if they dislike the work they do or the person they work for, say) they are likely to rely more and more heavily on an informal organisation at work to satisfy their personal needs in their work situation. When this happens, it has been argued (by Argyris and others) that the informal organisation of the individual will act against the efficiency of the formal organisation. Informal organisations always exist within a formal organisation, and if employees are properly motivated, these informal organisations should operate to the advantage of the formal organisation's efficiency and effectiveness.

1.25 A conclusion might therefore be that management should seek to harness the informal organisation to operate to the benefit of the formal organisation. In practice, however, this will be difficult because unlike formal organisation, which does not change even when individual employees move into and out of jobs (by promotion, transfer, appointment, resignation or retirement) most informal organisations depend on individual personalities. If one member leaves, the informal organisation is no longer the same, and new informal organisations will emerge to take its place.

Exercise 1

Make a list of all the different organisations that you belong to: your employer, your college, your sports club or whatever.

(a) See if you can analyse each in terms of its objects and its objectives.

(b) As far as possible identify your personal objectives in being a member of each organisation.

(c) Concentrate on one organisation (your employer is probably the best choice) and consider what differences there are between your own and your colleagues' formal roles and any informal roles adopted

Exercise 2

According to Laurie Mullins, *Management and Organisational Behaviour* (1993), there are three common factors in any organisation - *people*, *objectives*, and *structure* - and in addition 'some process of *management* is required by which the activities of the organisation, and the efforts of its members, are directed and controlled towards the pursuit of objectives'.

Read through Section 1 again noting the incidence of these four factors.

Solution

You should have been able to find many instances. For example, Buchanan and Huczynski's concentration on 'social groupings' (people), the way *managers* have to try to reconcile personal and corporate objectives, the *structure* provided by communication channels, and so on.

Continue to perform this exercise throughout the remainder of this chapter.

---

2. BUREAUCRACY

Max Weber

2.1 To a layman, the word 'bureaucracy' has unpleasant associations; however, the German writer, Max Weber (1864-1920) who is the organisational theorist most closely associated with the analysis of bureaucracy, was inclined to regard bureaucracy as the ideal form of organisation, which is 'from a purely technical point of view, capable of attaining the highest degree of efficiency and is in this sense formally the most rational means of carrying out imperative control over human beings.' Weber argued that bureacracy is the most efficient structure available for large organisations in particular, because it is impersonal and rational, based on a set pattern of behaviour and work allocation, and not allowing personality conflicts to get in the way of achieving goals.

2.2 Weber regarded an organisation as an authority structure. He was interested in why individuals obeyed commands, and he identified three grounds on which legitimate authority could exist.

(a) *Charismatic leadership*: in such an organisation, a leader is regarded as having some special power or attribute; decision-making is centralised in him and delegation strictly limited. The leader expects personal, sycophantic devotion from his staff and followers. Decisions are frequently irrational.

(b) *Traditional, or patriarchal leadership*: in such organisations, authority is bestowed by virtue of hereditary entitlement, as in the family firm, the lord of the manor. Tradition is glorified. Decisions and actions are bound by precedent.

(c) *Bureaucracy*: authority is bestowed by dividing an organisation into jurisdictional areas (production, marketing, sales and so on) each with specified duties. Authority to carry them out is given to the officials in charge, and rules and regulations are established in order to ensure their achievement. Leadership is therefore of a 'rational-legal' nature: managers get things done because their orders are accepted as legitimate and justified.

2.3 Weber specified several general characteristics of bureaucracy, which he described as 'a continuous organisation of official functions bound by rules'.

(a) *Hierarchy*: each lower office is under the control and supervision of a higher one.

(b) *Specialisation and training*: there is a high degree of specialisation of labour. Employment is based on ability, not personal loyalty.

(c) *Impersonal nature*: employees work full time within the impersonal rules and regulations and act according to formal, impersonal procedures.

(d) *Professional nature of employment*: an organisation exists before it is filled with people. Officials are full-time employees, promotion is according to seniority and achievement; pay scales are prescribed according to the position or office held in the organisation structure.

(e) *Rationality*: the 'jurisdictional areas' of the organisation are determined rationally. The hierarchy of authority and office structure is clearly defined. Duties are established and measures of performance set.

(f) *Uniformity* in the performance of tasks is expected, regardless of whoever is engaged in carrying them out.

(g) *Technical competence* in officials, which is rarely questioned within the area of their expertise.

(h) *Stability*.

2.4 Compared with other types of organisation the potential advantages of bureaucracy may seem apparent. Weber was impressed with the development and accomplishments of bureaucracy, and especially with the role of technical knowledge in bureaucratic administration which he regarded as the primary source of the superiority of bureaucracy as an organisation. He was also ready, however, to acknowledge the 'deadening' effect of bureaucracy and deplored an organisation of 'little cogs, little men, clinging to little jobs and striving towards bigger ones.'

Alvin Gouldner

2.5 Alvin Gouldner set out to investigate the application of *rules* - a key feature of bureaucratic organisation, controlling the pattern of interpersonal relationships with rationality, according to Weber. Gouldner rejected Weber's view that people comply with rules and that rules create order and uniformity. Instead he considered what were the various political 'interests' in organisations, and for which (if any) of them rules were created and maintained. His study of the General Gypsum Company (GGC) identified three 'patterns of industrial bureaucracy' in the use of rules.

(a) 'Mock bureaucracy' is a surface rule system which is generally ignored by employees, and not, apparently, taken seriously even by management, for whom it is a token gesture. A 'No Smoking' rule was taken to be an example of this. Most personnel smoked, *despite* the proliferation of signs, and management were seen to collaborate in this by not enforcing the rule and even by warning employees of impending visits by the fire inspector. The rule is seen purely as a cosmetic measure for the benefit of the fire insurers and inspectors.

(b) 'Representative bureaucracy' is a rule system which may be detailed and complex, and conformity to which is generally expected and stressed. Safety rules at GGC's plant were an example of this. 'Specific agencies existed which strove energetically to bring about their observance. These agencies placed continual pressures upon both workers and management, and sought to orient the two groups to the safety rules during their daily activities.'

(c) 'Punishment centred bureaucracy' is a rule system enforced by 'grievances'. Gouldner identified the 'bidding system' as an example of this: job vacancies were to be posted in the department concerned and employee applications considered in order of seniority, and only if no one in the department bid for the job, was bidding opened to other employees.

Workers were determined that supervisors conform to this system, but supervisors resisted the rules - sometimes with covert tactics such as posting the vacancy at a falsely low (and therefore unattractive) rate, only telling a chosen 'favourite' the truth, appointing him and then upgrading him. The rules were thus enforced only by strong support from workers, against the resistance and even evasion of local management.

Exercise 3

(a) Does the head of your organisation hold authority by virtue of charisma, tradition or bureaucracy, or is this analysis not appropriate?

(b) If possible, give an example of mock bureaucracy in your own organisation.

Criticisms of bureaucracies

2.6 Many criticisms of bureaucracies may already be familiar. Weber acknowledged their existence but did not discuss them. However ideal bureaucracy may appear in theory, there are many inefficiencies in practice which need to be overcome. The very strength of some of the characteristics of bureaucracy may in some cases be turned into a cause of weakness.

2.7 Influences commonly found in bureaucracies are as follows.

(a) The complexity of decision-making (eg to obtain the go-ahead for new projects, say) slows down the decision-making process, causing unwanted delays.

(b) Conformity creates ritualism, formalism and 'organisation man'.

(c) Personal growth of individuals is inhibited - although bureaucracies tend to attract, select and retain individuals with a tolerance for such conditions.

(d) Innovation is discouraged; there is a school of management thinking which believes that too much bureaucracy represses creativity and initiative in moulding organisation man.

(e) Control systems are frequently out of date. According to Michael Crozier, the control mechanism (whereby feedback on errors is used to initiate corrective action) is hampered by rigidity: bureaucracies cannot learn from their mistakes!

2.8 Michael Crozier (*The Bureaucratic Phenomenon*) developed his ideas as a result of empirical studies in two large French organisations. He confirmed that 'a system of organisation whose main characteristic is its *rigidity* will not adjust easily to change and will tend to resist change as much as possible'. He pointed out the consequences of this rigidity for the control system of the organisation as well as for response to external pressures. The feedback of information on errors, which in other structures might be used to readjust behaviour, does not tend to initiate control action. Because bureaucracies are highly centralised, constant adjustment or control action is impossible: decisions have to be taken at the top, against the grain of unwieldy upward communication mechanisms, so that corrective action is only taken 'when serious dysfunctions develop and no other alternatives remain'. Crozier further suggests that a bureaucracy cannot adjust to inevitable environmental changes without 'deeply felt crisis', because of delays in taking action, the magnitude of change when it finally occurs and the cultural resistance caused by the personality types attracted and recruited to the system.

2.9 Merton also discussed rigidity of behaviour in a bureaucracy, due to:

(a) the reduction in the amount of personalised relationships (officials work with officials, not people with people);

(b) the internationalisation of the rules of the organisation. Rules originally designed for efficiency take on a significance which is totally unrelated to any organisational goal;

(c) the increase in the use of categorisation of situations in order to reach decisions; decision-making is simplified (based on precedents);

(d) the development of an 'esprit de corps'. In spite of impersonalised relationships, there is a propensity amongst members of an organisation to defend each other against outside pressures.

2.10 Rigidity of behaviour, Merton argued, has three consequences.

(a) It creates reliability, thus meeting an important maintenance need of the system.

(b) It creates defensibility of individual action (due to categorisation for decision-making purposes).

(c) It increases difficulties with clients.

Client pressure (for example complaints from members of the public) creates a felt need for defensibility of individual actions, which in turn encourages further rigidity of behaviour.

2.11 Selznick suggested a further dysfunction in bureaucracy. As a result of increasing specialisations and technical competence, delegation of authority has increased. With delegation, however, comes the 'bifurcation of interests'. What one expert sees as the organisational goals are different from the views of another expert. This in turn raises conflict between sub-units of the organisation and there is a tendency for individuals to identify with the sub-goals of their sub-unit to the detriment of the organisation as a whole. Selznick concluded that delegation achieves a necessary purpose, as specialisation increases, but that it carries problems with it: it is both functional and dysfunctional at the same time.

2.12 Gouldner argued that rules are both functional and dysfunctional in a bureaucracy.

(a) *Rules are functional*: they take away from a subordinate the feeling that his superior, in issuing orders, holds power over him. This in turn reduces the interpersonal tensions which otherwise exist between superiors and subordinates, and for this reason the 'survival of the work group as an operating unit' is made possible by the creation of rules.

(b) *Rules are dysfunctional*: employees use rules to learn what is the minimum level of behaviour expected from them, and there is a tendency for employees to work at this minimum level of behaviour. This, in turn, suggested Gouldner, creates a requirement for close supervision. Greater pressure from supervisors will make subordinates more aware of the power the supervisor holds over him, thereby increasing tension within the work group.

Overcoming the disadvantages

2.13 The financial and technical advantages of bureaucratic organisations usually outweigh the disadvantages, especially in circumstances of slow change and a large customer/client base, to which bureaucracy is well suited. The dysfunctions, however, still need to be reduced to acceptable proportions. Arguably therefore:

(a) the organisation should be arranged:

- (i) into small working groups (to promote group loyalty and purpose);
- (ii) into small working establishments (for the same reasons);
- (iii) with as little centralisation as possible (to give junior management more scope);
- (iv) with a highly developed two-way communication system;

(b) culture may be used to increase the flexibility and humanity of the organisation. A more responsive structure/culture may be developed in particular units of the bureaucracy, such as those with direct customer/client contact;

(c) opportunities should be created for individualism and innovation - again, if only in certain units of the organisation. The bureaucracy may then be able to attract individuals of a less conforming type, better able to handle change, ambiguity and flexibility.

Bureaucracy in the 1990s

2.14 A bureaucracy, on balance, promotes efficiency in larger organisations, but it must achieve a balance whereby rules are impersonal but not inhuman, the structure is organised but not rigid, and employees are loyal without being unthinking conformists.

2.15 There are indications that bureaucracy in some forms and in some environments has suffered the consequences of its dysfunctions.

(a) In the UK, the recessions of the early 1980s and early 1990s necessitated cost-cutting which, together with factors such as improvements in information provision through technology, effectively wiped out middle management in many firms: thus the tall, highly 'tiered' vertical structure characteristic of bureaucracy is less common in the private sector.

(b) Market pressures have also created a need for client responsiveness - at least at the interface between organisation and client/customer. Sales and service units within bureaucracies have therefore had to undergo cultural and even structural change, so that thorough-going bureaucracies are rarer. Banks, for example, are beginning to decentralise authority to branches, to carry market segmentation throughout their organisation structure and to gear their cultures towards sales and service. Government organisations have come under the influence of a variety of client-based 'charters' (the Patient's Charter, the Passenger's Charter and so on).

(c) Management theory has in recent years emphasised the importance of flexibility and adaptability in the face of rapid environmental change. The 'fashion' is for a contingency approach to management and organisation, for differentiation of cultures and structures, for smaller task-centred units.

2.16 However, it has to be said that bureaucracy self evidently has not died. It is still much in evidence in the public sector and government, and elements of it survive in many large private sector organisations, in the formality of communication channels, the extent of specialisation and delegation, and the number of individuals who as administrative 'knowledge workers' are physically separated from the 'real work' of the organisation.

2.17 This survivability of bureaucracy can be explained by a number of factors.

(a) In certain slow-change environments, or those in which market pressures have little short-term effect, bureaucracy can still be a perfectly acceptable structure/culture.

(b) Bureaucratic systems tend to be self-defensive, in:

(i) the type of people who are attracted to, and selected for, the culture. Change to the organisation itself is therefore resisted, and the self-correcting mechanism is not highly developed;

(ii) problems and events being categorised for decision-making purposes. This, according to Merton, creates defensibility of individual action in the face of client/customer pressure;

(iii) the development of esprit de corps. In spite of the impersonality of the relationships, Merton suggests, there is a propensity for members of the bureaucracy to defend each other against outside pressures.

(c) The dysfunctions of bureaucracy are often also functional at the same time. Rigidity of behaviour also creates reliability. Rules help to control interpersonal tension and aid the survival of the work group.

Exercise 4

'Thirty years ago most people thought that change would mean more of the same, only better. That was incremental change and to be welcomed. Today we know that in many areas of life we cannot guarantee more of the same, be it work or money, peace or freedom, health or happiness, and cannot even predict with confidence what will be happening in our own lives. Change is now more chancy, but also more exciting if we want to see it that way.' (Handy, *The Age of Unreason*)

If Handy is correct, what do you think is the future for bureaucracies?

Solution

You must form your own view, and it will depend upon whether or not you are the sort of person who feels comfortable in a bureaucracy. Handy's own views may be readily deduced from the continuation of the above passage:

'a frog if put in cold water will not bestir itself if that water is heated up slowly and gradually and will in the end let itself be boiled alive, too comfortable with continuity to realise that continuous change at some point becomes discontinuous and demands a change in behaviour.'

3. LARGE AND SMALL ORGANISATIONS

3.1 It is wrong to confuse 'bureaucracy' with 'large' organisations, because the features of bureaucracy can be found in many small and medium-sized organisations. It might be useful, however, to consider the advantages and disadvantages of *large* size in this chapter.

3.2 The advantages of a large organisation are as follows:

(a) A large-scale organisation should have access to sufficient resources to command a significant market share. This in turn will enable it to influence prices in the market so as to ensure continuing profitability.

(b) A large organisation can provide for greater division of work and specialisation. Specialisation, and the development of a wide range of products or customer services, should enable the organisation to attract continuing customer support and market shares. In contrast, a small or medium-sized business will require greater competence and versatility from its top management, because they will not have the benefits of support from functional specialists which are available to the top managers of large organisations.

(c) A large organisation with a wide variety of products or customer services should be able to offer an attractive career to prospective employees, and it is therefore likely to receive job application requests from very talented people. This in turn should enable the large organisation to recruit and develop high-quality personnel for future top management positions.

(d) Specialisation brings with it the ability to provide expert services at a relatively low cost to the customer. A large organisation is also able to make use of the advantages of efficient 'large-scale' equipment such as advanced computer systems or manufacturing equipment. For these (and other) reasons, large organisations are able to achieve *economies of scale* in the use of resources. Cheaper costs in turn mean either lower prices for customers or higher profits for the organisation.

(e) A large organisation is more likely to provide continuity of goods or services, management philosophy, customer relations and so on than a smaller organisation. A smaller organisation might be prone to sudden policy changes or changes of product when a new management team takes over.

3.3 The disadvantages of a large organisation are as follows.

(a) There is a tendency for the management hierarchy to develop too many levels. The more management levels there are, the greater the problem of communication between top and bottom, and the greater the problems of control and direction by management at the top.

(b) An organisation might become so widely diversified in the range of products or services it offers that it becomes difficult, if not impossible, for management to integrate all of the organisation under a common objective and within a single 'management philosophy' and culture.

(c) Top management might spend too much time in maintenance of the organisation (that is, with problems of administration) and lose sight of their primary tasks of setting objectives and planning for the future.

(d) There is a tendency of top management in large organisations to become 'ingrown and inbred, smug and self-satisfied'. The tendency towards 'group-think' - an acceptance by all managers of a common attitude towards problems - might introduce an unconscious resistance to necessary changes and developments.

(e) The sheer size of an organisation may provide management with problems of co-ordination, planning policy and effective control. For example, a junior manager might find the organisation so large that he has relatively little influence. Decisions which he regards as important must be continually referred up the line to his superiors, for inter-departmental consultations. At the same time, the top management might find the organisation so large and complex, and changes in policy and procedures so difficult and time-consuming to implement, that they also feel unable to give direction to the organisation. The organisation is therefore a 'monster' which operates of its own accord, with neither senior nor junior managers able to manage it effectively.

(f) In a large organisation, many of the tasks of junior management are routine and boring. Even middle management might be frustrated by the restrictions on their authority, the impersonal nature of their organisation, the inability to earn a just reward for their special efforts owing to the standardisation of pay and promotion procedures and the lack of information about aspects of the organisation which should influence their work.

3.4 These difficulties of large organisations can, to some extent, be overcome by:

(a) *decentralisation and delegation of authority*. The aim of decentralisation should be to encourage decision-making at lower levels of management 'closer to the action'. Management motivation, but also management efficiency in target-setting, planning and control should improve;

(b) *pay policies* which provide for just rewards (individual or team bonuses) for outstanding effort, achievements and innovation;

(c) the introduction of comprehensive management and employee *information systems* which enable all managers and employees:

(i) to understand their planned contribution towards achieving organisational objectives;
(ii) to compare their actual achievements against their targets;

(d) *a task structure* within the organisation which stimulates employee commitment. A feature of the 1980s and 1990s has been what Peters and Waterman called 'chunking' - breaking organisational structures down into small, task-centred units.

Exercise 5

In the light of the above, from your own experience and from common sense, what features do you think characterise a *small* organisation as opposed to a large one? How are these differences reflected in the structure of the organisation and the way it is controlled, its responsiveness to change, the motivation of its staff and its use of resources?

Solution

Here are some suggestions.

(a) *Organisation structure, planning and control*

(i) Less rigid definition of jobs than in large organisations, and less specialisation.

(ii) Authority often centralised, although the management hierarchy will be small and all managers should feel quite close to the decision-making process.

(iii) Planning, control and communication (management information systems) will be less formal than in larger organisations.

(iv) Fewer people than in large organisations, and so lesser problems of co-ordination. Greater sense of teamwork but greater risk of disaster through personality clash.

(v) Not so much bound by formal rules and procedures as large organisations.

(vi) Decision-making procedures are relatively fast. Decisions can be taken quickly.

(b) *Adapting to change*

(i) Small organisations are usually staffed by individuals who are more innovative in their ideas, and more responsive to change. Ideas for innovation and adapting to change are more readily accepted by the small organisation culture.

(ii) However, resources are likely to be limited for research and development, personnel, marketing and so on.

(c) *Motivation*

(i) Individuals have a wider range of duties, and are often able to contribute more to the achievements of the organisation, compared with the achievements of individuals in large organisations.

(ii) Individuals might develop a closer sense of identity with the organisation - for example through personal association with the owner-managers.

(iii) Pay and reward systems are usually more personalised than in large organisations.

(iv) Management training and development are not encouraged as much as in large organisations.

(v) The 'culture' in small organisations is unlikely to be bureaucratic.

(d) *Resources*

(i) Small businesses are considered risky investments and a small organisation's borrowing capability might be limited.

(ii) Limited resources may restrict capacity for developing further new product ideas such as those on which the organisation may have been founded.

---

4. AUTHORITY, RESPONSIBILITY AND POWER IN ORGANISATIONS

4.1 *Authority* is the right to do something. In an organisation, it refers to the scope and amount of discretion given to a person to make decisions, by virtue of the position he or she holds in the organisation: it is conferred 'from the top down' because each manager can pass on some of his or her authority to subordinates in assigning tasks to them. However, authority is *not only* bestowed by the organisation (that is, by more senior people in it). Authority can be bestowed 'from the bottom up', when it is conferred on a leader figure by people at *lower* levels in the hierarchy for example by election: trade union officials have this kind of authority.

4.2 *Responsibility* is the liability of a person to be called to account for exercising his authority, actions and results. It is therefore a formal obligation to do something. *Accountability* is the duty of an individual to report to his superior how he has fulfilled his responsibilities.

4.3 It is important to realise that accountability cannot be 'delegated' or passed on to a subordinate. A manager may delegate *authority* and *responsibility* to a subordinate (if that authority is within his *own* power to delegate) - but he remains *accountable* to his own boss for seeing that the work gets done, albeit by the subordinate rather than by himself personally.

4.4 *Delegation* thus refers to the process whereby a superior gives a subordinate the authority and responsibility to carry out a given aspect of the superior's own task. It is delegation that shapes the hierarchical structure of the formal organisation, where authority and responsibility are passed down the 'chain of command' and accountability passes back up.

(Note, incidentally, that the words 'superior' and 'subordinate' are used for convenience only. Some organisations would prefer vaguer but less emotive terms.)

4.5 There should be a careful balance between delegated authority and responsibility in the organisation.

(a) A manager who is not held responsible for any area of his authority is free to exercise it in a capricious way: he is not bound to do otherwise.

(b) A manager who is held responsible for aspects of performance, but has not been given authority to control them, is in an impossible position. Sufficient authority must be given to enable the individual to do what is expected of him.

4.6 It should be evident that the boundaries of each manager's authority should be clear, to avoid ambiguity and overlap. This is an argument for 'unity of command': that each individual should report to only one superior, so as to avoid conflicts and doubts created by dual command. Dual command systems have, however, proved workable in practice, for example where there is project/product management: either clarified boundaries of authority or consensus management is required.

Power and influence

4.7 Influence is the process by which one person in an organisation, A, directs or modifies the behaviour or attitudes of another person, B. Influence can only be exerted by A on B if A has some kind of power from which the influence emanates. Power is therefore the ability to influence, whereas influence is an active process. Note that *power* is not the same as *authority*. A manager may have the right to expect his subordinates to carry out his instructions, but may lack the ability to make them do it. On the other hand, an individual may have the ability to make others act in a certain way, without having the organisational authority to do so: informal 'leaders' are frequently in this position.

4.8 Power and influence are clearly important factors in the structure and operations of an organisation. They help to explain how work gets done. In addition, it has also been suggested that:

(a) an individual who believes he exerts some influence is likely to show greater interest in his work. The research of writers, such as Likert, who support the principle of management by participation suggests that employees may be more productive when they consider that they have some influence over planning decisions which affect their work;

(b) some individuals are motivated by the need for power, and show great concern for exercising influence and control, and for being leaders.

4.9 Charles Handy (*Understanding Organisations,* 1993) identifies six types of power from different sources.

(a) *Physical power* - the power of superior force. Physical power is absent from most organisations (except the prison service and the armed forces), but it is sometimes evident in poor industrial relations (for example shop floor intimidation, or the use of riot police against workers).

(b) *Resource power* - the control over resources which are valued by the individual or group to be influenced. Senior managers may have the resource power to grant promotion or pay increases to subordinates; trade unions possess the resource power to take their members out on strike. The amount of power a person has then depends on how far he controls the resource, how much the resource is valued by others, and how scarce it is.

(c) *Position power* - the power which is associated with a particular job in an organisation. Handy noted that position power has certain 'hidden' benefits:

(i) access to information;

(ii) the right of access: for example entitlement to membership of committees and contact with other 'powerful' individuals in the organisation;

(iii) the right to organise conditions of working and methods of decision-making.

(d) *Expert power* - the power which belongs to an individual because of his expertise, although it only works if others acknowledge him to be an expert. Many staff jobs in an organisation (computer systems analysts, organisation and methods analysts, accountants, lawyers or personnel department managers) rely on expert power to influence line management. If the expert is seen to be incompetent (if an accountant, say, does not seem to provide sensible information) or if his area of expertise is not widely acknowledged (which is often the case with personnel department staff) he will have little or no expert power.

(e) *Personal power*, or *charisma* - the popularity of the individual. Personal power is capable of influencing the behaviour of others, and helps to explain the strength of informal organisations.

(f) *Negative power* is the use of disruptive attitudes and behaviour to stop things from happening. It is associated with low morale, latent conflict or frustration at work. A subordinate might refuse to communicate openly with his superior, and might provide false information; a colleague might refuse to co-operate; a typist might refuse to type an urgent letter because she is too busy; a worker might deliberately cause his machine to break down; a manager might refuse to co-operate with his colleagues, if an agreed policy adversely affects his position. Negative power is destructive and potentially very damaging to organisational efficiency.

4.10 Influence, the act of directing or modifying the behaviour of others, may then be achieved through:

(a) the application of force, eg physical or economic power;

(b) the establishment of rules and procedures - enforced through position and/or resource power;

(c) bargaining and negotiation - depending on the relative strengths of each party's position, (expert, resource or personal power and so on);

(d) persuasion, again associated with various sources of power.

4.11 Handy identifies two further, 'unseen' methods of influence.

(a) *Ecology*, or the environment in which behaviour takes place. The physical environment can be altered by a manager, who may be able to regulate noise levels at work, comfort and security of working conditions, seating arrangements, the use of open-plan offices or segregation into many small offices, the physical proximity of departments as well as individuals.

> 'The design of work, the work, the structure of reward and control systems, the structure of the organisation, the management of groups and the control of conflict are all ways of managing the environment in order to influence behaviour. Let us never forget that although the environment is all around us, it is not unalterable, that to change it is to influence people, that ecology is potent, the more so because it is often unnoticed.' *(Handy)*

(b) *Magnetism*, the unseen application of personal power. 'Trust, respect, charm, infectious enthusiasm, these attributes all allow us to influence people without apparently imposing on them.'

4.12 According to Amitai Etzioni, authority and power are exercised differently according to the environment, relationships and type of subordinates involved. He suggests that there are three forms of power that might be used, depending on the situation:

(a) *coercive power* (for example that used in prisons): power based on fear of physical punishment;

(b) *remunerative power* (for example that used in most work organisations): power based on control and administration of the reward system; and

(c) *normative power* (for example in professional or religious organisations): power based on the application of norms and standards.

Consultative management and the exercise of authority

4.13 Jaques (the 'Glacier investigations') found that even a well-defined role in the organisation posed problems for the person expected to fill it with regard to the exercise of authority. In particular, he found that in an organisation committed to consultative management, a superior may become more and more reluctant to exercise his authority. Mechanisms for avoiding responsibility and authority included:

(a) the exercise of a purely consultative relationship with others while ignoring roles involving line authority;

(b) misuse of formal joint consultation processes - using contact between higher management and workers' representatives as an excuse to ignore responsibility for immediate subordinates. Consultation must not by-pass intervening roles in the chain of command;

(c) pseudo-democracy - for example a superior, senior member on a committee asserting 'I'm just an ordinary member'.

One of the most important conclusions of the Glacier studies was that there is a distinctive leadership role in groups, and that members (in their roles as participants) *expect* it to be properly filled.

Delegation

4.14 It is generally recognised that in any large complex organisation, management must delegate some authority and responsibility because:

(a) there are physical and mental limitations to the work load of any individual or group in authority;

(b) routine or less important decisions can be passed 'down the line' to subordinates, freeing the superior to concentrate on the more important aspects of the work, like planning, and allowing the organisation to respond more quickly to the demands made on it;

(c) the increasing size and complexity of organisations calls for specialisation, both managerial and technical;

(d) employees in today's organisations have high expectations with regard to job satisfaction - including discretion and participation in decision-making;

(e) the managerial succession plan depends on junior managers gaining same experience of management processes in order to be 'groomed' for promotion.

4.15 In practice many managers are reluctant to delegate and attempt to do many routine matters themselves in addition to their more important duties. Amongst the reasons for this reluctance one can commonly identify:

(a) low confidence and trust in the abilities of the subordinates: the suspicion that 'if you want it done well, you have to do it yourself';

(b) the burden of responsibility and accountability for the mistakes of subordinates;

(c) a desire to 'stay in touch' with the department or team - both in terms of workload and staff - particularly if the manager does not feel 'at home' in a management role, and/or misses aspects of the subordinate job or the camaraderie;

(d) an unwillingness to admit that subordinates have developed to the extent that they could perform some of the manager's duties. The manager may feel threatened by this sense of 'redundancy';

(e) poor control and communication systems in the organisation, so that the manager feels he has to do everything himself, if he is to retain real control and responsibility for a task, and if he wants to know what is going on;

(f) an organisational culture that has failed to reward or recognise effective delegation by superiors, so that the manager may not realise that delegation is regarded positively (rather than as 'shirking responsibility');

(g) lack of understanding of what delegation involves - that is, realising that it does *not* mean giving the subordinates total control, making the manager himself redundant.

(h) a desire to continue to operate within their personal 'comfort zone', doing things that are familiar to them and then claiming that they are too busy to tackle the jobs they really should be doing, which are risky because they involve thought, forward planning, performance improvement and the initiation of change.

4.16 Handy writes of a 'trust-control dilemma' in a superior-subordinate relationship, in which the sum of trust + control is a constant amount.

$$T + C = Y$$

where T = the trust the superior has in the subordinate, and the trust which the subordinate feels the superior has in him;

C = the degree of control exercised by the superior over the subordinate;

Y = a constant, unchanging value;

Any increase in C leads to an equal decrease in T: if the superior retains more 'control' or authority, the subordinate will immediately recognise that he is being trusted less. If the superior wishes to show more trust in the subordinate, he can only do so by reducing C, in other words by delegating more authority.

4.17 To overcome the reluctance of managers to delegate, it is necessary:

(a) to provide a system of selecting subordinates who will be capable of handling delegated authority in a responsible way. If subordinates are of the right 'quality', superiors will be prepared to trust them more;

(b) to have a system of open communications, in which the superior and subordinates freely interchange ideas and information. If the subordinate is given all the information he needs to do his job, and if the superior is aware of what the subordinate is doing:

(i) the subordinate will make better-informed decisions;
(ii) the superior will not 'panic' because he does not know what is going on.

Although open lines of communication are important, they should not be used by the superior to command the subordinate in a matter where authority has been delegated to the subordinate; in other words, communication links must not be used by superiors as a means of reclaiming authority;

(c) to ensure that a system of control is established. Superiors are reluctant to delegate authority because they retain absolute responsibility for the performance of their subordinates. If an efficient control system is in operation, responsibility and accountability will be monitored at all levels of the management hierarchy, and the 'dangers' of relinquishing authority and control to subordinates are significantly lessened;

(d) to reward effective delegation by superiors and the efficient assumption of authority by subordinates. Rewards may be given in terms of pay, promotion, status, or official approval;

(e) to set the example from the very top of the organisation: an MD who has a hands-on approach to every aspect of the organisation's activities is unlikely to be able to encourage his managerial team to let go.

Exercise 6

The quality newspapers often carry profiles of leading business figures (the *Financial Times* every Monday, for example, the *Times* on Saturday, and often the quality Sunday press). It is worth reading articles such as these to see what is said about the style of management. From what sources do such figures derive their power? How do they exercise their authority? Are they good delegators? Do you notice common characteristics or are you struck by the differences between such figures?

Chapter roundup

- Do not worry if many of the terms and concepts you have encountered in this chapter are new to you. They will all be explained and developed in the chapters that follow. Use the index if you want to find out more about a particular topic straight away.
- An organisation is a social arrangement for the controlled performance of collective goals.
- Organisations have goals and objectives. These are not necessarily the same as the goals and objectives of the individuals within the organisation.
- Organisations can be formal or informal.
- Bureaucracies involve hierarchy, specialisation and rules. This can make them slow and inflexible.
- In organisations, authority refers to the right to make decisions. To be responsible is to be held accountable for those decisions.
- Power comes from a variety of sources, not just the right to make decisions: make sure you understand terms like 'expert power', 'normative power' and so on.
- We will discuss further aspects of authority and delegation in our chapter on the structures and cultures of organisations. First, however, we will get a brief overview of organisation theory.

Test your knowledge

1 Give four reasons why an organisation might be formed. (see para 1.6)

2 Informal organisations should not be allowed to develop. True or false? (1.23)

3 What are the eight general characteristics of a bureaucracy, according to Weber? (2.3)

4 What are the dysfunctions of bureaucracy? (2.8-2.12)

5 What are the advantages and disadvantages of large organisations? (3.2-3.3)

6 What types of power can be identified? Give examples of each. (4.9, 4.12)

7 Why might managers be reluctant to delegate? (4.15, 4.16)

Now try question 1 at the end of the text

Chapter 2

INTRODUCTION TO ORGANISATION THEORY

This chapter covers the following topics.

Signpost

- This chapter provides further background, but it must not be neglected because an understanding of the ideas and terms introduced here will be essential as you read on.
- Scientific management is referred to, for example, in the context of job design in Chapter 4. The human relations approach is important to the discussion of leadership in Chapter 6. Contingency theory is often mentioned, for example in Chapter 13 when discussing process theories of motivation.

1. WHAT ARE ORGANISATION THEORIES?

1.1 It is worth noting that ideas of what organisations are, and how they should be analysed, designed and managed, have changed over the years. Organisations have been likened to 'tribes', 'organisms', 'systems', 'kingdoms' and even 'shamrocks'.

1.2 Theories about organisations tend not to be 'theories' at all, strictly speaking, but 'approaches' offering ways of looking at issues such as organisational structure, management functions or motivation. None of them can be used to predict with certainty what the 'behaviour' of an organisation, manager or employee will be in any given situation. Nor can they guarantee that application of the principles they put forward will result in effectiveness and efficiency for the organisation.

1.3 They can however provide helpful and/or thought-provoking ways of analysing organisational phenomena, and 'frameworks' within which practical problems and situations can be tackled, or 'models' which can be used to isolate the basic characterisation of organisations.

2. SCIENTIFIC MANAGEMENT

2.1 The main early theorists who put forward ways of understanding organisations were practising managers, who analysed their own experience in management to produce a set of what they saw as 'principles' of organisation, applicable in a wide range of situations. Their approach was essentially prescriptive: it attempted to suggest what is good - or even best - for organisations, and contributed techniques for studying the nature of work more systematically than ever before, and solving problems of how it could be organised more efficiently.

F W Taylor

2.2 Frederick W Taylor (1856 - 1915) pioneered the *scientific management* movement. He argued that management should be based on 'well-recognised, clearly defined and fixed principles, instead of depending on more or less hazy ideas'. This involved the development of a 'true science of work', where all the knowledge gathered and applied in the work should be investigated and reduced to 'law' and techniques.

2.3 Taylor's famous four principles of scientific management were as follows.

(a) *The development of a true science of work*: all knowledge which had hitherto been kept in the heads of workmen should be gathered and recorded by management. 'Every single subject, large and small, becomes the question for scientific investigation, for reduction to law.' Very simply, he argued that management should apply techniques to the solution of problems and should not rely on experience and 'seat-of-the-pants' judgements.

(b) *The scientific selection and progressive development of workmen*: workmen should be carefully trained and given jobs to which they are best suited. Although 'training' is an important element in his principles of management, 'nurturing' might be a more apt description of his ideas of worker development.

(c) *The bringing together of the science and the scientifically selected and trained men*. The application of techniques to decide what should be done and how, using workmen who are both properly trained and willing to maximise output, should result in maximum productivity.

(d) *The constant and intimate co-operation between management and workers*: 'the relations between employers and men form without question the most important part of this art.'

2.4 The practical application of the approach was the use of work study techniques to break each job down into its smallest and simplest component parts: these single elements became the newly-designed 'job'. Workers were selected and trained to perform their single task in the most efficient way possible, as determined by techniques such as time and motion study to eliminate 'wasted motions' or unnecessary physical movement. Workers were paid incentives on the basis of acceptance of the new methods and output norms.

2.5 The *financial* implications of this were important to the approach.

> 'The great mental revolution that takes place in the mental attitude of the two parties under scientific management is that both parties take their eyes off the division of the surplus as the all-important matter, and together turn their attention toward increasing the size of the surplus until this surplus becomes so large that it is unnecessary to quarrel about how it should be divided. They come to see that when they stop pulling against one another, and instead both turn and push shoulder to shoulder in the same direction, the size of the surplus created by their joint effort is truly astounding. They both realise that when they substitute friendly co-operation and mutual helpfulness for antagonism and strife, they are together able to make this surplus so enormously greater than it was in the past that there is ample room for a large increase in wages for the workmen and an equally great increase in profits for the manufacturer.' *F W Taylor*

2.6 It is useful to consider an application of Taylor's principles. In testimony to the House of Representatives Committee in 1912, Taylor used as an example the application of scientific management methods to shovelling work at the Bethlehem Steel Works.

(a) Facts were first gathered by management as to the number of shovel loads handled by each man each day, with particular attention paid to the relationship between weight of the average shovel load and the total load shifted per day. From these facts, management was able to decide on the ideal shovel size for each type of material handled in order to optimise the speed of shovelling work done. Thus, scientific technique was applied to deciding how work should be organised.

(b) By organising work a day in advance, it was possible to minimise the idle time and the moving of men from one place in the shovelling yard to another. Once again, scientific method replaces 'seat-of-the-pants' decisions by supervisors.

(c) Workers were paid for accepting the new methods and 'norms' and received 60% higher wages than those given to similar workers in other companies in the area.

(d) Workers were carefully selected and trained in the art of shovelling properly; anyone falling below the required norms consistently was given special teaching to improve his performance.

(e) 'The new way is to teach and help your men as you would a brother; to try to teach him the best way and to show him the easiest way to do his work. This is the new mental attitude of the management towards the men....'

(f) At the Bethlehem Steel Works, Taylor said, the costs of implementing this method were more than repaid by the benefits. The labour force required fell from 400 - 600 men to 140 men for the same work.

2.7 A summary of scientific management, in Taylor's own words, might be:

(a) 'The man who is fit to work at any particular trade is unable to understand the science of that trade without the kindly help and co-operation of men of a totally different type of education.'

(b) 'It is one of the principles of scientific management to ask men to do things in the right way, to learn something new, to change their ways in accordance with the science and in return to receive an increase of from 30% to 100% in pay....'

Exercise 1

How well received do you think Taylor's two comments would be by the workers in a modern factory?

An appraisal of scientific management

2.8 Peter Drucker made some useful comments about scientific management, as follows.

(a) Scientific management has contributed a philosophy of worker and work. 'As long as industrial society endures, we shall never lose again the insight that human work can be studied systematically, can be analysed, can be improved by work on its elementary parts. Like all great insights, it was simplicity itself'.

(b) However, it is capable of providing solutions to management problems only up to a certain point, and it seems incapable of providing significant further developments in future. 'Scientific management....has been stagnant for a long time....During the last thirty years, it has given us little but pedestrian and wearisome tomes on the techniques, if not on the gadgets, of narrower and narrower specialities....'

(c) One major weakness of scientific management is that by breaking work down into its elementary parts, and analysing a job as a series of consecutive 'motions', the solution to management problems often provided is that each separate 'motion' within the entire job should be done by a separate worker. This is profoundly un-satisfying to workers, and treats them like (poorly designed) 'machine tools': operations *should* be analysed in this way, but then *reintegrated* into a whole job.

(d) A further criticism of scientific management is that it divorces planning work from doing the work. 'The divorce of planning from doing reflects a dubious and dangerous philosophical concept of an elite which has a monopoly on esoteric knowledge entitling it to manipulate the unwashed peasantry'.

2.9 In spite of this it is notable that scientific management has enjoyed something of a resurgence in the late 1980s and early 1990s when high unemployment has meant that companies can more easily attract and retain staff. Enlightened organisations may persist in attempting to create 'satisfying' jobs for their workers, but others (for example Nissan in Washington) have returned to molecurised work design, involving minimal training for workers, close supervision and tightly-timed work activities.

3. THE CLASSICAL SCHOOL

Henri Fayol

3.1 Fayol (1841-1925) was a French industrialist, who was associated with the *classical* school of management thought, which was primarily concerned with the structure and activities of the formal organisation, and the rational principles by which they could be directed most effectively.

3.2 Applying the principles of management is a 'difficult art requiring intelligence, experience, decision and proportion'. 'Seldom do we have to apply the same principle twice in identical conditions; allowance must be made for different changing circumstances'.

Among his fourteen principles of management, Fayol listed the following.

(a) *Division of work,* ie specialisation. The object of specialisation is to produce more and obtain better results.

(b) *Authority and responsibility*: Fayol distinguished between a manager's official authority (deriving from his office) and personal authority (deriving from his experience, moral worth, intelligence and so on).

'Authority should be commensurate with responsibility', in other words the holder of an office should have enough authority to carry out all the responsibilities assigned to him.

He also suggested that 'generally speaking, responsibility is feared as much as authority is sought after, and fear of responsibility paralyses much initiative and destroys many good qualities'.

A good leader should encourage those around him to accept responsibility.

(c) *Discipline*: 'the state of discipline of any group of people depends essentially on the worthiness of its leaders'. A fair disciplinary system, with penalties judiciously applied by worthy superiors, can be a chief strength of an organisation.

(d) *Unity of command*: for any action, a subordinate should receive orders from one boss only. 'This rule seems fundamental to me...' Fayol saw dual command as a disease, whether it is caused by imperfect demarcation between departments, or by a superior S2 giving orders to an employee, E, without going via the intermediate superior, S1.

(e) *Unity of direction*: there should be one head and one plan for each activity. Unity of direction relates to the organisation itself, whereas unity of command relates to the personnel in the organisation.

(f) *Subordination of individual interests*: the interest of one employee or group of employees should not prevail over that of the general interest of the organisation.

(g) *Remuneration*: it should be 'fair', satisfying both employer and employee alike.

(h) *Scalar chain*: the scalar chain is the term used to describe the chain of superiors from lowest to highest rank. Formal communication is up and down the lines of authority, eg E to D to C to B to A. If, however, communication between different branches of the chain is necessary (eg D to H) the use of a 'gangplank' of horizontal communication saves time and is likely to be more accurate.

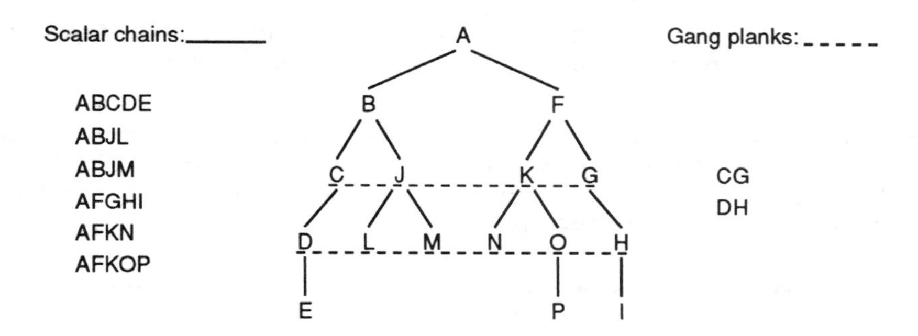

If C has a problem which affects G, instead of referring it up to B (who might then refer it to A before it could be discussed with G's boss, F) C could cross the 'gangplank' - and communicate directly and horizontally with G. The problem might then be solved jointly by C and G.

(i) *Stability of tenure of personnel*: 'It has often been recorded that a mediocre manager who stays is infinitely preferable to outstanding managers who merely come and go'.

(j) *Esprit de corps*: personnel should not be isolated: cohesion should be encouraged. 'In union, there is strength'.

(k) *Initiative*: 'it is essential to encourage and develop this capacity to the full'.

Mary Parker Follett

3.3 Mary Parker Follett (1868 - 1933) saw management as a continuous process rather than a series of discrete events. She suggested that in any industrial organisation everyone in the organisation should have a rational appreciation of what is required by the situation. When applying the idea to one aspect of management, the giving of orders, she said that in order to avoid the extremes of being too bossy on the one hand and not giving enough orders on the other, 'My solution is to depersonalise the giving of orders, to unite all concerned in a study of the situation....to discover the law of the situation and to obey that....*one person* should not give orders to another *person*, but both should agree to take their orders from the situation.'

3.4 One of the largest contributions of the scientific management method, she added, was to depersonalise orders, because it makes the job of management not one of giving orders, but of devising methods of discovering what orders are appropriate to a particular situation - and both managers and employees must then follow the dictates of the situation they find revealed. She postulated four fundamental principles of organisation.

(a) Co-ordination by direct contact
(b) Co-ordination in the early stages
(c) Co-ordination as the 'reciprocal relating' of all factors in a situation
(d) Co-ordination as a continuing process

Exercise 2

(a) Attempt to explain the following terms in your own words.

(i) Unity of command
(ii) Scalar chain

(b) Can you spot any links between Follett's four principles of organisation and Fayol's principles?

Solution

(a) Compare your explanation with ours given above.

(b) Division of work is what leads to the need for co-ordination; unity of command is in keeping with Follet's suggestion that co-ordination should be by direct contact. You can probably spot other links.

It is often useful to try and see connections between the arguments of different writers, or to try and identify differences. Comparing one section of this text with another is a more effective and stimulating way of studying than simply trying to learn the material off by heart - even if you conclude that there are *no* connections.

An appraisal of the classical school

3.5 'What Fayol did achieve was the first real attempt to produce a theory of management based on a number of principles which could be passed on to others. Many of these principles have been absorbed into modern organisations. Their effect on organisational effectiveness has been subject to increasing criticism over the last decade, mainly because such principles were not designed to cope with conditions of rapid change and issues such as employee participation in the decision-making processes of organisations.' (GA Cole, *Management: theory and practice*, 1993)

4. THE HUMAN RELATIONS APPROACH

4.1 In the 1930s, a critical perception of scientific management, in particular, emerged.

'By the end of the scientific management period, the worker has been reduced to the role of an impersonal cog in the machine of production. His work became more and more narrowly specialised until he had little appreciation for his contribution to the total product... Although very significant technological advances were made....the serious weakness of the scientific approach to management was that it de-humanised the organisational member who became a person without emotion and capable of being scientifically manipulated, just like machines'.

(*Hicks*)

Elton Mayo

4.2 Mayo (1880-1949) was pioneer of a new approach, which emphasised the importance of human attitudes, values and relationships for the efficient and effective functioning of work organisations.

This was called the 'human relations' approach. It was developed mainly by social scientists - rather than practising managers - and based on research into human behaviour, with the intention of describing and thereafter predicting behaviour in organisations. Like classical theory, it was essentially prescriptive in nature.

> 'We have failed to train students in the study of social situations; we have thought that first-class technical training was sufficient in a modern and mechanical age. As a consequence we are technically competent as no other age in history has been; and we combine this with utter social incompetence. This defect of education and administration has of recent years become a menace to the whole future of civilisation. The administrator of the future must be able to understand the human-social facts for what they actually are, unfettered by his own emotion or prejudice. He cannot achieve this ability except by careful training - a training that must include knowledge of relevant technical skills, of the systematic ordering of operations, and of the organisation of co-operation.' (*Mayo*)

4.3 The human relations approach concentrated mainly on the concept of 'Social Man' (*Schein*): man is motivated by 'social' or 'belonging' needs, which are satisfied by the social relationships he forms at work.

This emphasis resulted from a famous set of experiments (the 'Hawthorne Studies') carried out by Mayo and colleagues for the Western Electric Company in the USA. The company was using a group of girls as 'guinea pigs' to assess the affect of lighting on productivity: they were astonished to find that productivity shot up *whatever* they did with the lighting. The conclusion was that the girls' sense of being a group singled out for attention raised their morale. The next stage involved interviews, and revealed that work relationships were considered very important. A later stage of the research studied group behaviour specifically, and noted how a powerful and self-protecting informal organisation, with its own goals, rules and norms, was operating - even to the detriment of the company (for example by restricting output, falsifying reports and 'freezing out' disliked supervisors). The Hawthorne Studies are discussed in more detail later in this text.

4.4 Mayo's ideas were followed up by various social psychologists - eg Maslow, Herzberg, Likert and McGregor, whose theories on motivation and leadership we discuss elsewhere - but with a change of emphasis. People were still considered to be the crucial factor in determining organisational effectiveness, but were recognised as having more than merely physical and social needs. Attention shifted towards man's 'higher' psychological needs for growth, challenge, responsibility and self-fulfilment. Herzberg suggested that only these things could positively motivate employees to improved performance: work relationships and supervisory style, along with pay and conditions, merely ward off *dis*satisfaction (and then only temporarily).

This phase was known as the 'neo-Human Relations' school.

An appraisal of the human relations approach

4.5 The human relations approaches contributed an important awareness of the influence of the human factor at work (and particularly in the work group) on organisational performance. Most of its theorists attempted to offer guidelines to enable practising managers to satisfy and motivate employees and so (theoretically) to obtain the benefits of improved productivity.

4.6 However, the approach tends to emphasise the importance of work to the workers without really addressing the economic issues: there is still no proven link between job satisfaction and motivation, or either of these and productivity or the achievement of organisational goals.

Employee counselling (prescribed by Mayo) and, for example, job enrichment (prescribed by Herzberg) have both proved at best of unpredictable benefit to organisations applying them in practice.

5. THE SYSTEMS APPROACH

Systems theory

5.1 There is no universally accepted definition of a 'system', although Ludvig von Bertalanffy, a pioneer of general system theory in the 1930s, said it is 'an organised or complex whole' and 'organised complexity'. Alternatively, it is 'an entity which consists of interdependent parts', so that system theory is concerned with the attributes and relationships of these inter-acting parts.

5.2 General system theory makes a distinction between open and closed systems.

(a) A *closed system* is a system which is isolated from its environment and independent of it, so that no environmental influences affect the behaviour of the system (the way it operates) nor does the system exert any influence on its environment.

(b) An *open system* is a system connected to and interacting with its environment. It takes in as inputs influences (or 'energy') from its environment (and outputs from other systems) and through a series of activities, converts these inputs into outputs (or inputs into other systems). In other words it influences its environment by its behaviour. An open system is a stable system which is nevertheless continually changing or evolving. All social systems are open systems.

Inputs to the organisation include labour, finance, raw materials, components, equipment and information. Outputs include information, services provided and goods produced.

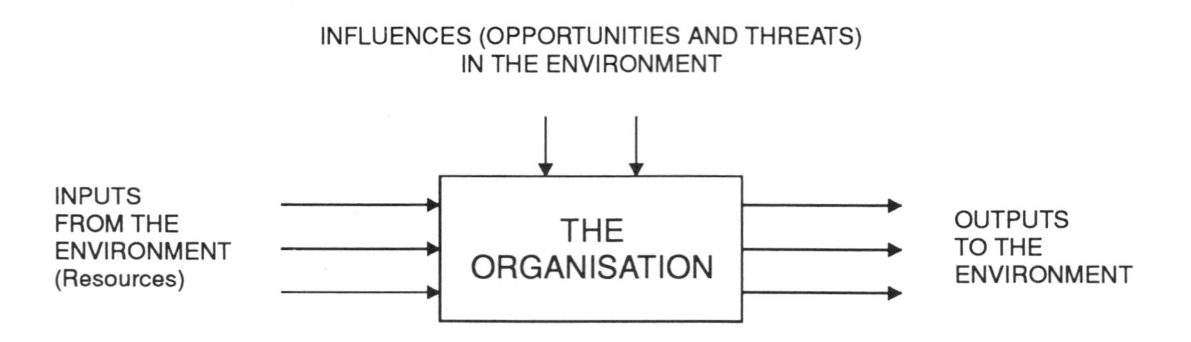

5.3 The organisation 'open system' must remain sensitive to its external environment, with which it is in constant interaction: it must respond to threats and opportunities, restrictions and challenges posed by markets, consumer trends, competitors, the government and so on. Changes in input will influence output.

5.4 A system might keep an unchanging state, or it might change. A 'homeostatic' system is one which remains static, but in order to do so, has to react to its own dynamic elements and also to a dynamic environment. It must make internal adjustments so as to remain the same. A 'dynamic' open system is one which transforms inputs from the environment so as to be continually changing (growing or shrinking).

5.5 For a business, homeostasis would not mean keeping an absolutely steady state, but would mean that the business has a 'dynamic or moving equilibrium', so that it is continually adjusting (for example to changes in customer demand or raw material supply), without necessarily growing in size or changing radically in character.

Example: socio-technical systems

5.6 An organisation is not simply a structure: the organisation chart reflects only one sub-system of the overall organisation. Trist and his associates at the Tavistock Institute have suggested that an organisation is a 'structured *sociotechnical* system', that is, it consists of at least three sub-systems:

(a) a structure;

(b) a technological system (concerning the work to be done, and the machines, tools and other facilities available to do it); and

(c) a social system (concerning the people within the organisation, the ways they think and the ways they interact with each other).

5.7 As is explained in greater detail in the next chapter, the practical application of socio-technical systems theory can be illustrated by the Durham coal-field research of Trist and Bamforth, which argued that work organisation is not wholly determined by technology but by organisational choices: the social system has properties independent of the technical system, and can be designed so as to meet technical demands and human needs. In other words, any given technical system can be operated by different social systems. The problem is to find a 'fit' that will meet technical demands and human needs. The socio-technical systems school advocated *composite autonomous group* working - 'composite' in terms of the range of skills in the group as a whole, and 'autonomous' in terms of self-determination of shifts and rotas. This proved effective in the organisational redesign of the Durham pits, and the introduction of the 'composite longwall' method of working.

The contribution of the systems approach

5.8 General systems theory can contribute to the principles and practice of management in several ways, not least by enabling managers to learn from the experience of experts and researchers in other disciplines.

(a) It draws attention to the *dynamic* aspects of organisation, and the factors influencing the growth and development of all its sub-systems.

(b) It creates an awareness of sub-systems, each with potentially conflicting goals which must be integrated. *Sub-optimisation* (where sub-systems pursue their own goals to the detriment of the system as a whole) is a feature of organisational behaviour.

(c) It focuses attention on interrelationships between aspects of the organisation, and between it and its environment, that is, on the needs of the system as a whole: management should not get so bogged down in detail and small political arenas that they lose sight of the overall objectives and processes.

(d) It teaches managers to reject the deterministic idea that A will always cause B to happen. 'Linear causality' may occur, but only rarely, because of the unpredictability and uncontrollability of many inputs.

(e) The importance of the *environment* on a system is acknowledged. One product of this may be customer orientation, which Peters and Waterman note is an important cultural element of successful, adaptive companies.

5.9 Like any other approach, managers should take what they find useful in practice in the systems view, without making a 'religion' of it.

5.10 The theory has an imaginative appeal. It is possible to draw the analogy between living systems and organisations (the 'organic analogy') and make some assumptions about how organisations are likely to behave on that basis. The analogy provides a framework for thinking about organisations and designing their structure.

5.11 However, it is only an analogy, and as such cannot be stretched too far, or provide a basis for devising testable hypotheses. It is therefore not a 'theory' at all in scientific terms. As an *approach*, it offers a useful, accessible language for discussing organisations, but - as with many behavioural frameworks - its scientific status cannot be reckoned in the same way as theories in the natural sciences, for example.

Exercise 3

Pick *one* feature of *each* of the organisation theories discussed so far in this chapter and provide an example of how it applies in your own organisation or in your own job (that is, take whatever is useful for *you* from each theory).

Solution

You were probably already doing this as you read through the sections above - for example you may have thought of an activity that you perform that used to involve 18 different tasks but which you have refined to a 5 stage process (the scientific approach).

---

6. THE CONTINGENCY APPROACH

5.10 The contingency approach to organisation developed as a reaction to prescriptive ideas of the classical and human relations schools, which claimed to offer a universal 'best way' to design organisations, to motivate staff, to introduce technology and so on. Research by Burns and Stalker, Joan Woodward, Lawrence and Lorsch and others indicated that different forms of organisational structure could be equally successful, that there was no inevitable correlation between classical organisational structures and effectiveness, and that there were a number of variables to be considered in the design of organisations. Essentially, 'it all depends'.

6.2 The emerging contingency school rejected the universal 'one-best-way' approach, in favour of analysis of the internal factors and external environment of each organisation and the design of organisational structure as a best fit between the tasks, people and environment *in the particular situation*. As Buchanan and Huczynski put it: 'With the coming of contingency theory, organisational design ceased to be "off-the-shelf", but became tailored to the particular and specific needs of an organisation.'

6.3 The reaction against the universality of management/organisation principles was founded on sound research evidence. Research by Blain in 1964 showed that the principles advanced by the classical management school - for example, small span of control and unity of command - did not necessarily correlate with organisational effectiveness.

6.4 Lawrence and Lorsch compared the structural characteristics of a 'high-performing' container firm, which existed in a relatively stable environment, and a 'high-performing' plastics firm which existed in a rapidly changing environment. They concluded that in a stable environment the most efficient structure was one in which the influence and authority of senior managers was high and of middle managers low: the converse was true of the dynamic environment firm.

6.5 Joan Woodward's research with firms in Essex highlighted the importance of technology as a major factor contributing to variances in organisation structure: 'It appeared that different technologies imposed different kinds of demands on individuals and organisations and that these demands have to be met through an appropriate form of organisation.'

6.6 However, two points should be noted.

(a) John Child, among others, suggests that 'One major limitation of the contemporary contingency approach lies in the lack of conclusive evidence to demonstrate that matching organisational designs to prevailing contingencies contributes *importantly* to performance.'

(b) 'Contingency theory' is again not properly a 'theory' at all. It is more of a 'philosophy' or 'approach' - a way of thinking about organisation design - than a theory, which implies causal elements about which hypotheses can be derived and tested empirically.

6.7 According to Tom Lupton: 'It is of great practical significance whether one kind of managerial "style" or procedure for arriving at decisions, or one kind of organisational structure, is suitable for all organisations, or whether the managers in each organisation have to find that expedient that will best meet the particular circumstances of size, technology, competitive situation and so on.'

Awareness of the contingency approach will therefore be of value in:

(a) encouraging managers to identify and define the particular circumstances of the situation they need to manage, and to devise appropriate ways of handling them. A belief in universal principles and prescriptive theories can hinder problem-solving and decision-making by obscuring some of the available alternatives. It can also dull the ability to evaluate and choose between alternatives that *are* clearly open, by preventing the manager from developing relevant criteria for judgement;

(b) encouraging responsiveness and flexibility to changes in environmental factors through organisational structure and culture. Task performance and individual/group satisfaction are more important design criteria than permanence and unity of design type. Within a single organisation, there may be bureaucratic units side by side with task-centred 'matrix' units (for example in the research and development function) which can respond to particular pressures and environmental volatility.

Exercise 4

In an article in the *Administrator* in April 1992 Colin Coulson-Thomas draws upon a series of surveys of larger organisations to support his argument that managers are not being equipped with the skills needed to bring about the transformations that are required in organisations operating in the 1990s. The article includes a series of tables which very effectively summarise current organisational trends and these are reproduced below.

Table 1: Business environment statements (in order of 'strongly agree' replies).

	%
Human resource is a critical success factor	74
Customers are becoming more demanding	50
The rate of change is speeding up	47
Environmental and social pressures are increasing	43
Information can be processed and transferred more quickly	41
Markets are becoming more open and competitive	40
More markets are becoming global	31
Human resource is a limiting constraint	24
There is greater sensitivity to demographic trends	9

Table 2: What respondents' organisations are doing to better respond to challenges and opportunities within the business environment.

	%
Creating a slimmer and flatter organisation	88
More work is being undertaken in teams	79
Creating a more responsive network organisation	78
Functions are becoming more inter-dependent	71
Procedures and permanency are giving way to flexibility and temporary arrangements	67
Organisations are becoming more inter-dependent	55

Table 3: Change requirements (in order of 'very important' replies).

	%
Clear vision and strategy	86
Top management commitment	86
Sharing the vision	71
Employee involvement and commitment	65
Communicating the purpose of change	65
An effective communications network	54
Communicating the expected results of change	44
Understanding the contributions required to the achievement of change	42
Communicating the timing of change	38
Linking a company's systems strategy with its management of change	38
Project management of change	27
Ongoing management education and development programmes	23
One off management education and development programmes	8

Table 4: Management qualities (in order of 'very important' and 'important' replies).

	%
Understanding of business environment	100
Adaptability	99
Ability to communicate	98
Flexibility	98
A balanced perspective	98
Broad perspective on the organisation's goals	95
Ability to handle uncertainty and surprise	94
Commitment to on-going learning	92
Ability to contribute to teams	91
Awareness of ethics and values	90
Ability to assume greater responsibility	90
Mobility	62
Multi-skills	60
Specialist expertise	55
Tolerance of ambiguity	50

To what extent do you think that the organisational issues outlined in this chapter are adequate to explain the current trends?

(*Note.* This is a difficult exercise to do well and thoroughly at this early stage, but it is worth attempting because it will help you to become familiar with the contents of this chapter, it anticipates matters dealt with in more detail in later chapters and it will enhance your awareness of current issues.)

Chapter roundup

- Organisational theories are models of the way organisations actually behave. They arose out of the study of management for its own sake.
- Scientific management divided an organisation into a managerial elite and a workforce whose activities were planned to rigid detail.
- The classical school (Fayol) prescribed certain basic rules for the rational behaviour of organisations.
- The human relations approach emphasised the value of individuals and human motivation in the workplace.
- The systems school regards the organisation as a dynamic entity.
- The contingency school would hold out there is no golden rule by which organisations should be run.

Test your knowledge

1 What were Taylor's four principles of scientific management? (see para 2.3)

2 What are the main drawbacks of scientific management? (2.9, 4.1)

3 List some of the principles of the classical school. (3.2)

4 Describe the human relations approach (4.2 - 4.3)

5 What is the contribution of the systems approach? (5.11)

6 What is the rationale behind the contingency school? (6.3 -6.5)

7 What approach was propounded by Elton Mayo? (4.2)

Now try question 2 at the end of the text

Chapter 3

THE STRUCTURE AND CULTURES OF ORGANISATIONS

This chapter covers the following topics.

1. The design of organisations
2. Centralisation and decentralisation
3. Line and staff management
4. Span of control
5. Departmentation
6. Matrix organisation
7. Social and technological factors
8. Change and the adaptive organisation
9. Culture

Signpost

- The distinction between line and staff management is further developed in Chapter 7. The impact of technology is also described in the next chapter. Labour flexibility is a topic in Chapter 8.
- The themes of change and of organisation culture are relevant throughout this Study Text.

1. THE DESIGN OF ORGANISATIONS

1.1 In this chapter we shall briefly look at *formal* organisation structure and consider a variety of views of how this structure might be established so as to optimise the efficiency of the organisation.

1.2 Organisational design or structure implies a framework or mechanism intended to:

(a) link individuals in an established network of relationships so that authority, responsibility and communications can be controlled;

(b) group together (in any appropriate way) the tasks required to fulfil the objectives of the organisation, and allocate them to suitable individuals or groups;

(c) give each individual or group the authority required to perform the allocated functions, while controlling behaviour and resources in the interests of the organisation as a whole;

(d) co-ordinate the objectives and activities of separate units, so that overall aims are achieved without gaps or overlaps in the flow of work required;

(e) facilitate the flow of work, information and other resources required, through planning, control and other systems.

1.3 Many factors influence the structural design of the organisation.

(a) Its *size*. As an organisation gets larger, its structure gets more complex: specialisation and subdivision are required. The process of controlling and co-ordinating performance, and communication between individuals, also grows more difficult as the 'top' of the organisation gets further from the 'bottom', with more intervening levels. The more members there are, the more potential there is for interpersonal relationships and the development of the *informal* organisation.

(b) Its *task*, that is the nature of its work. Structure is shaped by the division of work into functions and individual tasks, and how these tasks relate to each other. Depending on the nature of the work, this can be done in a number of ways. The complexity and importance of tasks will affect the amount of supervision required, and so the ratio of supervisors to workers. The nature of the market will dictate the way in which tasks are grouped together: into functions, or sales territories, or types of customer;

(c) Its *staff*. The skills and abilities of staff will determine how the work is structured and the degree of autonomy or supervision required. Staff aspirations and expectations may also influence job design, and the amount of delegation in the organisation, in order to provide job satisfaction.

(d) Its legal, commercial, technical and social *environment*. Examples include: economic recession necessitating staff streamlining especially at middle management level; market pressures in the financial services sector encouraging a greater concentration of staff in specialised areas and at the bank/customer interface; new technology reducing staff requirements but increasing specialisation.

(e) Its *age*, that is the time it has had to develop and grow, or decline, whether it is very set in its ways and traditional, or experimenting with new ways of doing things and making decisions.

(f) Its *culture and management style* - how willing management is to delegate authority at all levels, how skilled they are in organisation and communication (for example in handling a wider span of control), whether teamwork is favoured, or large, impersonal structures are accepted by the staff.

Exercise 1

Consider how each of the factors listed in paragraph 1.3 might affect the structural design of a service organisation (for example, a bank) and a manufacturing organisation (for example, a cement manufacturer).

Solution

Just taking elements of the first two factors as an example, a small bank may have just one office located in the area where it does its business - probably the City of London - and employ fairly specialised autonomous staff. A large bank would need a network of branches and regional offices as well as a central HQ and would employ a larger proportion of relatively unskilled workers, with greater supervision.

A cement manufacturer would need to have its production facilities located on top of the natural resources used. Larger organisations might have administrative offices and distribution depots elsewhere. The distinction between 'productive' workers and administrative workers would be much more marked than in a bank, and these differences would be accentuated the larger the organisation was.

---

2. CENTRALISATION AND DECENTRALISATION

2.1 Centralisation and decentralisation refer to the degree to which authority is delegated in an organisation - and therefore the *level* at which decisions are taken in the management hierarchy.

2.2 'From decentralisation we get initiative, responsibility, development of personnel, decisions close to the facts, flexibility - in short, all the qualities necessary for an organisation to adapt to new conditions. From co-ordination we get efficiencies and economies. It must be apparent that co-ordinated decentralisation is not an easy concept to apply'. (*A. Sloan*)

2.3 Whatever system is set up, it is of paramount importance that all managers at all levels should clearly know where they fit into the organisation. They should know the nature and extent of their authority and responsibility and that of fellow managers at all levels. Managers can then exercise as much authority and carry as much responsibility as possible within the constraints of the policies set by the organisation and the commitments they have made to their own superior executive.

2.4 Arguments in favour of centralisation and decentralisation

	Pro centralisation		*Pro decentralisation/delegation*
1	Decisions are made at one point and so easier to co-ordinate.	1	Avoids overburdening top managers, in terms of workload and stress.
2	Senior managers in an organisation can take a wider view of problems and consequences.	2	Improves motivation of more junior managers who are given responsibility since job challenge and entrepreneurial skills are highly valued in today's work environment.
3	Senior management can keep a proper balance between different departments or functions - eg by deciding on the resources to allocate to each.	3	Greater awareness of local problems by decision makers. Geographically dispersed organisations should often be decentralised on a regional/area basis.
4	Quality of decisions is (theoretically) higher due to senior managers' skills and experience.	4	Greater speed of decision making, and response to changing events, since no need to refer decisions upwards This is valued by customers and is particularly important in rapidly changing markets.
5	Possibly cheaper, by reducing number of managers needed and so lower cost of overheads.	5	Helps junior managers to develop and helps the process of transition from functional to general management. Centralised organisations tend to fill senior positions from external sources.
6	Crisis decisions are taken more quickly at the centre, without need to refer back, get authority etc.	6	Separate spheres of responsibility can be identified: controls, performance measurement and accountability are better.
7	Policies, procedures and documentation can be standardised organisation-wide.	7	Communication technology allows decisions to be made locally, with information and input from head office if required.

Exercise 2

In December 1992 Christopher Lorenz wrote in the *Financial Times* about the modern management trend of 'federalism', which according to Charles Handy implies that power in an organisation should reside fundamentally with the constituent parts, even though some is ceded to the centre for the benefit of all. Lorenz finds the 'federal' metaphor limited for a number of reasons, including the following.

'as business conditions alter, the balance of power [can] shift informally between the central authority and the units. It must be able to do so, within outline parameters, if a company is to be capable consistently of striking the optimum balance between integration and autonomy.

In business, power-sharing with the centre will need to vary not only over time, but also as between business units, depending on their changing competitive, technological and financial circumstances.

Even within a business unit, different value-adding activities - such as design, development, purchasing, production and marketing - will need, over time, to have different degrees of international integration or national differentiation (and semi-independence)'.

Does this remind you of a theory encountered in an earlier chapter?

Solution

The contingency approach argues that there is no universally appropriate system for all organisations: systems should be matched to circumstances. Lorenz's stance on decentralisation is similar.

3. LINE AND STAFF MANAGEMENT

3.1 'Line' and 'staff' can be used to denote functions in the organisation.

(a) *Line* management consists of those managers directly involved in achieving the objectives of an organisation (for example all production and sales managers).

(b) Every other manager is a *staff* manager (for example accounting, personnel, research and development). Staff activities are those which primarily exist to provide advice and service.'

The terms are also used to denote authority relationships. A manager (even a personnel - staff - manager) has 'line' authority over his own subordinates: authority passed down the chain of command. 'Staff' authority depends on persuasion and 'expert' power, and crosses departmental boundaries.

3.2 Accountants, personnel administrators, economists, data processing experts and statisticians are all experts in a specialised field of work. Where this expertise is 'syphoned off' into a separate department, the problem naturally arises as to whether:

(a) the experts *advise* line managers, who may accept or reject the advice given; or
(b) the experts can step in to *direct* the line managers in what to do.

3.3 Staff authority is now usually considered to be purely advisory: the organisation and methods department, for example, or the 'personal assistant' are such roles.

The term *functional authority* is a step further in the recognition that some 'staff' management has become highly specialised in areas of work which form a fundamental part of line management. This expert power becomes formally recognised in the organisation/management structure, merging line and staff authority: the expert has formally delegated to him the authority to influence specific areas (those in which he is a specialist) of the work of other departments and their managers.

3.4 This is a move towards dual authority, since the line manager retains ultimate authority for the functioning of his department. For this reason, and to avoid complex political problems, functional authority is usually exercised through the establishment of systems, procedures and standards, rather than by on-going direct intervention on the part of functional specialists.

3.5 There are drawbacks to using staff departments.

(a) There is a danger that staff experts may, intentionally or not, undermine the authority of line managers.

(b) Friction may also occur when staff managers report on the line departments' performance, over the heads of line managers, to the managing director, say.

(c) Staff managers have no line authority and are therefore not accountable for the effect of their advice.

3.6 The solutions to these problems are easily stated, but not easy to implement in practice.

(a) Authority must be clearly defined, and distinctions between line, staff and functional authority clearly set out (eg in job descriptions).

(b) The use of experts should become part of the organisational culture - with emphasis on the building of multi-disciplinary teams if possible.

(c) Staff managers must be fully informed about the operational aspects of the business, so they will be less likely to offer impractical advice.

4. SPAN OF CONTROL

4.1 Span of control or 'span of management', refers to the number of subordinates responsible to a superior. In other words, if a manager has five subordinates, the span of control is five.

4.2 Various classical theorists suggest that:

(a) there are physical and mental limitations to any given manager's ability to control people, relationships and activities; but that

(b) there needs to be tight managerial control from the top of an organisation downward.

The span of control should therefore, they argued, be restricted, to allow maximum control consistent with the manager's capabilities: usually between three and six.

If the span of control is too wide, too much of the manager's time will be taken up with routine problems and supervision, leaving less time for planning: even so, subordinates may not get the supervision, control, communication and so on that they require.

4.3 On the other hand, if the span is too narrow, the manager may fail to delegate, keeping too much routine work to himself and depriving subordinates of decision-making authority and responsibility. There may be a tendency to interfere in or over-supervise the work that is delegated to subordinates - and the relative costs of supervision will thus be unnecessarily high. Subordinates tend to be dissatisfied in such situations, having too little challenge and responsibility and perhaps feeling that the superior does not trust them.

4.4 *James Worthy* reported in 1950 that the policy of the American Sears Roebuck company was to have as wide a span of control as possible between stores managers and their subordinates, the merchandising managers. A wide span of control forced stores managers to delegate authority: the consequences, Worthy claimed, were improved morale and greater efficiency of merchandising management.

4.5 *Lyndall Urwick* put forward a counter-argument that a wide span of control had been possible in this example because the work of the merchandising managers did not interlock, and so the need for co-ordination and integration was not present: this reduced the burdens of supervision and made a wider span of control feasible. Urwick concluded that the maximum management span of control should be six, when the work of subordinates interlocks.

4.6 The appropriate span of control will depend on various factors.

(a) *Ability of the manager*. A good organiser and communicator will be able to control a larger number. The manager's work-load will also be relevant, as will be his ability to handle interruptions to his own work.

(b) *Ability of the subordinates*. The more able, intelligent, trustworthy and well-trained subordinates are, the easier it is to control large numbers, as they need less support and advice to work on their own.

(c) *Nature of the task*. It is easier for a supervisor to control a large number of people if they are all doing routine, repetitive or similar tasks. Where closer control is necessary, the span would have to be smaller. The technology of the task may also make control easier (if the task is highly automated) or more necessary (if it is complex and specialised).

(d) The *geographical dispersal* or grouping of the subordinates, and the communication system of the organisation. These may or may not facilitate control.

Tall and flat organisations

4.7 The span of control has implications for the 'shape' of the organisation. An organisation with a narrow span of control will have more levels in its management hierarchy than an organisation of the same size with a wide span of control: the first organisation will be narrow and *tall*, while the second will be wide and *flat*.

In fact, it is the flat organisation that reflects a greater degree of delegation in the structure: the more a manager delegates, the wider his span of control can be, and authority is shared among more people, further down the organisation. A tall organisation reflects tighter supervision and control, and lengthy chains of command and communication.

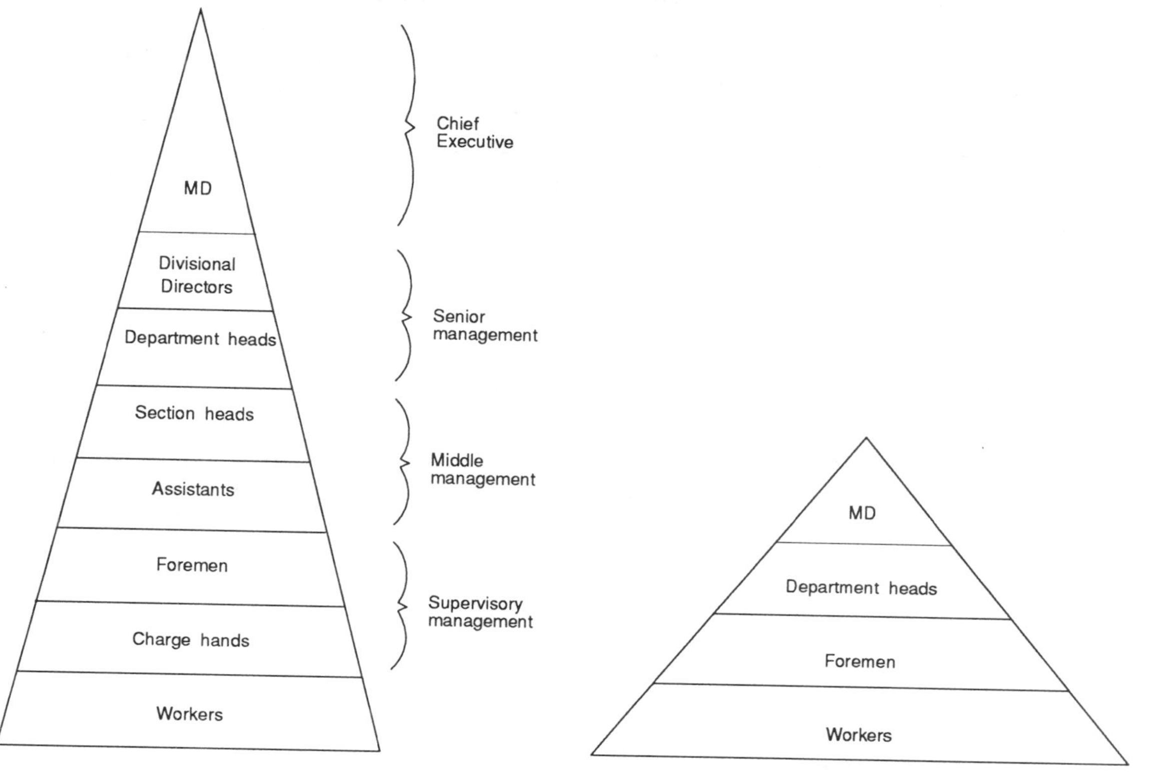

4.8 Some classical theorists accepted that a tall organisation structure is inefficient, because:

(a) it increases overhead costs;

(b) it creates extra communication problems, since top management is more remote from the work done at the bottom end of the organisation, and information tends to get distorted or blocked on its way up or down through the organisation hierarchy;

(c) management responsibilities tend to overlap and become confused as the size of the management structure gets larger. Different sections or departments may seek authority over the same territory of operations, and superiors find it difficult to delegate sufficient authority to satisfy subordinates;

(d) the same work passes through too many hands;

(e) planning is more difficult because it must be organised at more levels in the organisation.

4.9 Behavioural theorists add that tall structures impose rigid supervision and control and therefore block initiative and ruin the motivation of subordinates. There may be more 'rungs' available in the promotional ladder, but there are unlikely to be real increases in responsibility between one and another.

4.10 Nevertheless, not all researchers favour flat organisation structures. Carzo and Yanouzas suggested that if work is organised on the basis of small groups or project teams (therefore narrow spans of control and a tall organisation structure) group members would be able to plan their work in an orderly manner, encourage participation by all group members in decision-making and monitor the consequences of their decisions better, so that their performance will be more efficient than the work of groups in a flat structure with a wide span of control. As mentioned earlier, Peters and Waterman also observe the success of 'chunking' - breaking the organisation into small, task-centred units.

Empowerment

4.11 Flat organisations are becoming more common. This is partly because the information processing traditionally done by middle management can be just as effectively carried out by computer, and partly due to the modern emphasis on quality. An article by Max Hand (*Management accounting*, January 1991) refers to the place of *empowerment* in the quality control process: 'the people lower down the organisation possess the knowledge of what is going wrong within a process but lack the authority to make changes. Those further up the structure have the authority to make changes but lack the profound knowledge required to identify the right solutions. The only solution is to change the culture of the organisation so that everyone can become involved in the process of improvement and work together to make the changes'. This approach risks conflicts and requires discipline. It may mean the removal of whole layers of middle management. If it can be achieved, however, the results can be dramatic.

4.12 Empowerment has two key aspects.

(a) Allowing workers to have the freedom to decide how to do the necessary work, using the skills they possess and acquiring new skills as necessary to be an effective team member.

(b) Making those workers personally responsible for achieving production targets and for quality control. (The French word for empowerment is 'responsibilisation'.)

Exercise 3

(a) What areas of your own organisation's work do you think would benefit from centralisation or decentralisation?

(b) Identify the line managers and the staff managers in your own organisation.

(c) Is you own organisation tall or flat? Do you have a span of control? If so do you think it is too wide, too narrow or about right?

5. DEPARTMENTATION

5.1 The grouping of organisational activities (usually into 'departments' or larger 'divisions') can be done in different ways.

(a) *By numbers*: when menial tasks are carried out by large numbers of workers, supervision can be divided by organising the labourers into gangs of equal size. Departmentation by numbers alone is rare; an example might be the organisation of a conscript army of infantrymen into divisions, and battalions.

(b) *By shifts*: with shift-working employees organised on the basis of 'time of day'.

(c) *By function*: this is a widely-used method of organisation. Primary functions in a manufacturing company might be production, sales, finance, and general administration. Sub-departments of the production function might be manufacturing (machining, finishing, assembly etc), production control, quality control, servicing and purchasing. Sub-departments of sales might be selling, marketing, distribution and warehousing.

Functional organisation is logical and traditional and accommodates the division of work into specialist areas. Apart from the problems which may arise when 'line' management resents interference by 'staff' advisors in their functional area, the drawback to functional organisation is simply that more efficient structures might exist which would be more appropriate in a particular situation.

(d) *By territory*: this method of organisation occurs when similar activities are carried out in widely different locations. Water and electricity services, for example, operate region by region. Within many sales departments, the sales staff are organised territorially.

The *advantage of territorial departmentation* is better local decision-making at the point of contact between the organisation (eg a salesman) and its customers. Localised knowledge is put to better use, 'relationship' management is easier at local level in personal service sectors (like branch banking) and in the right circumstances it may be less costly to establish area factories/offices than to control everything through Head Office (costs of transportation and travelling may be less).

The *disadvantage of territorial departmentation* might be the duplication of management effort, increasing overhead costs and the risk of disintegration. For example, a national organisation divided into ten regions might have a customer liaison department at each regional office.

(e) *By product*: some organisations group activities on the basis of products or product lines. Some functional division of responsibility remains, but under the overall control of a manager with responsibility for the product, product line or brand, with authority over the personnel of different functions involved in its production, marketing and so on.

Advantages of product departmentation are that:

(i) individual managers can be held accountable for the *profitability* of individual products;

(ii) specialisation can be developed. For example, salesmen and/or engineers will be trained to sell and/or service a specific product in which they may develop technical expertise and thereby offer a better sales and after-sales service to customers;

(iii) the different functional activities and efforts required to make and sell each product can be co-ordinated and integrated by the divisional/product manager.

The disadvantage of product departmentation is that it increases the overhead costs and managerial complexity of the organisation and involves a certain amount of the ambiguity of 'dual authority'.

(f) *By customer or market segment*: a manufacturing organisation may sell goods through wholesalers, export agents and by direct mail. It may therefore organise its functions (particularly marketing) on the basis of types of customer, market segment or distribution channels. Banks, for example, are starting to reorganise around market segments - domestic and foreign, personal, small business, corporate, 'high net worth' and so on.

The *advantages of market-oriented organisations* are that:

(i) they encourage efficient marketing techniques (especially 'cross-selling' of products to appropriate customers of other products) and PR;

(ii) they promote a culture whose values are heavily customer-oriented, which may be a strong motivating force (Peters and Waterman).

The *disadvantages of market-orientation* may be that it requires special leadership (a 'bureaucratic' management would fail to achieve the required inter-relationships between the organisation and its customers) and is likely to be costly in terms of staffing and other overheads.

(g) *By equipment specialisation*: the most obvious example of departmentation based on equipment specialisation is provided by the data processing departments of large organisations. Batch processing operations are conducted for other departments at a computer centre (where it is controlled by DP staff) because it would be uneconomical to provide each functional department with its own large mainframe computer.

Example: departmentation

5.2 Say you work for an organisation which manufactures a range of cars and buses. These are both manufactured and sold in Europe and Asia. The company has its head office, and most of its research facilities, in the UK. It is a UK registered company, and has no shareholders outside the UK. The company employs 20,000 people. If you work for a large organisation, you will be familiar with the fact that some people work in marketing and selling, and others work in production, and still more in finance and administration.

5.3 Such a company is likely to be quite complex. There is bound to be a central committee of top executives in charge of the overall running of the business, but how about the rest of the business? There are a number of ways by which it could be organised.

(a) Would the business be structured geographically, like this?

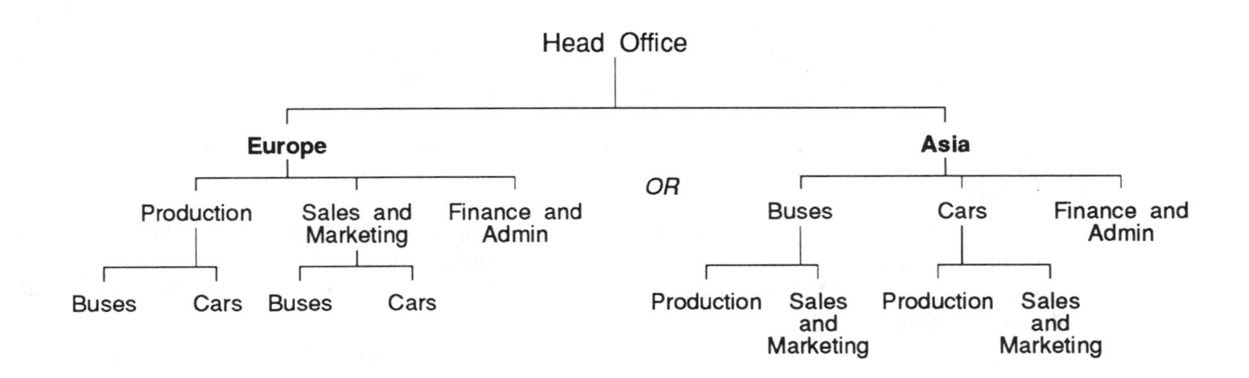

In this *geographical organisation* structure the responsibility for all the activities of the company is divided on an area basis. The manager for Asia is in charge of producing and selling products in that area.

(b) Would the business be structured on a product divisional basis like this?

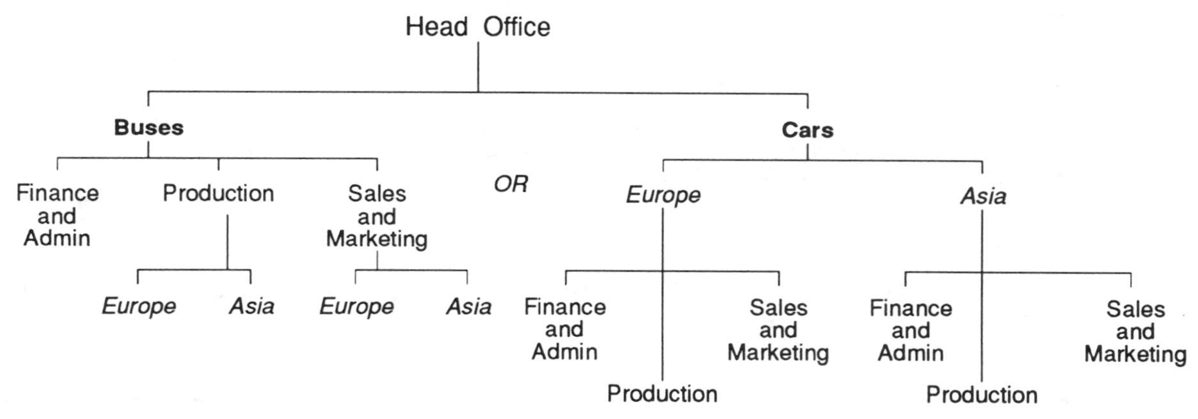

In this *product division structure* the responsibilities for buses worldwide and cars worldwide are given to one individual.

(c) Or would the business be structured on a functional basis like this?

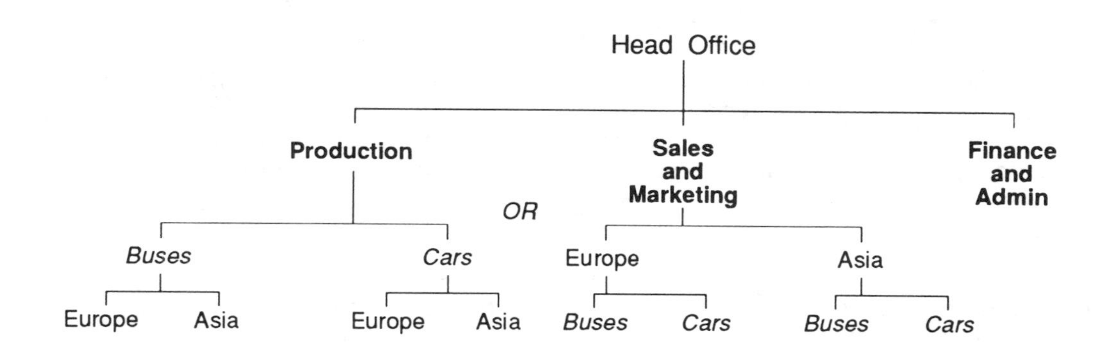

In this *functional organisation structure*, worldwide control of production is vested in one person, as is the case with sales and marketing.

5.4 Note that in each case there can be different variations in the structure lower down the organisation. Production, for example, in a functional organisation structure (diagram (c)) could *either* be structured on a production division basis *or* a geographical basis.

Exercise 4

'A year ago Paul Allaire, [Xerox's] chairman, unveiled a radical redesign of the company's management structure, or what he calls its "organisational architecture", in an attempt to solve one of the central problems facing large companies everywhere: how to retain the entrepreneurial flair and flexibility of a small business.

The document processing business was organised as a classic monolithic company, divided into departments such as sales and manufacture and with a large central staff. But this, he says, means that, "people get too accustomed to sitting on their hands and waiting for a decision to come down from above."

Under the new system, Xerox consists of nine separate stand-alone businesses, each organised around particular products and markets and each with total responsibility for their profit and loss account.

Allaire likes to describe this as turning the traditional, vertically organised business on its side: at one end, there is an integrated research and development arm, which feeds ideas into the nine businesses.

These in turn supply products to a unified customer operations division, which sells, installs and services equipment and is organised by geographic region.

The sales force was not divided up among the nine businesses because Xerox felt it important to keep a single face to the customer, and did not want to run the risk of salespeople from different operations competing with each other.'

(*Financial Times,* March 1993)

Required

Draw organisation charts showing Xerox's old structure and new structure, as you interpret them from this extract.

---

6. MATRIX ORGANISATION

6.1 In recent years, the awareness of internal and external influences on organisational structure and operation has contributed to a new emphasis on flexibility and adaptability in organisational design, particularly since the pace of the change in the technological and competitive environment has put pressure on businesses to innovate, to adopt a market orientation.

Part of this shift in emphasis has been a trend towards task-centred structures, such as multi-disciplinary project teams, which draw experience, knowledge and expertise together from different functions to facilitate flexibility and innovation. In particular, the concept of 'matrix' organisation has emerged, dividing authority between functional managers and product or project managers or co-ordinators - thus challenging classical assumptions about 'one man one boss' and the line/staff dilemma.

6.2 Matrix management first developed in the 1950s in the USA in the aerospace industry. Lockheed-California, the aircraft manufacturers, were organised in a functional hierarchy. Customers were unable to find a manager in Lockheed to whom they could take their problems and queries about their particular orders, and Lockheed found it necessary to employ 'project expediters' as customer liaison officials. From this developed 'project co-ordinators', responsible for co-ordinating line managers into solving a customer's problems. Up to this point, these new officials had no functional responsibilities.

With increasingly heavy customer demand, Lockheed eventually created 'programme managers', with full authority for project budgets and programme design and scheduling. This dual authority structure may be shown diagramatically as a management *grid*; for example:

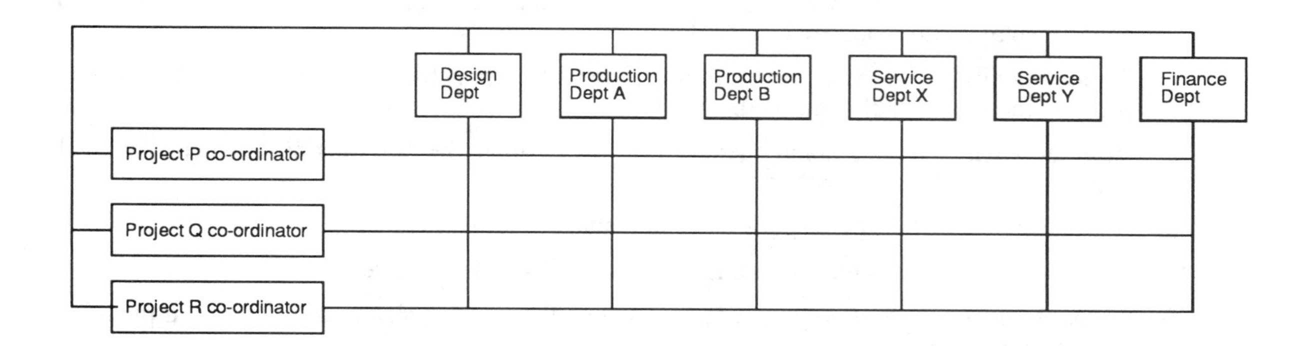

Functional department heads are responsible for the internal organisation of their departments, but project co-ordinators are responsible for the aspects of all departmental activity that affects their particular project.

6.3 The authority of product or project managers may vary from organisation to organisation. J K Galbraith drew up a range of alternative situations, as shown.

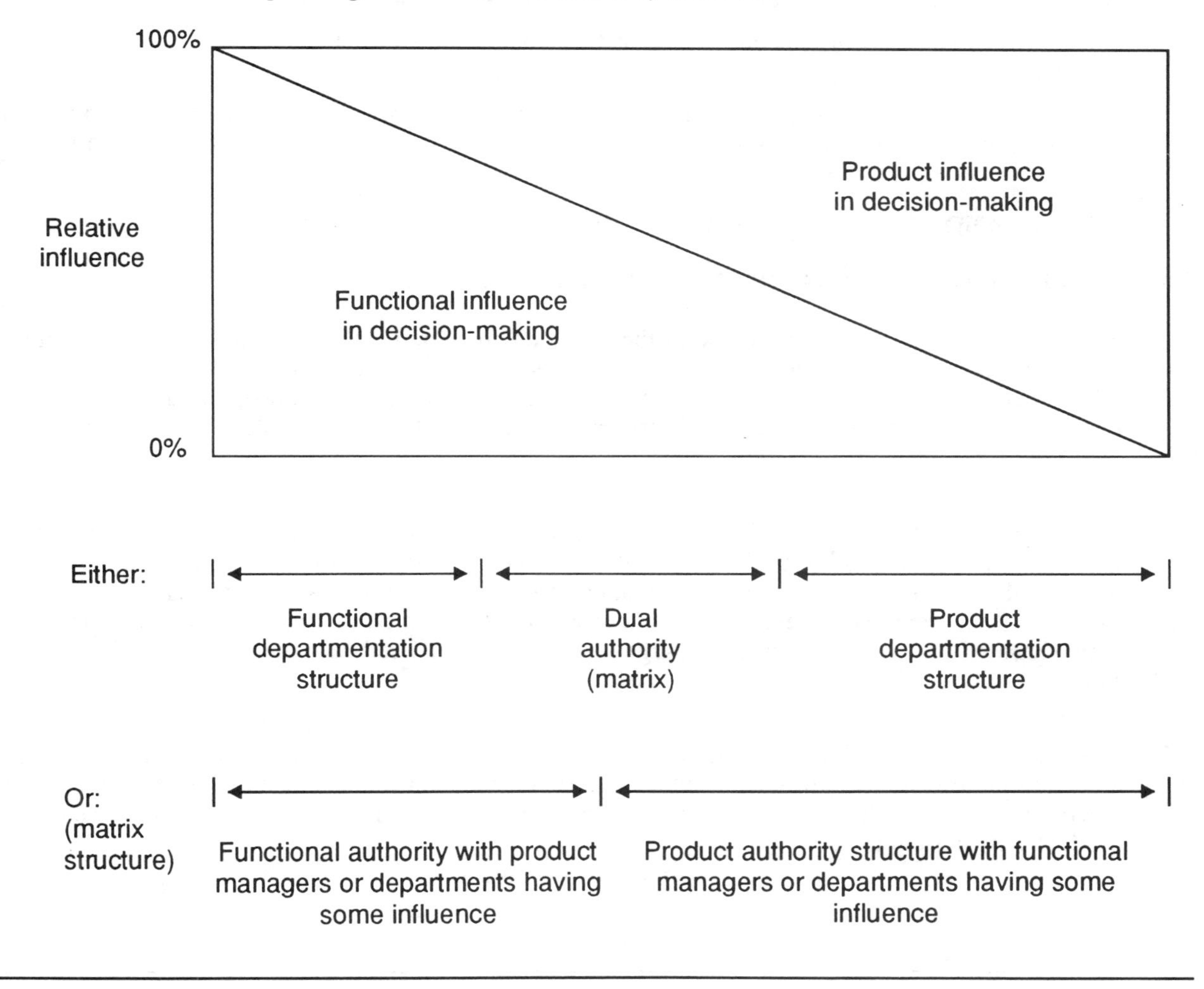

6.4 Matrix management thus challenges classical ideas about organisation in two ways:

(a) it rejects the idea of one man, one boss;
(b) its subverts the bureaucratic ethic of authority based on status in the formal hierarchy.

6.5 The *advantages* of a matrix structure are said to be as follows.

(a) Greater flexibility of people. Employees develop an attitude geared to accepting change, and departmental monopolies are broken down

(b) Greater flexibility of tasks and structure. The matrix structure may be short-term (as with project teams) or readily amended.

(c) Re-orientation. A functional department will often be production-oriented: product management will create a market orientation.

(d) A structure for allocating responsibility to managers for end-results.

(e) Inter-disciplinary co-operation and a mixing of skills and expertise.

(f) Arguably, motivation of employees by providing them with greater participation in planning and control decisions.

6.6 The *disadvantages* of matrix organisation are said to be as follows.

(a) Dual authority threatens a conflict between functional managers and product/project managers. Where matrix structure exists it is important that the authority of superiors should not overlap and areas of authority must be clearly defined. A subordinate must know to which superior he is responsible for a particular aspect of his duties.

(b) One individual with two or more bosses is more likely to suffer stress at work.

(c) Matrix management can be more costly - product management posts are added, meetings have to be held, and so on.

(d) It may be difficult for the management of an organisation to accept a matrix structure and the culture of participation, to share authority and live with the ambiguity that it fosters.

7. SOCIAL AND TECHNOLOGICAL FACTORS

The socio-technical systems approach to organisation

7.1 This approach was discussed briefly in the previous chapter. It takes the view that organisation structure (along with the work done by an individual, and the way in which he can relate to fellow-workers) tends to be influenced by the technology of the operations. Organisations are complex social systems, but must be seen in the context of their technology - they are *socio-technical systems*.

7.2 The originators of socio-technical systems theory are reputedly Eric Trist and his colleagues at the Tavistock Institute who (in the early 1950s) studied the effects of technological change on the morale of workers. They based their studies on the coal mines where, owing to increased mechanisation, the miners were merged into larger and larger groups, found it difficult to identify their share of the work within the enlarged groups and suffered low morale: productivity was falling.

7.3 Trist realised that productivity was linked to job satisfaction and the ability of the miner to associate his own extra effort with extra reward. When he introduced a method of organisation where the large shifts were sub-divided into small, identifiable units, morale improved and productivity increased.

7.4 Trist also believed that the primary task of management was to relate the organisation to its environment. Management must realise that an organisation is a conversion process that converts units of input from the environment into units of output desired by the consumer in that environment. The market is constantly changing and management should monitor these changes and react to them.

7.5 Trist has consistently developed a systems approach to organisations, in which task requirements and individuals' needs are inter-related as an interdependent 'socio-technical' system.

'It is difficult to see how these (organisation) problems can be solved efficiently without restoring responsible autonomy to *primary groups* throughout the system and ensuring that each of these groups has a satisfying sub-whole as its work task, and some scope for flexibility in work-pace. Only if this is done will the stress of the deputy's role be reduced and his task of maintaining the cycle receive spontaneous support from the primary work groups.' (*Trist and Bamforth*)

The research work of Trist

7.6 The Coal Board had been trying to introduce new mechanical processes into coal mining in order to increase productivity, but the innovation provoked severe industrial unrest. Trist and his colleagues were invited to study the problem and to come up with a solution.

Prior to the technical innovations by the Coal Board, miners had been used to working in small autonomous groups or teams. Each team had its own place at the coal seam and was responsible for hewing coal with a pick or drill, loading it into tubs for transportation out of the mine and propping up the roof as they advanced. Each miner in the team was an all-rounder and did not have to specialise in any single activity within the overall task of the group. Each team was paid as a group and the pay was shared out equally between its members.

7.7 The Coal Board decided to change the work organisation in order to introduce new coal-cutting equipment capable of cutting a long stretch of wall at a time. Their new organisation (known as the *conventional longwall* system) divided the mining work into three separate tasks or shifts.

(a) One group of miners did the cutting.

(b) A second group of miners loaded the loose coal onto a moving conveyor which took it away from the coal face.

(c) A third group moved the coal cutting equipment and conveyor forward, and propped up the roof.

7.8 The new arrangements proved unsuccessful. Within each specialised task there were some miners more willing and more able than others to carry out the specialised work. There were also problems in co-ordinating the work of the three different groups or shifts. As a result, it was found that closer supervision was required by management to ensure that the work was done properly and that every individual did his fair share.

7.9 Trist and his colleagues suggested that close managerial supervision was unsuitable to mining work, which was carried out in dangerous conditions.

They agreed that the technical equipment used in operations must influence the type of work organisation for employees: in the case of coal mining, the new cutting equipment and conveyor belts made working in small groups no longer practicable. However, a work group has social and psychological properties and the work organisation should not be arranged in such a way that the advantages of technological improvements are offset by employee resistance and unrest.

7.10 Trist et al argued that an organisation is a socio-technological system which must attempt to balance the work organisation's:

(a) economic advantages;
(b) technological advantages; and
(c) social and psychological advantages.

It is not sufficient to introduce the most up-to-date technology if a cheaper older technology is available or if the new work organisation will create serious employee unrest. Nor is it sufficient to create worker satisfaction if this entails inefficiency and uneconomic working which could be improved by better technological equipment.

7.11 Their solution to the coal mining problem was to recommend a *composite longwall* method of working. Under this method the new technology was retained, but the workforce was no longer divided for three separate tasks. The team as a whole was given the responsibility for the whole task and for assigning individuals to particular jobs. By this means, the work group was given autonomy, self-regulation, multi-skilled roles and a complete task to perform. As far as possible, the social conditions of the traditional system of mining were restored, while the 'three-task' group was defined as the new primary work group for coal mining operations.

Joan Woodward (1916 - 1971)

7.12 Joan Woodward developed the socio-technical systems approach with extensive research into differing types of organisations. She discovered that the structure of organisations varied very widely and that technology was a major factor contributing to the variances.

7.13 Woodward categorised the levels of technology into:

(a) unit production, or small batch production;
(b) mass production, or large batch production;
(c) process production, or continuous flow production.

This categorisation also describes a rising scale of *technical complexity*: process production is more complex than mass production, which is more complex than unit production. By 'technical complexity' she meant the extent to which the production process is controllable and its results predictable.

Woodward described the findings of a survey of firms in Essex.

'When the firms were grouped according to similarity of objectives and techniques of production, and classified in order of the technical complexity of their production systems, each production system was found to be associated with a characteristic pattern of organisation. It appeared that technical methods were the most important factor in determining organisational structure and in setting the tone of human relationships inside the firm.'

7.14 Elaborating further on the survey, Woodward noted the following.

(a) Different objectives of different firms controlled and limited the techniques of production they could use (for example a firm developing prototypes of electronic equipment cannot go in for mass production).

(b) Analysing the firms into a continuum of ten levels of technical complexity (sub-divisions, slightly overlapping, of the three main levels described above) firms using similar technical methods also had similar organisational structures.

> 'It appeared that different technologies imposed different kinds of demands on individuals and organisations and that these demands have to be met through an appropriate form of organisation.'
>
> *(Woodward)*

7.15 Specific findings were that:

(a) the number of levels in the management hierarchy increases with technical complexity: complex technologies lead to tall organisation structures, while simpler technologies can operate with a flat structure;

(b) the span of control of first-line supervisors was at its highest in mass production, and then decreased in process production;

(c) labour costs decreased, and the ratio of indirect labour increased as technology advanced;

(d) the span of control of the chief executive widened with technical advance;

(e) the proportion of graduates in supervisory positions increased with technical advance;

(f) the organisation was more flexible at both ends of the scale of complexity, that is, duties and responsibilities were less clearly defined. In mass production, duties and responsibilities are clearly set out, largely owing to the nature of the technology, and this favours a formal, authoritarian structure. As a consequence:

  (i) the amount of written, as opposed to verbal, communication peaked with mass production, being a feature of bureaucracy and formal structures;

  (ii) specialisation between the functions of management was most frequent in large batch and mass production companies. The clear-cut demarcation (and resulting conflicts) between line and staff management was also most frequent here;

  (iii) the administration of production (the 'brainwork') and the actual supervision of production work are the most widely separated in large-batch and mass production companies;

(g) industrial and human relations were better at both extremes of the scale than in large-batch and mass production companies, possibly owing to heavier pressure on all individuals in this type of organisation;

(h) the size of the firm was not related to its technical complexity, in other words it is not possible to attribute the 'faults' of mass production to the size of the firm rather than to the nature of its technology;

8. CHANGE AND THE ADAPTIVE ORGANISATION

8.1 Most organisations exist in a changing environment and must adapt in order to survive. Although formalisation and bureaucratic organisation helps a small company to develop into a large one, it may be insufficient to enable the organisation to survive continuing environmental changes.

8.2 Handy suggests that an organisation adapts to change in one of three ways.

(a) *By deliberation*: the organisation 'seeks to reinforce the formal structure by more formal structures'. Companies or governments might establish committees with powers to investigate, recommend or even to make decisions. Special project teams might be created, or new departments established (for example the corporate planning department or economic advisory section).

(b) *By reproduction*: large national organisations might delegate authority ('decentralise') to regional headquarters. Unfortunately, decentralisation of this sort usually results in regional structures which duplicate the former national structure: bureaucracy in the same form, but on a smaller scale. Unless the environment is fairly stable, such adaptation is likely to be inefficient.

(c) *By differentiation*: the organisation employs different structures with different cultures, in separate parts of the organisation, using a contingency approach - choosing the most suitable structure for each particular situation:

  (i) stable, routine work will be performed in a formalised bureaucratic manner (*role culture*);

(ii) adaptation to change (development of new products and new markets, or meeting environmental 'threats') should be organised on a task basis *(task culture)*;

(iii) any sudden crisis might have to be dealt with by key individuals with emergency powers (*power culture*);

(iv) overall policy decisions of the organisation should be set by a ruling body of key individuals (board of directors, the Cabinet of government ministers, or the supreme policy-making councils of other organisations) (*power culture*).

'One culture should not be allowed to swamp the organisation' (Handy). However, where differentiation, on a contingency basis, is applied in an organisation structure, there is a potential for conflict. Project teams might resent policy decisions of senior managers because they believe them to be inappropriate to the problems of the organisation; line managers might resent 'free-wheeling' 'undisciplined' members of project teams. The management of an organisation must be capable of reconciling differences and integrating the work of all employees towards a common aim.

8.3 Lawrence and Lorsch compared the structural characteristics of a 'high-performing' container firm, which existed in a relatively stable environment, and a 'high-performing' plastics firm which existed in a rapidly changing environment. They concluded that:

(a) in a stable environment (the container firm) the most efficient structure was one in which the influence and authority of senior managers were relatively high and of middle managers low;

(b) in a dynamic environment (the plastics firm) the most efficient structure was one in which the influence and authority of senior managers were somewhat less, and of middle managers correspondingly greater.

Organic and mechanistic organisations

8.4 Burns and Stalker contributed significant ideas about managing organisation growth and change. They identified the need for a different organisation structure when the technology of the market is changing; innovation is crucial to the continuing success of any organisation operating in the market.

8.5 They recommended an *organic structure* (also called an 'organismic structure') which has the following characteristics.

(a) There is a 'contributive nature' where specialised knowledge and experience are contributed to the common task of the organisation.

(b) Each individual has a realistic task which can be understood in terms of the common task of the organisation.

(c) There is a continual re-definition of an individual's task, through interaction between the individual and others.

(d) There is a spread of commitment to the concern and its tasks.

(e) There is a *network* structure of authority and communication.

(f) Communication tends to be *lateral* rather than vertical.

(g) Communication takes the form of information and advice rather than instructions and decisions (in keeping with the notion of empowerment).

8.6 Burns and Stalker contrasted the organic structure of management, which is more suitable to conditions of change (and more in sympathy with the needs of customers and suppliers), with a *mechanistic* system of management, which is more suited to stable conditions. A mechanistic structure has the following characteristics.

(a) Authority is delegated through a hierarchical, formal scalar chain, and 'position power' is used in decision-making.

(b) Communication is *vertical* rather than lateral.

(c) Individuals regard their own tasks as specialised and not directly related to the goals of the organisation as a whole.

(d) There is a precise definition of duties in each individual job (rules, procedures, job definitions).

8.7 Mechanistic systems are unsuitable in conditions of change because they tend to deal with change by cumbersome methods. For example:

(a) the *ambiguous figure system*: in dealing with unfamiliar problems authority lines are not clear, matters are referred 'higher-up' and the top of the organisation becomes over-burdened by decisions;

(b) *mechanistic jungle*: jobs and departments are created to deal with the new problems, creating further and greater problems;

(c) *committee system*: committees are set up to cope with the problems. The committees can only be a temporary problem-solving device, but the situations which create the problems are not temporary.

8.8 Burns and Stalker recognised that organic systems would only suit individuals with a high tolerance for ambiguity and personal stresses involved in being part of such an organisation - but the freedom of manoeuvre is considered worth this personal cost, for individuals who prize autonomy and flexibility.

(a) 'Growth is defined as change in an organisation's size, when size is measured by the organisation's membership or employment; development is defined as change in an organisation's age.' (*Starbuck 1965*)

(b) 'Development involves policy decisions that change organisational objectives. Growth, on the other hand, involves technical or administrative improvement by which it is possible more effectively to accomplish old objectives.' (*Hicks 1967*)

(c) Organisation development 'is a complex educational strategy intended to change the beliefs, attitudes, values and structure of organisations so that they can better adapt to new technologies, markets and challenges and to the dizzying rate of change itself'. (*Bennis 1969*)

9. CULTURE

9.1 At this point, it is worth discussing in more detail the concept of 'culture' in organisations. It may be defined as the complex body of shared values and beliefs of an organisation.

Peters and Waterman, in their study (*In Search of Excellence*) found that the 'dominance and coherence of culture' was an essential feature of the 'excellent' companies they observed. A 'handful of guiding values' was more powerful than manuals, rule books, norms and controls formally imposed (and resisted). They commented: 'If companies do not have strong notions of themselves, as reflected in their values, stories, myths and legends, people's only security comes from where they live on the organisation chart.'

9.2 Handy sums up 'culture' as 'that's the way we do things round here'. For Schein, it is 'the pattern of basic assumptions that a given group has invented, discovered, or developed, in learning to cope with its problems of external adaptation and internal integration, and that have worked well enough to be considered valid and, therefore, to be taught to new members as the correct way to perceive, think and feel in relation to these problems.'

> 'I believe that the real difference between success and failure in a corporation can very often be traced to the question of how well the organisation brings out the great energies and talents of its people. What does it do to help these people find common cause with each other? And how can it sustain this common cause and sense of direction through the many changes which take place from one generation to another?...I think you will find that it owes its resiliency not to its form of organisation or administrative skills, but to the power of what we call *beliefs* and the appeal these beliefs have for its people.' (Watson (IBM), quoted by Peters and Waterman)

9.3 All organisations will generate their own cultures, whether spontaneously or under the guidance of positive managerial strategy. The culture will consist of:

(a) the *basic, underlying assumptions* which guide the behaviour of the individuals and groups in the organisation, for example customer orientation, or belief in quality, trust in the organisation to provide rewards, freedom to make decisions, freedom to make mistakes, and the value of innovation and initiative at all levels. Assumptions will be reflected in the kind of people employed (their age, education or personality), the degree of delegation and communication, whether decisions are made by committees or individuals etc;

(b) *overt beliefs* expressed by the organisation and its members, which can be used to condition (a) above. These beliefs and values may emerge as sayings, slogans, mottos such as 'we're getting there', 'the customer is always right', or 'the winning team'. They may emerge in a richer mythology - in jokes and stories about past successes , heroic failures or breakthroughs, legends about the 'early days', or about 'the time the boss...'. Organisations with strong cultures often centre themselves around almost legendary figures in their history. Management can encourage this by 'selling' a sense of the corporate 'mission', or by promoting the company's 'image'; it can reward the 'right' attitudes and punish (or simply not employ) those who aren't prepared to commit themselves to the culture;

(c) *visible artefacts* - the style of the offices or other premises, dress 'rules', display of 'trophies', the degree of informality between superiors and subordinates.

9.4 'Positive' organisational culture may therefore be important in its influence on:

(a) the motivation and satisfaction of employees (and possibly therefore their performance) by encouraging commitment to the organisation's values and objectives, making employees feel valued and trusted, fostering satisfying team relationships, and using 'guiding values' instead of rules and controls;

(b) the adaptability of the organisation, by encouraging innovation, risk-taking, sensitivity to the environment, customer care, willingness to embrace new methods and technologies etc; and

(c) the image of the organisation. The cultural attributes of an organisation (attractive or unattractive) will affect its appeal to potential employees and customers. For example, the moves of banks to modernise and beautify branch design are meant to convey a 'style' that is up-to-date, welcoming, friendly but business-like, with open-plan welcome areas, helpful signposting and lots of light and plants.

Cultural problems and how to change culture

9.5 Not all organisation cultures are so 'positive' in their nature and effect, however. The symptoms of a negative, unhealthy or failing culture (and possibly organisation as a whole) might be as follows.

(a) No 'visionary' element: no articulated beliefs or values widely shared, nor any sense of the future.

(b) No sense of unity - because no central driving force. Hostility and lack of co-ordination may be evident.

(c) No shared norms of dress, habits or ways of addressing others. Sub-cultures may compete with each other.

(d) Political conflict and rivalry, as individuals and groups vie for power and resources and their own interests.

(e) Focus on the internal workings of the organisation rather than opportunities and changes in the environment. In particular, disinterest in the customer.

(f) Preoccupation with the short term.

(g) Low employee morale, expressed in low productivity, high absenteeism and labour turnover, 'grumbling'.

(h) Abdication by management of the responsibility for doing anything about the above - perhaps because of apathy or hopelessness.

(i) No innovation or welcoming of change: change is a threat and a problem.

(j) Rigorous control and disciplinary systems have to be applied, because nothing else brings employees into line with the aims of the business.

(k) Lacklustre marketing, company literature and so on.

9.6 A pretty depressing picture. So what can be done about it? There are many factors which influence the organisational culture, including the following.

(a) *Economic conditions*

In prosperous times organisations will either be complacent or adventurous, full of new ideas and initiatives. In recession they may be depressed, or challenged. The struggle against a main competitor may take on 'heroic' dimensions.

(b) *The nature of the business and its tasks*

The types of technology used in different forms of business create the pace and priorities associated with different forms of work, for example the hustle and frantic conditions for people dealing in the international money market compared with the studious life of a research officer. Task also influences work environment to an extent, and this is an important visual cultural indicator.

(c) *Leadership style*

The approach used in exercising authority will determine the extent to which subordinates feel alienated and uninterested or involved and important. Leaders are also the creators and 'sellers' of organisational culture: it is up to them to put across the vision.

(d) *Policies and practices*

The level of trust and understanding which exists between members of an organisation can often be seen in the way policies and objectives are achieved, for example the extent to which they are imposed by tight written rules and procedures or implied through custom and understanding.

(e) *Structure*

The way in which work is organised, authority exercised and people rewarded will reflect an emphasis on freedom or control, flexibility or rigidity.

(f) *Characteristics of the work force*

Organisation culture will be affected by the demographic nature of the workforce, for example its typical manual/clerical division, age, sex and personality.

9.7 It is possible to 'turn round' a negative culture, or to change the culture into a new direction.

(a) The overt beliefs expressed by managers and staff can be used to 'condition' people, to sell a new culture to the organisation by promoting a new sense of corporate mission, or a new image. Slogans, mottos ('we're getting there'), myths and so on can be used to energise people and to promote particular values which the organisation wishes to instil in its members.

(b) Leadership provides an impetus for cultural change: attitudes to trust, control, formality or informality, participation, innovation and so on will have to come from the top - especially where changes in structure, authority relationships or work methods are also involved. The first step in deliberate cultural change will need to be a 'vision' and a sense of 'mission' on the part of a powerful individual or group in the organisation.

(c) The reward system can be used to encourage and reinforce new attitudes and behaviour, while those who do not commit themselves to the change miss out or are punished, or pressured to 'buy in or get out'.

(d) The recruitment and selection policies should reflect the qualities desired of employees in the new culture. To an extent these qualities may also be encouraged through induction and training.

(e) Visible emblems of the culture - for example design of the work place and public areas, dress code, status symbols - can be used to reflect the new 'style'.

Example: 'culturefit'

9.8 'In order to gain competitive edge Heron Distribution looked at how well it served the needs of its customers. It drew up a diagram of each of its client companies' cultures, then looked at how closely its own attitudes and practices fitted in to the machine. Where there was a mismatch it geared up the recruitment and training of its own employees accordingly', a process it calls 'culturefit'. (*Personnel Management*, March 1993)

Exercise 5

(a) You may remember that British Airways ran a series of corporate advertisements in the late 1980s based on the idea that their cabin staff possessed superhuman powers. Who do you think the advertising was aimed at - cabin staff or customers?

(b) Market research indicates that British Airways really is 'The World's Favourite Airline'. Which do you think came first, the slogan or the fact?

(c) Watch out in the press for other examples of organisations' attempts to change their culture. Local government should be a good source, as should organisations due for privatisation like British Rail and the Post Office.

Types of culture

9.9 Different writers have identified different types of culture, based on particular aspects of organisation and management.

Charles Handy discusses four cultures and their related structures. He recognises that while an organisation might reflect a single culture, it may also have elements of different cultures appropriate to the structure and circumstances of different units in the organisation. (The customer service or marketing units may, for example, have a more flexible, and dynamic 'style' than administrative units, which tend to be more bureaucratic.)

(a) The *power culture*. Mainly in smaller organisations, where power and influence stem from a central source, through whom all communication, decisions and control are channelled. The organisation, since it is not rigidly structured, is capable of adapting quickly to meet change; however, the success in adapting will depend on the luck or judgement of the key individuals who make the decisions. Political competition for a share of power is rife, and emotional behaviour is encouraged by the 'personality' cult surrounding the leader.

(b) The *role culture* or bureaucracy, discussed earlier.

(c) The *task culture*, reflected in a matrix organisation, in project teams and task forces. The principal concern in a task culture is to get the job done; therefore the individuals who are important are the experts with the ability to accomplish a particular aspect of the task. Such organisations are flexible and constantly changing as tasks are accomplished and new needs arise. Innovation and creativity are highly prized. Job satisfaction tends to be high owing to the degree of individual participation, communication and group identity.

(d) The *person culture*, in an organisation whose purpose is to serve the interests of individuals within it. Organisations designed on these lines are rare, but some individuals may use any organisation to suit their own purposes; to gain experience, further their careers or express themselves.

9.10 Dale and Kennedy (*Corporate Cultures*) consider cultures to be a function of the willingness of employees to take risks, and how quickly they get feedback on whether they got it right or wrong.

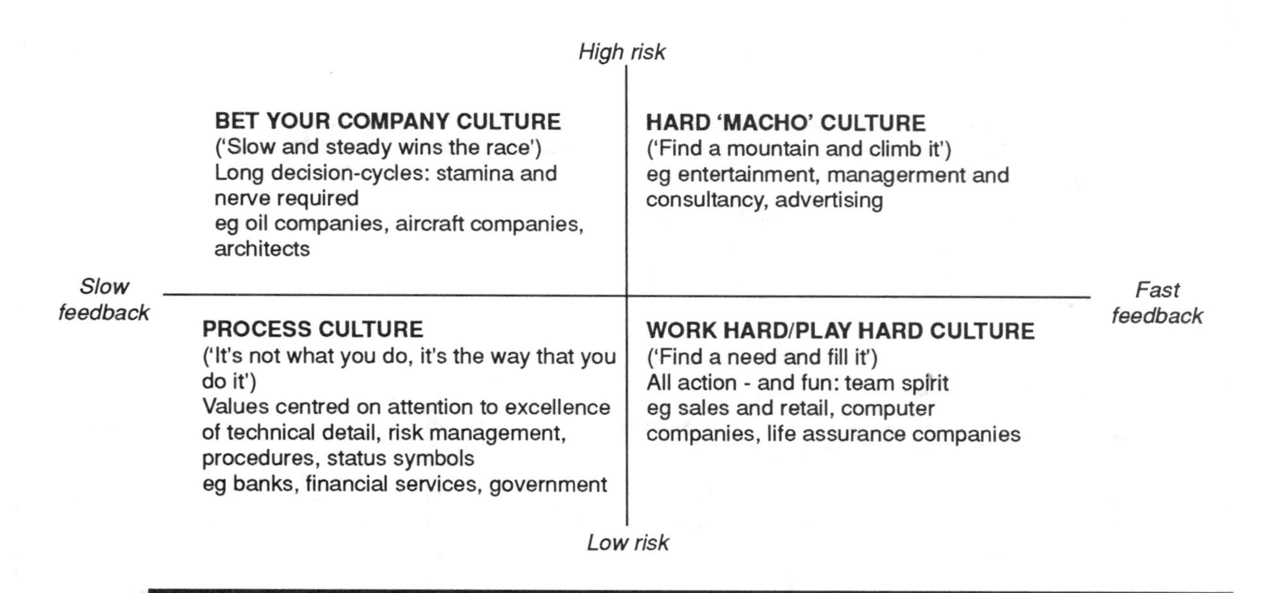

Exercise 6

Read the following passage (slightly adapted) from Tom Peters' book *Liberation Management: Necessary Disorganisation for the Nanosecond Nineties* (1992).

'IBM decentralizes. Then decentralizes again. Then decentralizes one more time. And joins in alliances with this company, then that company. But it's still too sluggish by a mile. (And can't gather the nerve to deconstruct its monster Sales organization). The long 80-year-old shadow cast by even a diminished corporate headquarters thwarts initiative after initiative. Like nearly every other clever product from IBM the innovative database language SQL had been developed in secret. The development group lied about it, then finally showed it to the big shots who were too impressed to turn the product down. *The Wall Street Journal* filed an eerily similar report on IBM's 1990 Nobel prize-winners for superconductivity, J Georg Bednorz and K Alex Mueller: "The two scientists hunted quietly, telling a supervisor a half-truth, and steering a curious visitor off track ... The issue is a sore point now for IBM brass."'

Now diagnose what was going wrong at IBM using as many as possible of the new terms and concepts you have encountered in this Study Text so far.

Solution

Here are some suggestions.

IBM seems to have attempted to decentralise without obtaining any of the benefits of decentralisation. There appears to be something of a cultural mess because the organisation cannot let go of its past: bureaucracy smothers initiative; a task culture seems to be what is called for (and what seems to exist, to some extent, informally) but the organisation is unable to adapt to change.

There are a number of other points you could make. For example, what parts of the organisation seems to have the greatest power and influence? How effective are the communication channels? Are the supervisors' spans of control too narrow?

Chapter roundup

- In this chapter we have considered how a variety of formal organisation structures can be established to optimise the efficiency of the organisation.
- There are advantages and disadvantages of centralisation. The modern trend favours decentralisation.
- Line managers are directly involved in achieving the objectives of an organisation while staff managers provide advice and service. This can cause conflict.
- A tall organisation is one with narrow spans of control and many levels in its management hierarchy. The modern trend is towards flat organisations and empowerment of employees lower down the organisation.
- There are numerous ways in which an organisation's activities can be grouped, for example by function, by product or geographically.
- A matrix organisation *divides* authority between functional managers and project or product managers.
- The interrelation between social and technological factors can have important implications for organisation structure.
- Organisations must be able to respond to change (or discontinuity) in order to survive. This requires an organic structure rather than a mechanistic one.
- Organisational culture and how to change it have been key issues in the 1980s and 1990s. Culture consists of basic underlying assumptions like belief in quality, overt beliefs which emerge in slogans and organisational mythology, and visible artefacts such as buildings and dress.

Test your knowledge

1. What are the advantages of centralisation? (see para 2.4)
2. How can the conflict between line and staff be resolved? (3.6)
3. Are tall organisation structures inefficient? (4.8 - 4.10)
4. Give some examples of departmentation (5.1)
5. Describe a matrix organisation. (6.2)
6. What is composite autonomous group working? What was Trist's argument for introducing it? (7.5 - 7.11)
7. Distinguish between 'mechanistic' and 'organic' systems. (8.3, 8.4)
8. What is 'culture' in an organisation? (9.1 - 9.3)
9. How might a negative culture be changed? (9.7)
10. What are Handy's four types of culture? (9.9)

Now try question 3 at the end of the text

Chapter 4

THE WORK ENVIRONMENT AND JOB DESIGN

This chapter covers the following topics.

Signpost

- Technology was also discussed in Chapter 3; hours of work are a topic in Chapter 8.
- The views of Herzberg and others are described in Chapter 13 on motivation and rewards.

1. ENVIRONMENT AND MOTIVATION

1.1 Organisations are composed of individuals. While external environment and internal systems shape and direct the organisation as a whole, it is also relevant to consider the impact of the working conditions under which individuals are called upon to function daily. We will look at general and specific aspects of this, and then go on to consider particular features of the work environment and design of jobs; the human effects of operating systems, conditions and the organisation of work.

1.2 A recent survey from America suggested that the office environment is of importance to workers but is no longer closely linked in their minds with increased productivity.

As we shall see in a later chapter these findings fit in with Herzberg's view of working conditions as a 'hygiene' factor. A good working environment will not be able to motivate employees to enhanced performance (beyond the short-term effects of a fresh improvement, perhaps) but shortcomings will be important to them, as a source of *dis*satisfaction.

1.3 The physical environment's main effect on worker morale and/or productivity will therefore be the product of the following.

(a) *The health and safety of workers*. A hazardous or unhygienic environment is - humanly and legally - unacceptable, and will not give workers the physical support they need to put in consistent, sustained effort.

(b) *Enabling of performance*. Certain conditions physically and psychologically enable the worker to perform tasks efficiently, without stress, fatigue and other forms of 'interference'.

In addition, the planning and layout of space in the office or factory can contribute to the efficiency of work flow, the movement of people, documents and work in progress.

(c) *Contribution to organisational culture*. The physical environment is an expression of the organisation's self-image to customers/clients and to employees. It can alter the way employees feel about their work and their organisation (is it smart or 'shabby'? high-tech or antiquated? a sign of caring or indifference on the part of the organisation?) It can also affect the amount of social contact and interaction available to workers: consider the difference between cubicles and open plan offices, for example, in encouraging informal communication ('networking').

1.4 In addition, the environment will contribute to the performance of the *business*, if it:

(a) projects a positive and attractive *image* to potential customers, employees and suppliers;

(b) is *flexible* enough to allow for change and growth, should that be required;

(c) makes *economical* use of space, equipment and work flows;

(d) encourages the informal *communication* network of the organisation for the exchange of information and ideas.

Exercise 1

A question often asked in examinations for the ICSA's old *Office Administration* examination concerned 'telecommuting', whereby 'office' workers work at home and are linked to their organisation's central premises by telephone and computer. Telecommuting has been technically feasible for perhaps most office workers for some time but it is not yet widespread. Why do you think this is so?

Solution

Surveys indicate that the main opposition comes from *employers* (in spite of the potential savings in premises costs) because they fear that they will lose control of their workers. Look also at the paragraphs above, and consider the following from Handy's *Age of Unreason*: 'For some it is too lonely, or the temptations of home become too great; one American lady complained that she put on two stone because of frequent trips to the fridge, and one man attributed his divorce to the fact that he was at home all the time.'

2. WORK ENVIRONMENT

External environment

2.1 The 'surroundings' of a work place include not only the immediate space in which employees carry out their duties, but also the external environment in which they may shop, bank, eat, commute and park their cars and the overall design and construction of the complex within which their particular office may be placed.

2.2 The *siting* of buildings is probably the first point to be considered in the acquisition or construction of offices or factories.

(a) Particularly, does the business want to be located in an urban area, or out of town? In which case, how far out of town?

(b) Are available transport and communication links sufficient to keep up the in- and out-flow of materials/goods/information/people?

(c) What about *cost*: could the business take advantage of lower land prices, lower rates and insurance costs or development grants from the government outside urban areas?

(d) Where do most of the *employees* live? Could they commute easily? Would they rather move out of town? Or is there an existing 'pool' of sufficient suitable labour in the area?

(e) Are *customers* all based in town? How would the company keep in touch with them? Could it attract new business without an office in town to present a public 'profile'?

2.3 One consideration in locating the site must be the facilities in the surrounding area. From the point of view of employees and the organisation, the proximity of banks, postal services and transport will be very important. For the employees' well-being, shops, restaurants and perhaps recreation/sports facilities would also be welcome.

2.4 As ever, financial considerations largely determine the organisation's attitudes to such factors. The site, size and age of buildings acquired, and whether they are rented, bought or built, will be determined by their cost and will in turn determine such operating costs as insurance and rates, heating and lighting, maintenance and renovation.

Office layout

2.5 In considering the layout of a new or adapting office, flexibility will obviously be needed to cope with the needs of different individuals and activities, the shapes and sizes of available rooms. But there are some basic features which any office should possess:

(a) economical use of space. This usually means flexibility - and room for expansion where this is anticipated;

(b) arrangement for efficient work flow;

(c) accessibility of shared equipment and facilities;

(d) arrangement for ease of supervision;

(e) provision for security where necessary;

(f) safety of occupants;

(g) the required 'image' of the premises, particularly in areas accessible to potential customer/clients.

2.6 There are four general types of office layout.

(a) *Small closed offices* linked by corridors. These have the advantage of privacy, peace and security, and are desirable for 'status conscious' managers for whom a separate office is an important symbol of authority, but supervision and communication is hindered, and this system is generally not favoured in new office designs.

(b) *Open plan offices*. These are arranged like classrooms and lecture halls, to do away with the maze of walls and doors to make better use of the space than small closed offices. It has been estimated that a 33% space saving is possible under this system. Advantages include:

(i) easier supervision;
(ii) freer communication;
(iii) flexible arrangement of furniture and equipment in the available space;
(iv) economies on heating and lighting;
(v) sharing of equipment, such as photocopiers.

Disadvantages include:

(i) the 'soulless' arrangement of the desks in ranks;
(ii) the lack of privacy;
(iii) the distraction from noise and movement;
(iv) the loss for managers of the status of a separate office, possibly lowering morale;
(v) the tendency for managers to become unnecessarily involved in routine matters;
(vi) the difficulty of satisfying every individual's needs and preferences.

(c) *Landscaped offices* are a variation of the open plan system, overcoming some of the latter's problems. Desks and equipment are 'clustered' as appropriate to the work. Movable screens of variable height, shape and colour, usually 'acoustic' to absorb noise, and flame-proof for safety, are used to break up the office space. Filing cabinets and large pot plants may also be placed to give staff more privacy in their own areas and to cut down on noise and distractions, while still preserving the communication advantages of the open plan. Equipment and furnishings are generally of good quality, but more space and better furnishings may be given to management to enhance their status and morale.

Open plan designs, and layout that facilitate frequent deliveries of components and sub-assemblies are equally relevant in modern factories.

2.7 Work study might be used to determine the best layout for sections in an open plan system, based on the amount of contact usually necessary between each section. Within each section, attention should be given to the *proximity* of:

(a) people regularly working together (such as the manager and his/her secretary); and
(b) supervisors and those under their control.

Attention should also be given to the *accessibility* of:

(a) people whose advice or services are required by the section as a whole (supervisors, typists);

(b) equipment and facilities regularly used and shared by the section (files, photocopier, coffee machine).

The same principles will apply to work flow between different departments on the premises. The flow of work could be looked at from the point of view both of staff movements (free and 'straight' passage from one place to another), and the movement of documents.

Internal factors

2.8 There are various factors which combine to affect the physical comfort and (arguably) performance of employees.

(a) *Heating, lighting and ventilation*

The effect of these is fairly obvious, particularly when one or more of them is inadequate. Wherever possible lighting should be natural rather than artificial and strip lighting rather than bulbs because fluorescent tubes emit a light closer to natural light. Ventilation should be provided without causing draughts, and if necessary, steps should be taken to filter smoke and dust out of the air for the sake of equipment and personnel alike.

(b) *Noise*

Strangely enough, we do need noise at work - or at least a certain amount of it. Constant loud noise, or intermittent noise at any volume, is distracting and should be reduced by carpeting, sound-absorbent walls and ceilings; 'acoustic hoods' are used to cover noisy machinery like some computer printers, to muffle the noise; soft bleep phones are increasingly used instead of bell-ringing ones. A total absence of any noise, however, causes uneasiness. Some large office blocks have solved this problem by using speakers around the building to give out 'white noise', a low background hiss which is sufficient and suitable noise; some prefer background music.

(c) *Decor*

A well-planned colour scheme can be a surprisingly important factor in the psychology of work. Brown, yellow and magnolia are considered good for work areas as they are warm colours; white and cream are light colours which can give an impression of space in small rooms; green is widely used, for example on VDU screens, 'black' boards etc as a 'restful' colour, easy on the eyes. Surprisingly, however, this seems to be a cultural factor: green is

(d) *Furniture*

Ergonomics (discussed later) includes the design of office equipment so that, for example, chairs and desks are designed to be comfortable and supportive. Desks and tables should be of a convenient height, and chairs in particular should be adjustable to give proper back support.

Office furniture design is being taken very seriously, with increasing emphasis on the attractive-and-functional rather than the beautiful-but-impractical. Systems or module furniture is being designed, with light, easy to move, linkable desks, cabinets, printer tables and so on for total flexibility and use of available space. Easily cleaned surfaces, and colour/style co-ordination are also taken into account. A 'hi-tech' or traditional *image* may be an important part of the culture of the organisation.

(e) *Cost*

This is not a 'physical factor' in itself, of course, but must be considered together with all of them. The costs of purchasing and maintaining office buildings, furniture and equipment, and the costs of heating, lighting, cleaning must be weighed against all possible benefits to the organisation from employee satisfaction and motivation, enhanced work flow and smart office image. The 'perfect' ergonomically-designed office would still have to have questions of cost, durability, and justification asked of it.

Exercise 2

Consider your own work environment. How well is your workplace sited? Do you sometimes *not* seek the help of others simply because of where they are located in relation to you? What impact does noise have on your work? Does the way you organise and perform your work depend to any extent on the physical facilities you have for filing it?

Solution

This is one of those exercises that prompts you to do what (hopefully) you have been doing already: study *actively*. You are unlikely to be asked questions like this in the exam! Their purpose is to encourage you to assimilate information by applying it to your own experience, or to translate your own experience into the terms that are being introduced to you.

---

3. HEALTH AND SAFETY

3.1 In 1972, the Royal Commission on Safety and Health at Work reported that unnecessarily large numbers of days are lost each year through industrial accidents, injuries and diseases, because of the 'attitudes, capabilities and performance of people and the efficiency of the organisational systems within which they work'.

Since then, major legislation has been brought into effect in the UK, most notably:

(a) Health and Safety at Work Act 1974;

(b) Safety Representatives and Safety Committees Regulations 1977 (under the authority of the 1974 Act).

(c) Six sets of regulations implementing EC directives on health and safety at work. These were introduced in January 1993.

3.2 It is arguable, too, that since 1972, society as a whole has become more aware of health and safety, through:

(a) legislation requiring health warnings and descriptions of contents of goods;

(b) the raising of 'issues' such as unsafe toys, additives in food, highly flammable materials in furniture, asbestos poisoning and so on;

(c) experience of notable disasters such as Chernobyl, the Kings Cross fire and so on.

Moreover, in the UK the general quality of life and environments has become highly modern and sophisticated. Employees are less likely to accept poorly lit or ventilated, cramped and hazardous environments at work. In some sectors of business, extremely 'high-tech', designer environments are accepted as the norm for the sake of corporate culture and image.

3.3 However, it would be wrong to paint too 'rosy' a picture at present, since:

(a) legislation sets bare minimum standards for (and levels of commitment to) health and safety. It does not represent satisfactory practice for socially responsible organisations;

(b) healthy and safety are still regarded with a negative attitude by many managers. Provisions are costly, and have no immediately quantifiable benefit. Like other forms of risk management - eg insurance - they are perceived as a regrettable necessity, which would be avoided if it were 'safe' to do so, given the constraints of legal obligations, trade union pressure and the threat of adverse publicity (from a disaster such as that on the Piper Alpha oil rig, or the Bhopal chemical plant);

(c) positive discipline only goes so far, and irresponsible or ignorant behaviour can still cause accidents - eg where operators disobey safety instructions on machinery;

(d) new health and safety concerns are constantly emerging, as old ones are eradicated. For example:

    (i) new technology and ergonomics may make physical labour less stressful, but it creates new hazards and health risks - for example a sedentary, isolated lifestyle and problems associated with working long hours at VDUs;

    (ii) new 'issues' in health are constantly arising, such as smoking in the workplace, and alcohol abuse with the increasing stress of work in highly competitive sectors.

One of the main provisions of the EC directives is to force *all* employers to undertake a continuous risk assessment programme to identify health hazards.

3.4 Health and safety at work is important:

(a) primarily, to protect employees from pain and suffering;

(b) because an employer has legal obligations for the health and safety of employees;

(c) because accidents cost the employer money (not just legal damages, but operating costs as well); and

(d) because the company's image in the marketplace (to which it sells goods and services, and from which it recruits labour and buys in other resources) will suffer if its health and safety record is bad.

Accidents

3.5 Apart from obviously dangerous equipment in offices, there are many hazards to be found in the modern working environment. Many accidents could be avoided by the simple application of common sense and consideration by employer and employee, and by greater safety consciousness - encouraged or enforced by a widely acceptable and well-publicised safety policy, advising and warning about safe and unsafe procedures, activities and attitudes.

Common causes of injury include:

(a) slippery or poorly maintained floors;

(b) frayed carpets;

(c) trailing electric leads;
(d) obstacles (files, books, open drawers etc.) in gangways;
(e) standing on chairs (particularly swivel chairs!) to reach high shelving;
(f) staircases used as storage facilities; and
(g) lifting heavy items without bending properly;

3.6 The *cost* of accidents to the employer consists of:

(a) time lost by the injured employee;

(b) time lost by other employees who choose to or must of necessity stop work at the time of or following the accident;

(c) time lost by supervision, management and technical staff following the accident;

(d) a proportion of the cost of employing first aid, medical staff, etc;

(e) the cost of disruption to operations at work;

(f) the cost of any damage to the equipment or any cost associated with the subsequent modification of the equipment;

(g) the cost of any compensation payments or fines resulting from legal action;

(h) the costs associated with increased insurance premiums;

(i) reduced output from the injured employee on return to work;

(j) the cost of possible reduced morale, increased absenteeism, increased labour turnover among employees;

(k) the cost of recruiting and training a replacement for the injured worker.

3.7 You should be aware, however, that occupational health programmes may deal with many aspects of the working environment - not just hazardous chemicals or processes, unsafe methods and so on. We will now look in more detail at particular aspects of working conditions, and their effects on work behaviour.

Fatigue and monotony

3.8 You may not think that 'fatigue' is a product of the working environment. It is more commonly associated with the depletion of an individual's energy resources, that is, 'tiredness'.

'Fatigue', however, is merely a useful term to describe a variety of phenomena:

(a) the depletion of 'fuel reserves' or 'blood sugars';

(b) exhaustion and increased heart rate from over-heating and lack of oxygen in hot, stuffy conditions;

(c) temporary disability, resulting from the build-up of lactic acid in the blood during short bursts of intense activity with insufficient oxygen.

3.9 Studies of 'industrial fatigue', moreover, have shown that 'fatigue' is part of the conflict between the individual and his environment that we mentioned in an earlier chapter. Performance is at its optimum when the worker is in a 'steady state': interference may create a condition of 'dis-equilibrium', which might render him temporarily unable to continue at the same level of activity, or indeed at all.

Studies of hours of work, rest pauses, lighting, heating, ventilation, individual suitability to particular tasks, posture, physique and so on have been aimed at identifying sources of this dis-equilibrium - whether within the individual himself, or in the relationship between the individual and his job or work environment.

3.10 Elton Mayo (1880-1949) led the famous investigation of the Hawthorne works of the Western Electric Company in Chicago. His initial interests were in fatigue, accidents and labour turnover, and the effect on them of rest pauses and physical conditions of work. (A previous study had concentrated on lighting conditions). In the end (as we shall see later), Mayo's attention was diverted from purely physical working conditions; the effects of the informal organisation (working attitudes and so on) on performance became the central part of his thesis. But his observations on fatigue are interesting - given his early assumption that physical conditions are responsible for falling production.

Mayo argued that from a purely physical aspect, we are entitled to infer a 'steady state' where organic processes are kept in balance with the required expenditure of energy: in such a state, work could continue indefinitely. Even under such 'internal' biological conditions, however, workers do slow down or stop - so external conditions must be affecting performance.

3.11 *Monotony* - and the experience of boredom - was early on identified as another form of 'interference' in worker performance.

(a) Specialisation, and the breaking of the job into smaller and smaller tasks, such as were advocated by 'scientific management' theorists, created conditions that gave the worker no mental stimulation: inattention, daydreaming or social conversation were usually the result, with accidents and errors following. It might be argued that repetitive and routine work on assembly lines, in largely automated processes, or in the typing pool, have such an effect. The strain on the worker is even worse, if the social outlet is not available to him - the typist working in an isolated booth for example, or the assembly worker separated from the group by the layout of machinery.

(b) The extent to which workers suffer boredom - or role 'underload' - depends on the mental capacity (and self-concept) of the individuals. The more intelligent a person is, the more keenly he will *perceive* the monotony of repetitive tasks: some workers will derive satisfaction from a long uninterrupted run of work (however undemanding) - especially if there are good piece rates for the job.

3.12 Robert Karasek, in studies of Swedish and American workers, found that *stress* was related to work overload and level of discretion in work methods: high workload/ low discretion jobs are the most highly stressful. (If both factors are low, the job will require little mental or physical activity, and as long as discretion is high, there will be some challenge and opportunity.)

Handy reports on the following situations.

A group of inmates from a mental institution was given jobs on an assembly line. The inmates proved to be highly productive workers in the plant, never getting bored or needing breaks, despite long hours on the job. Their job satisfaction was reckoned to be very high.

A pigeon was trained as an inspector in a drug factory. Human inspectors had been employed to watch thousands of pills pass on a conveyor belt, and to discard rejects: they worked only short spells, and there were high absenteeism and turnover rates. The pigeons, however, worked longer hours - and made fewer errors.

3.13 The general conclusions of studies of monotony have been that it is less likely to arise if:

(a) the activity is changed from time to time, and rest pauses allowed; workers might be trained to alternate jobs, or might simply be allowed to move around a bit, for example by fetching their own materials or tools;

(b) work is grouped into whole, self-contained tasks - rather than the repetition of a single part of a job by each individual; completely autonomous work groups have even been set up, with responsibility for the production of whole units - and even in competition with each other;

(c) workers are permitted the outlet of social interaction, ie are allowed to form groups, rather than being isolated by the way the work place is designed;

(d) payment by results is used - depending on the significance of pay as a motivator to the individual.

One of the supposed benefits of IT is that it enables computers to take over monotonous tasks.

3.14 The application of the above conclusions to the organisation of work, and job design, is discussed later.

Stress

3.15 Stress is a term which is often loosely used to describe feelings of tension or exhaustion - usually associated with too much, or overly demanding, work. In fact, stress is the product of demands made on an individual's physical *and mental* energies: monotony and feelings of failure or insecurity are sources of stress, as much as the conventionally-considered factors of pressure, overwork and so on.

3.16 It is worth remembering, too, that demands on an individual's energies may be stimulating as well as harmful: many people, especially those suited to managerial jobs, work well under pressure, and even require some form of stress to bring out their best performance. (It is excessive stress that can be damaging: this may be called *strain.*) This is why we talk about the management of stress, not about its elimination: it is a question of keeping stress to helpful proportions and avenues.

3.17 Harmful stress, or strain, can be identified by its effects on the individual and his performance. Symptoms usually include:

(a) *nervous tension.* This may manifest itself in various ways: irritability and increased sensitivity, preoccupation with details, a polarised perspective on the issues at hand, or sleeplessness. Various physical symptoms - such as skin and digestive disorders - are also believed to be stress-related;

(b) *withdrawal.* This is essentially a defence mechanism which may manifest itself as unusual quietness and reluctance to communicate, or as physical withdrawal in the form of absenteeism, poor time-keeping, or even leaving the organisation;

(c) *low morale:* low confidence, dissatisfaction, expression of frustration or hopelessness;

(d) signs that the individual is repressing the problem trying to deny it. Forced cheerfulness, boisterous playfulness or excessive drinking may indicate this.

3.18 It is worth noting that some of these symptoms - say, absenteeism - may or may not be *correctly* identified with stress: there are many other possible causes of such problems, both at work (lack of motivation) and outside (personal problems). The same is true of physical symptoms such as headaches and stomach pains: these are not invariably correlated with personal stress.

All these things can, however, adversely affect performance, which is why stress management has become a major workplace issue. Considerable research effort has been directed to:

(a) investigating the causes of stress;
(b) increasing awareness of stress in organisations; and
(c) designing techniques and programmes for stress control.

3.19 Stress can be caused or aggravated by:

(a) *personality.* Competitive, sensitive and insecure people feel stress more acutely;

(b) ambiguity or conflict in the *roles* required of an individual. If a person is unsure what is expected of him at work, or finds conflict between two incompatible roles (employee and mother of small children, say), role stress may be a problem;

(c) *insecurity, risk and change*. A manager with a high sense of responsibility who has to initiate a risky change, and most people facing career change, end or uncertainty, will feel this kind of stress;

(d) *management style*. A recent American report pointed out particular management traits that were held responsible by workshop interviewees for causing stress and health problems (high blood pressure, insomnia, coronary heart disease and alcoholism). These included:

    (i) unpredictability. Staff work under constant threat of an outburst;
    (ii) destruction of worker's self esteem - making them feel helpless and insecure;
    (iii) setting up win/lose situations - turning work relationships into a battle for control;
    (iv) providing too much - or too little - stimulation.

3.20 Greater *awareness* of the nature and control of stress is a feature of the modern work environment. *Stress management techniques* are increasingly taught and encouraged by organisations, and include:

(a) counselling
(b) time off or regular rest breaks
(c) relaxation techniques (breathing exercises, meditation)
(d) physical exercise and self-expression as a safety valve for tension
(e) delegation and planning (to avoid work-load related stress)
(f) assertiveness (to control stress related to insecurity in personal relations).

In addition, *job* training can increase the individual's sense of competence and security and *ecological* control can be brought to bear on the problem of stress, creating conditions in which stress is less likely to be a problem: well designed jobs and environments, and an organisation culture built on meaningful work, mutual support, communication and teamwork.

Noise

3.21 Noise can be a particularly stressful form of 'interference'. Continuous, loud, random noise may have to be accepted by workers in some situations - although sound-proofing is increasingly a feature of offices and factories - and perceptual selectivity usually filters it out after a while: you get used to it, and it becomes part of the 'background' to activity.

Elusive but potentially meaningful sounds (which you strain to hear and make sense of), sharp intermittent noises or unexplained variations in noise level can, however, be irritating, distracting or hard on the nerves. Concentration can easily lapse if there is a sound competing with other more relevant sensory information: 'mental blinking' (the shifting of attention from one sensory input to another), can cause significant lapses. Any work which demands alertness and concentration may suffer.

Ergonomics

3.22 Ergonomics is usually described as the scientific study of the relationship between man and his working environment. This sphere of scientific research explores the demands that can arise from a working environment and the capabilities of people to meet these demands.

Through this research, data is made available to establish machines and working conditions which, apart from functioning well, are best suited to the capacities and health requirements of the human body. In old people's homes and hospitals, for example, switches are placed according to measurements of chair height and arm reach. In the same way, computer consoles and controls, office furniture, factory layout and so on can be designed so that the individual expends minimal energy and experiences minimal physical strain in any given task.

3.23 Both work study and ergonomics are concerned with 'fitting the job to the worker' and ergonomic data is used in establishing workplace layout. The operator's comfort will depend on many factors connected with the particular job, but there are certain general considerations which ensure a comfortable position.

(a) The operator should be allowed to sit where possible.

(b) The chair should permit alternate sitting and standing.

(c) In either case the elbows should be 2-3 inches above the working surface. This often calls for benches to be higher than those normally found.

(d) There should be room for the operator to put both legs under the bench.

(e) Work, tools and equipment should be within easy reach.

(f) Movements should be natural, rhythmical, and symmetrical.

If possible, the workplace should be tailored to the requirements of the individual operator.

3.24 Industrial psychology increasingly enters into this field. Apart from purely mechanical considerations - in what position should a worker be sitting in order to exert maximum force over a long period of time without physical strain or fatigue? - the ergonomist must now take into account the increasing problems of the worker as information processor. The perceptual limitations of the worker can also be measured, and systems designed which do not make unreasonable demands on the worker's attention span or capacity to absorb information - for example, the use of sound signals to attract attention to visual displays or equipment.

Exercise 3

'The US Government estimates that between 30 and 75 million American workers are at risk of becoming ill because of the buildings in which they work. It sees indoor air pollution as one of the five most urgent environmental issues in the US, and its occupational health agency has studied more than 1,000 cases of suspected sick building syndrome.

In more than 50 percent of cases it identified inadequate ventilation, followed by chemical contamination and problems related to microbiological agents such as moulds, bacteria and fungi.'

Personnel Management Plus, March 1993

In your role as office manager a number of employees have recently complained to you that they feel below par after an hour or two in the building, but seem to recover when they pop out at lunch time and on the way home in the evening. What do you do?

Solution

Dr Aric Sigman, the author of the above piece in *Personnel Management Plus*, suggests the following.

'If employees suspect they are working in a sick building and they have eliminated their boss and their job as potential causes of their condition, they should act as follows.

- Document the symptoms - who gets what, where and when. All doctor's visits should be recorded. Liaise with the occupational health unit.
- Check out the building - staff must take responsibility for looking around for sources of concern. Contact the Health and Safety Executive for advice on what to look for.
- Suggest action - clearly and in writing.
- Seek expert assistance.'

4. HOURS OF WORK

4.1 It is generally recognised that personal, family and social commitments must be allowed for in the demands made on workers' time. Role conflict may be caused or aggravated if, for example, an individual works late or at weekends when he/she should be with his/her family. A shorter working week, 'preserved' weekends, long weekends, and more holidays are developments towards the promotion of leisure time as the benefit of increased prosperity.

4.2 Overtime, however, has become an established feature of working life: it is taken for granted in planning and scheduling by many organisations, even though in most spheres it is still considered voluntary, and paid at a premium. Despite common expectations, it has been shown that individuals who regularly work overtime have reasonably good morale, attendance records and so on. Pay is considered to be the major incentive, where:

(a) low basic rates necessitate 'earning a bit extra'; or

(b) more highly-paid workers also work overtime to preserve differentials between themselves and others.

High achievement needs may also lead an individual to work extra hours - 'to get the job done', 'tie up loose ends' etc. There may even be political 'points' to be scored with superiors being seen at work beyond the call of duty. Overtime may even be an 'addictive', self-sustaining activity, whereby the worker compulsively puts his time and energy into his work, for its own sake, as a substitute for other activities which the individual feels insecure in tackling, or as a means of acquiring more and more luxury goods from the overtime payments.

4.3 The number of hours in the 'standard' working week have fallen from around fifty to forty, while the hours actually worked have also fallen - though not by so much. Since the Second World War, average hours worked have hovered around forty-seven. As well as overtime hours, some individuals may seek supplementary jobs elsewhere - preferring increased earnings to increased leisure time.

In fact, average hours worked may well continue to fall, with the influence of new technology reducing man-hours required, unemployment, and a higher proportion of the working population (notably skilled workers and married women) now tending to work part-time or shorter hours. An EC directive is likely to introduce a 48-hour week in 1994, although the UK has secured a 10 year grace period before this could be introduced.

4.4 The hours worked by any given individual will depend on many factors, including employment law, domestic and social demands, sex and age.

Rates of pay determine that labourers tend to work longer hours than technicians or craftsmen, and manual workers longer hours than clerical workers - although this has a confusing effect on pay differentials between skilled and unskilled workers and may (albeit illogically) cause some resentment. *Methods* of payment also affect the hours worked by those who seek cash rewards: piece-rate workers can earn more by increased effort during standard hours, whereas time-paid workers have no way of increasing their take-home pay except through overtime

4.5 The value of longer hours worked by a minority is not significant in production terms. Overtime will obviously be desired by management if it enables essential, round-the-clock services to be kept running, or enables them to cope with occasional peaks of activity without the necessity of recruiting new staff.

Long working days, however, almost certainly involve a lower average effort over the hours worked. Also, if overtime is worked unit costs of production will rise - unless increased production results from the practice.

4.6 Making the best use of the hours worked is perhaps what 'organisational behaviour' is all about. In trial programmes, the institution of natural pauses (whenever the work allows, as opposed to fixed, 'off the clock' breaks), and kiosks/safe-smoking areas close to production units was shown to save around 25,000 productive man hours per year. Workers can (generally) be trusted not to abuse the facilities and flexibility - and are able to maintain a steady rate of working more easily under the more congenial conditions.

> 'Most types of continuous or sustained work may lead to a build-up of fatigue. In many jobs there are natural breaks where you can move about to do something different. Jobs should be designed to allow such changes in activity, but if this is not possible, short frequent breaks seem to prevent fatigue. Being able to choose when to take a break is preferable to having fixed rest break schedules.' *Health & Safety Executive* booklet for VDU users

Shift work

4.7 Increasing numbers of workers are employed on a *shift basis*. The advantage to the organisation is that overhead costs can be spread over a longer productive 'day', and equipment and labour can be used more efficiently.

There are three main systems in operation.

(a) *Double-day system* - ie two standard working day (eight hour) shifts, say from 6am - 2pm, and 2pm - 10pm. The physical and social problems of such a system are much less acute than where night-work is required - although the hours may seem somewhat 'anti-social' and hard on the evening social life.

(b) *Three-shift system* - ie three eight-hour shifts covering the twenty-four hour 'day' (say 6am - 2pm, 2pm - 10pm, 10pm - 6am). The main problem here is the 'unnaturalness' of night-time work, on the third shift. There are physiological, psychological and social effects - which we will discuss below. Most complaints are directed at the so-called 'dead fortnight', when the pattern of afternoon and night shifts interfere most with normal social life.

(c) *Continental or 3-2-2 system*. This entails more frequent changes than the traditional system, enabling employees to have 'normal' leisure time at least two or three times per week! Over a four-week cycle, shifts rotate so that workers do 3 mornings, 2 afternoons, 2 nights, 3 rest days, 2 mornings, 2 afternoon etc (see table below). This gets away from the 'dead fortnight', but it may cause confusion initially, and also means that there are no entirely free weekends - which may be important to families.

3-2-2 system

Week	*Mon*	*Tue*	*Wed*	*Thur*	*Fri*	*Sat*	*Sun*
1	6-14	6-14	14-22	14-22	22-6	22-6	
2	-	-	6-14	6-14	14-22	14-22	22-6
3	22-6	22-6	-	-	6-14	6-14	14-22
4	14-22	14-22	22-6	22-6	-	-	6-14

Average 42 hours per week

4.8 The 3-2-2 system is becoming increasingly popular - and is already established in the chemical and iron and steel industries. ICI operate it in two factories, and surveys show 80% of the workforce in favour because:

(a) shorter, though more frequent, spells on each shift were found to be less fatiguing than longer periods on, say, the night shift;

(b) the variety was more enjoyable;

(c) employees felt that they had more time off for social and family life; and

(d) senior staff found it easier to keep in touch with the shiftworkers.

A study by Folkard and Monk suggests that the best way of adapting to hours which are not normal for the human body is to work permanently on one shift, allowing the body to develop a 'revised schedule'. This is, however, unlikely to be widely implemented because of the anti-social nature of evening and night work.

4.9 The effects of shiftwork have been well researched.

(a) *Physiological or medical effects* - a disruption of body-temperature, disturbance of digestion, inability to sleep during the day resulting from the disorientation of the body's 'clock', its regular cycle of meals, sleep and energy expenditure. Shiftwork tends to conflict with the body's 'circadian rhythms', or 24-hour body cycles. Stress-related ill-health may also be caused. Some people suffer more from the physical disruption than others: in particular, diabetics, epileptics and those prone to digestive disorders should be screened by management, and excluded from shiftwork for health reasons.

(b) *Psychological effects*. The experience of variety can be stimulating. On the other hand, the fatigue and sense of physical disorientation can be stressful. Those with strong security or structure needs may feel threatened by a lack of 'rhythm' in working life. A sense of isolation and lack of variety arising from the social problems of shiftwork may also be threatening - particularly if strain is being put on non-work relationships and roles.

(c) *Social effects*. Some forms of shiftwork involve high social costs, though others - in particular double-day working, very little. In some systems, the normal hours of socialising - afternoon and evening - are taken up at work, which may isolate the individual from his non-work social circle. Family problems may be acute - especially where weekends are lost: not only is the worker absent, leaving a role gap in the family, but the routine of the whole family will be disrupted by his sleeping and eating patterns.

(d) *Economic effects*. Overtime is not necessarily eliminated by shiftwork: premiums for double-shift and Sundays are common in practice. Shiftworking itself is inherently unpopular, and its appeal will largely depend on financial incentives.

4.10 The EC Commission Report on 'The Problems of Shiftworking', in 1977, proposed to reduce the extent of shiftworking, and to improve conditions by:

(a) giving older shiftworkers, or those with a certain length of shift service, the right to return to day work - to counter negative physical and psychological effects; and

(b) reducing the length of the working week, lengthening rest periods and offering earlier retirement to shiftworkers.

Flexitime

4.11 There are many 'flexitime' systems in operation providing freedom from the restriction of a '9 to 5' work-hours routine. The concept of flexitime is that predetermined fixed times of arrival and departure at work are replaced by a working day split into two different time zones.

(a) The main part of the day is called 'core time' and is the only period when employees must be at their job (this is commonly 10.00 to 16.00 hours).

(b) The flexible time is at the beginning and the end of each day and during this time it is up to the individual to choose when he arrives and leaves. Arrival and departure times would be recorded by some form of 'clocking in' system. The total working week or month for each employee must add up to the prescribed number of hours, though he may go into 'debit' or 'credit' for hours from day to day, in some systems.

4.12 A flexitime system can be as flexible as the company wishes.

(a) The most basic version involves flexibility only within the day, with no possibility of carrying forward debit or credit hours into other days, so if you arrive late, you work late. This type of system is used in some factories where transport difficulties make arrivals and departures difficult.

(b) Another system is flexible hours within the span of the week. Hours can be carried forward to the next day. This enables an employee to cope with a fluctuating work load without overtime and gives him some control over his work and leisure time.

(c) Finally, flexitime may be operated by the month. Each employee will have a coded key which can also serve as an identity card, and this will be used on arrival and departure by insertion into a key acceptor, which records hours worked on a 24 hour clock.

4.13 Flexitime enables workers to plan their lives on a more personal basis, as long as they work their contracted hours in the period. Where the employee works longer hours than necessary, he is usually able to save these hours for holiday or days off.

4.14 Morale is improved by allowing staff to arrange their days to fit particular needs, leisure pursuits, available transport and so on. Stress (related to frustration, monotony, isolation, fears associated with being 'late for work') is reduced. The temptation to take time off is also reduced: pioneer schemes showed that absenteeism dropped, because trips to the dentist, social visits and so on were allowed for, and the idea that 'I'm late for work; I may as well not go in at all' discounted.

'Flexible Working Hours not only bring order into this disorder [peak hour travel, traffic strikes etc]; within limits that are clearly defined and acceptable to management, it gives staff a new facility deliberately to vary arrival and departure times for personal reasons, and to do so with candour and dignity'. (M W Cuming, *Personnel Management*)

4.15 In *Office Magazine* (January 1987) it was reported how the Bedford Offices of the then Anglian Water Authority implemented a flexitime scheme. With offices on the edge of town, staff had been experiencing annoying traffic delays. However, the Authority was reluctant to incur extra clerical costs in implementing the system, and installed an electronic 'clock-in/clock-out' system to record staff hours. Staff could monitor their own hours, with flexibility up to four hours in debit and eight hours in credit in any one 'settlement' period of four weeks. The Authority found that attendance improved, and that staff were generally highly satisfied with the arrangement.

5. AUTOMATION AND TECHNOLOGY

5.1 It has been said that we are in the middle of a 'New Industrial Revolution' based on developments in the field of microchip technology. The effects of this revolution on jobs is largely a matter of the changing relationship between man and machine. Machines have progressively replaced human effort and skills, in the interests of increased efficiency, standardisation, speed and precision. Computers and micro-processors have changed the nature of production and clerical work alike, and we might expect continued changes in:

(a) the way we work (the electronic office or robotic factory);

(b) where we work (with communications bringing remote locations within reach);

(c) how we work (with new skills and materials);

(d) whether we work or not (the time a job takes and therefore the number of people required to do the work is altered by the new technology);

(e) how we live (because IT has extended into the home).

5.2 The progressive mechanism of production processes can be traced to demonstrate this.

5.3 Initially, engineering tasks were performed by hand, or by hand tool. The initiating source of the activity, the control of the 'machine response', and the power source all resided with the human operator. Towards the end of the 19th century it became possible to 'program' machine tools to carry out single functions, or even a sequence of functions, automatically. Initiation, control and power source resided within the machine: human skills were, however, required to strip and reset the machines.

5.4 With the beginning of computerisation in the 1950s, the work changed again. Even the task of setting and resetting the tool was performed by electronic - rather than mechanical - means.

5.5 During the 1970s cheap, reliable microprocessing technology was built into machine tools. Computers can initiate action according to variables in the environment, and can respond to signals (for error detection, for example) or can select from a range of programmed options (changing speed or position according to measurement data, or segregating/rejecting units according to size, or even identifying and initiating appropriate sequences of action). Computers are now even capable of 'adaptive control' - responding to working conditions as they arise (over-heating, dulling of tools) and modifying their behaviour over a wide range of variables - correcting after, during, or even before the event. Human intervention is reduced even further, to loading and removing workpieces, and replacing dulled tools although robotics is beginning to encroach on these areas also.

5.6 The work of several machines may be scheduled and controlled from a single central computer, which tells the operator, via a VDU (visual display unit), which workpieces to fit on which machines.

5.7 The systematic reduction of human intervention and control in work has made work less tiring and safer for the worker, and more efficient, less prone to human failings (carelessness, idleness or hostility) from the point of view of the organisation.

5.8 However, the American Marxist sociologist Harry Braverman suggests that 'the remarkable development of machinery becomes, for most of the working population, the source not of freedom but of enslavement, not of mastery but of helplessness, and not of the broadening of the horizon of labour, but of the confinement of the worker within a blind round of servile duties...'

5.9 Robert Blauner carried out a study in America of working conditions in various industries.

(a) Printing - dominated by *craft work* - allowed workers to set their own pace, choose their own methods, practice high-status skills and have strong social relationships at work.

(b) Cotton spinning - dominated by *machine minding* - offered textile workers simple, repetitive, non-discretionary work. However, they lived in close rural communities, whose 'culture' overcame the alienation experienced to a degree at work (see below).

(c) Car manufacture - then dominated by *process production* - gave assembly workers little control over their work, little perceived 'significance' in their tasks, social isolation in the work place, and no opportunity to develop skills. Alienation was very high.

(d) Chemicals manufacture was dominated by *process production* but gave the process workers an advanced work environment, where manual work was automated. The workers had team work and social contact, and control over their work pace and movements. They had opportunities to learn about the processes they monitored, and also derived satisfaction from a sense of responsibility, achievement and belonging.

5.10 Blauner identified feelings of *alienation* in workers, due to a sense of:

(a) powerlessness - loss of control over work and conditions;
(b) meaningless - loss of significance of work;

(c) isolation - loss of sense of belonging and relatedness; and
(d) self estrangement - loss of personal identity, of any sense of work as a central life activity.

His conclusion was that *advanced* technology - as employed only by the chemical processing plant at that time - actually eliminated alienation.

5.11 So is technology 'good' or 'bad' for the worker? Two sides have been argued.

(a) Process operators are victims of management's use of technology to create work that is unskilled - and unlikely to offer any learning opportunities - boring, repetitive, tightly controlled, lacking meaning, and socially isolating. Microelectronic extension of automation may alter the demand for operators' human skills, and may damage the quality of working life: it may replace the exercise of human mental capacity altogether.

(b) Process operators are skilled, knowledgeable decision makers, with responsibility, discretion and prosperous working conditions. Process automation eliminates dirt and hazard, and can offer a motivating work environment with task variety, meaning, learning opportunities and discretion. Electronic controls lack human flexibility and creativity: systems can enhance job skills and interest.

5.12 Obviously the motivational and political implications must be considered together with the nature of the technology - its capabilities and limitations, and the skills required to operate it effectively.

Computer

'It has a mathematical brain but it is not capable of exercising intuition, discretion or creativity; nor can it make moral or strategical judgements. If a problem can be mathematically posed or if it can be structured in a logical series of questions, then a computer will provide answers with remarkable speed and absolute accuracy. But real decisions, particularly those involving social implications, must be made by people. A mathematically viable solution is sometimes not acceptable when account has to be taken of such intangibles as moral attitudes, human and political responses and the vagaries of economic forces.' Ronald Pitfield

Attitudes to advanced technology

5.13 Buchanan and Huczynski comment that computers have in recent years moved from the 'background' functions of accounting and administration, and into 'foreground' tasks - text production, order processing, product and process design and equipment control.

5.14 Advanced technology therefore has a 'high-profile' image, which in the popular mind may have little basis in fact or experience. Common fears include the following.

(a) 'Computers/robots' are replacing people in industry.
(b) The only jobs left for people will be routine 'de-humanised' tasks.
(c) When the 'paperless office' arrives, what will happen to the clerical/administrative worker?
(d) Skills are becoming irrelevant, as is any desire for autonomy or control.
(e) We'll never learn to cope with all these new-fangled ways.

Let's look at some of these attitudes.

5.15 A principle fear is that of *replacement* - the substitution of 'intelligent' machines for people at work. (Information technology is in fact often *sold* on this basis - it will increase productivity because machines are more efficient than people.) Job opportunities may be lost and redundancies will result, in so far as machines *do* replace people. However, there are certain compensatory mechanisms, in that:

(a) technology generates new products and services (personal computers, computer games, compact disc players and so on) which encourage investment and expansion, with new employment opportunities;

(b) lower unit costs, arising from higher productivity, can be passed on to the consumer as lower or static prices. There may be increased demand for that - or other - product(s), arising from the extra spending power, again offering new employment opportunities;

(c) there will inevitably be a lapse of time, in any case, between the arrival of a new device or system, and its successful incorporation into existing facilities - which may have to be 'run down' over a lengthy period; 'natural' labour turnover may sufficiently reduce the workforce during this time to make redundancies unnecessary;

(d) the risks and costs involved in new technologies will make organisations slow to test and initiate changes. They also suggest that an organisation which 'takes the plunge' must be confident of expansion and increased demand - so there is a likelihood of new employment opportunities;

(e) new technology is not all-powerful - nor does it always live up to its advertising. Existing jobs, skills and systems may have to work alongside the new devices, for a while at least.

5.16 Technical change may even be seen as the way of *preserving* jobs in the face of technologically advanced foreign competitors who innovate faster and produce goods better and more cheaply than domestic technology-resistant firms. (The economic motive for technological change must not be underestimated: information gathering, storage, display, processing and transmission can be achieved with greater speed, accuracy and consistency through the use of information technology - and this must be desirable from the organisation's point of view.)

5.17 Another fear is that skills will die out, or become irrelevant and wasted. However, Buchanan and Huczynski suggest that: 'as machines do more, people do not necessarily do less. Computer technology may be tools that *complement* human skills and create more interesting and meaningful work.... The outcome depends to a large extent on *management decisions* on how to organise the work around the new devices.'

This is a very important concept. The introduction of technology *per se* cannot be seen in isolation from *managerial choices* about workplace layout, organisation of work, culture and so on. Look back at the work of Trist in the Durham coalfields described in the previous chapter. There was nothing wrong with the technology itself - but any technical system can be operated by a number of social systems, and faults in the work organisation and provision for the social system *rendered the technology ineffective*.

5.18 Computerised processes - for the 'process supervisor' in a computerised factory, as for the word processor operator in the office - *can* offer satisfaction.

(a) Retained discretion to control the process, and responsibility for its correct, safe operation (in the case of a word processor, responsibility for file management, since errors may cause files to be lost or erased).

(b) New 'user' skills in a high-status field - with opportunities to develop those skills as the technology progresses.

(c) Rapid feedback on performance.

(d) Satisfaction from more efficient and productive performance.

(e) Often, a modern and highly 'designed', working environment - the 'modern office'.

(f) Identification with the 'new revolution', innovation, high-tech and so on.

5.19 In a comparison by Buchanan and Boddy (1977) of the word processor typist's job with the copy typist's, it was found that the VDU operators:

(a) had more variety (associated with greater output), but fewer breaks in the rhythm of work - that is, more continuous time at the keyboard;

(b) felt they had more control over the quality and appearance of the end product - with text manipulation and corrections on screen, and not showing up on the printed version;

(c) had less to do in the way of preparation and other tasks; they did not have to handle or load paper, file finished documents or lay out pages (taking account of line ends, page ends etc) which was done automatically. Corrections were simple and worry-free (compare the copy typist reaching the end of an otherwise perfect page...). However, the VDU operators had other tasks of file management, data protection and so on and needed to acquire skills in the use of codes, menu selection and other operating mechanisms;

(d) found it physically easier to use - with a lighter touch, flatter keyboard, quieter operation (particularly with the new, silent laser printers) - though more demanding on the concentration.

In this study, jobs of the VDU operators were not found to have been enhanced overall, because of changes in work organisation that deprived the typists of author contact and involvement in whole, 'meaningful' tasks, but the advantages of the new *technology* - as opposed to the old - were admitted.

5.20 In some cases, the resistance to a new system might stem from a fear that it will result in a loss of status for the *department* concerned. For example, the management of the department might believe that a computer system will give 'control' over information gathering and dissemination to another group in the organisation. Dysfunctional behaviour might therefore find expression in:

(a) interdepartmental squabbling about access to information;

(b) a tendency to disregard the new sources of information, and to stick to old methods of collecting information instead.

5.21 New systems might also disrupt the established *social system* in the office. Individuals who are used to working together might be separated into different groups, and individuals used to working on their own might be expected to join a group. Office staff used to moving around and mixing with other people in the course of their work might be faced with the prospect of having to work much more in isolation at a keyboard, unable to move around the office as much. Where possible, new systems should be designed so as to leave the 'social fabric' of the workplace undamaged. Group attitudes to change should then be positive rather than negative.

5.22 Shoshana Zuboff has argued that IT can have a devastating affect on *middle management*. This is because the expertise required by workers to use IT is the type of expertise which was once the preserve of middle management. IT also enables senior management to delve far deeper into the tactical and operational behaviour of the organisation. Middle managers are thus threatened from above and below.

5.23 A new system will reveal weaknesses in the previous system, and so another fear of computerisation is that it will show up exactly how poor and inefficient previous methods of information gathering and use had been. If individuals feel that they are put under pressure by the revelation of any such deficiencies, they might try to find fault with the new system too. When fault-finding is not constructive - not aimed at improving the system - it will be dysfunctional in its consequences.

Exercise 4

Staffords was originally a traditional corner shop selling a range of groceries and other sundry goods and staffed by Mr Stafford and two shop assistants. On Mr Stafford's retirement the shop has been bought by a chain which is proposing to redesign the shop, expanding it and running it along the lines of a modern '7-11' mini-supermarket.

What sort of changes do you imagine will occur to the work environment and how will the jobs of the two assistants be affected?

6. JOB DESIGN

6.1 There is no particular mystique about 'job design': it is merely the way in which tasks are fragmented or grouped to form a given job, and what decisions are made about specialisation, discretion, autonomy, variety and other job elements.

Early job design

6.2 Frederick Taylor ('scientific management') was an early exponent of systematic job design. His technique was as follows.

(a) Decide on the optimum degree of task fragmentation, breaking down a complex job into its simplest component parts.

(b) Decide the most efficient way of performing each component.

(c) Train employees to carry out a single task in the 'best' way.

6.3 This micro-division of labour is based on a production line organisation of work, and offers some efficiencies.

(a) Each task is so simple and straightforward that it can be learned with very little training.

(b) Since skills required are low, the effects of absenteeism are minimised: workers can be shifted from one task to another very easily.

(c) Similarly, high labour turnover is not critical, because replacements can be found and trained without difficulty.

(d) Tasks are closely defined, standardised and timed, so output and quality are more easily predicted and controlled.

Problems of task fragmentation

6.4 The question of 'job design' acquired its prominence when human relations theorists became interested in the motivational aspects of the job itself, and the role of 'job satisfaction' in employee performance. It was recognised that jobs made up of low-skilled, repetitive tasks (of which there will inevitably be some, in any organisation's operations) could offer little satisfaction to the workers performing them, being socially isolating, meaningless and monotonous.

6.5 Studies of human behaviour at work also suggested that the existence of such tasks poses problems for management.

(a) Monotony, and the experience of boredom, is part of what may be called 'industrial *fatigue*', an element which interferes with the 'steady state' in which workers work best. Tasks which provide little mental stimulation for the worker may result in inattention, daydreaming or preoccupation with social interactions and diversions. Errors and even accidents may result from this. If the worker has *no* social outlet, however, the strain of monotony is even worse.

(b) Stress is related to high workload, low discretion jobs. Its symptoms - including nervous tension, withdrawal and low morale - will invariably affect performance.

(c) Motivation will suffer, unless particular efforts are made to compensate the workers for lack of satisfaction in the work itself.

(d) If such tasks are perceived to be the lot of the worker ('us'), under the control of management ('them'), interpersonal relations between manager and workers will be hampered.

6.6 We mentioned earlier how job design and work organisation might minimise the experience of monotony. Such areas of interest focused attention on job design as a factor in employee motivation, satisfaction and performance. A systematic approach to job satisfaction and its relationship to job design was first put forward by Frederick Herzberg in the 1950s.

6.7 Herzberg's theory suggested that the job itself can be a source of satisfaction, offering various ways of meeting the individual's needs for personal growth. A similar prescription was drawn up by other theorists. Concepts such as combining tasks, forming natural work units, establishing client relationships, vertical loading (ie increased delegation, reduced controls) and feedback were said to result in enhancement of work experience in five *core job dimensions*:

(a) skill variety;
(b) task identity;
(c) task significance;
(d) autonomy;
(e) feedback.

The experience of these dimensions was said to meet employee growth needs and lead to high motivation and satisfaction, high quality performance and low absenteeism/turnover rates.

6.8 Buchanan and Huczynski point out that 'the design of an individual's job determines both the kinds of rewards that are available and what the individual has to do to get those rewards'. Intrinsic rewards - for example satisfaction and accomplishment - are more directly linked to the job itself, and have a more direct influence on motivation (and - theoretically - therefore work performance). Moreover, job design affects an individual's experience of work, his expectations and valued outcomes.

6.9 Interest therefore came to be focused on 'job enrichment' as a job design technique. Job enrichment was popular throughout the 1970s and contributed significantly to the 'quality of working life movement', and the high-profile work organisation experiments of companies such as Volvo, Saab-Scania and Atlas Copco, the introduction of autonomous working groups and so on.

6.10 However, the world economic recession of the 1980s and 1990s diverted management theorists' attention away from such issues. As Buchanan and Huczynski note: 'The quality of working life is less important when there is little work to be had.'

6.11 Nevertheless, the theories have offered managers ideas about what their subordinates look for and get out of their work, and what variables can be manipulated to give them greater challenge and satisfaction in their work. Relatively simple managerial changes, for example giving more direct feedback on performance, or reducing the number of formal controls on employee behaviour, can affect the employee's experience of the core job dimensions (paragraph 6.7 above).

Chapter roundup

- The work environment can have positive and negative effects on productivity.
- Organisations are required to ensure that health and safety legislation is followed, and that fire regulations are observed.
- Routine and monotony, while maximising skills, can cause accidents as workers get bored.
- An *ergonomic* approach recognises that the work environment should be designed to fit the people that use it.
- Technology can remove some of the routine and boring work, but causes its own problems.
- Job design is a means of structuring a job to meet some of the employee's goals.

Test your knowledge

1. Why is the physical environment of work important to organisational effectiveness? (see paras 1.1-1.4)
2. What considerations should be taken into account in the layout of office space? (2.5-2.7)
3. List some of the costs of accidents to employers. (3.6)
4. What are four causes of stress? (3.19)
5. What is ergonomics? (3.22)
6. What are the advantages and disadvantages in human terms of the three main types of shift work. (4.9, 4.7-4.8, 4.10)
7. What are people's main fears about the introduction of IT? (5.14)
8. What is the main ingredient in making IT successful? (5.17)

Now try question 4 at the end of the text

PART B

THE WORK GROUP AND LEADERSHIP

Chapter 5

THE NATURE OF WORK GROUPS

This chapter covers the following topics.

1. What is a group?
2. The formation of groups
3. Group behaviour
4. Communicating in groups
5. Effective and ineffective work groups

Signpost

- Leadership of groups is discussed in the next chapter, which should be regarded as a continuation of this one.
- Other relevant chapters are Chapter 1 (Formal and informal organisations), Chapter 13 (Motivation and rewards) and Chapter 16 (Communication).

1. WHAT IS A GROUP?

1.1 Handy in *Understanding Organizations* (1993) defines a group as 'any collection of people who perceive themselves to be a group'. The point of this definition is the distinction it implies between a random collection of individuals and a 'group' of individuals who share a common sense of identity and belonging.

1.2 Groups have certain attributes that a random 'crowd' does not possess.

(a) *A sense of identity*. Whether the group is formal or informal, its existence is recognised by its members: there are acknowledged boundaries to the group which define who is 'in' and who is 'out', who is 'us' and who is 'them'. People generally need to feel that they 'belong', that they share something with others and are of value to others. Organisations try to establish a sense of corporate identity or company image among their employees as well as among outsiders through advertising and public relations.

(b) *Loyalty to the group*, and acceptance within the group. This generally expresses itself as conformity or the acceptance of the 'norms' of behaviour and attitude that bind the group together and exclude others from it. This type of 'solidarity' may formalise itself in entrance qualifications, rule books, oaths of allegiance or may be an unspoken acceptance of norms: if you have ever travelled on a commuter train you will know how a group can develop a 'style' of its own, without its existence ever having to be formalised. Again, organisations try to encourage employee loyalty and commitment to the rules and objectives of the organisation, and a sense of solidarity against competition and problems.

(c) *Purpose and leadership*. Most groups have an express purpose, aim or set of objectives, whatever field they are in: most will, spontaneously or formally, choose individuals or sub-groups to lead them towards the fulfilment of those goals. Strength of personality, a high level of expertise, seniority of age or status and other factors will determine who 'rises to the top' in any group. This hierarchy will be desirable so that the group can co-ordinate and control its members and their activities: it will of course be highly developed in formal organisations.

1.3 You should bear in mind that although an organisation as a whole may wish to project itself as a large group, with a single, united identity, loyalty and purpose, any organisation will in fact be composed of many sub-groups, with such attributes of their own. People in organisations will be drawn together into groups by:

(a) a preference for small groups, where closer relationships can develop;
(b) the need to belong and to make a contribution that will be noticed and appreciated;
(c) familiarity: a shared office or canteen;
(d) common rank, specialisms, objectives and interests;
(e) the attractiveness of a particular group activity (joining an interesting club, say);
(f) resources offered to groups (for example sports facilities);
(g) 'power' greater than the individuals could muster (trade union, pressure group).

Formal and informal groups

1.4 Some such groupings will be the result of *formal* directives from the organisation: for example, specialists may be 'thrown together' in a committee set up to investigate a particular issue or problem; a department may be split up into small work teams in order to facilitate supervision.

(a) *Informal* groups may spring up as a result of these formal arrangements (for example if the members of the committee become friends), and will invariably be present in any organisation. Informal groups include workplace 'cliques', and networks of people who regularly get together to exchange information, groups of 'mates' who socialise outside work and so on. They have a constantly fluctuating membership and structure, and leaders emerge usually through personal (rather than 'positional') power. The purposes of informal groups are usually related to group and individual member satisfaction, rather than to a task.

(b) *Formal* groups will have a formal structure; they will be consciously organised for a function allotted to them by the organisation, and for which they are held responsible - they are task oriented, and become *teams*. Leaders may be chosen within the group, but are typically given 'legal' authority by the organisation. Permanent formal groups include standing committees, management teams (for example the board of directors) or specialist services (an information technology support). *Temporary* formal groups include task forces, designed to work on a particular project, ad hoc committees and so on.

Primary working groups

1.5 It is worth looking again at Trist's ideas on the role of the *primary working group*. Primary groups are important in an industrial organisation because:

(a) they form the smallest units within it;

(b) they are the immediate social environment of the individual worker; and

(c) the methods and team-spirit which prevail there will determine the efficiency and inter-relationships of the whole company.

1.6 The optimum size of a primary working group is important; the intimate, face-to-face relationships on which a primary group depends cannot be formed among more than, say, a dozen people. Anthony Jay (*Corporation Man*) identifies a group of ten - a 'ten group' - as the linear descendant of the primaeval hunting-band, balancing the individuality which is necessary for *generating* new ideas with the support and comradeship necessary for *developing* them.

1.7 In the sense that most industrial work falls naturally into small groups of up to a dozen or so, primary groups are commonly found in industry. In the sense that these groups are provided for in the plan of organisation and given *official* leaders, primary groups have (until relatively recently) rarely been found. Too often, still, the official organisation of a company comes to an end well above the primary group level.

1.8 Whenever the formal organisation of a company fails to provide for primary groups, informal primary groups will spring up. These will usually have a self-protective purpose and will offer roles to unofficial leaders whose aims will not necessarily be in harmony with the official aims of the organisation.

1.9 There are many roles in the organisation (those of setters, overlookers and the like) which are in effect leadership roles. In many cases, however, these are recognised only for their technical content and not for their man-management content. If the people in such positions were held responsible for the whole of their leadership role - for relationships and team spirit among the group as well as for its technical working - management organisation would extend to the primary working groups.

Exercise 1

A major trend in modern industry and commerce is to dismantle traditional organisation structures and ways of doing things and introduce small multi-disciplinary teams.

For example, Asea Brown Boveri, the Swedish-Swiss engineering business, aims to halve all lead times in the company's activities by the end of 1993. It has already made significant progress through the creation of what it calls 'Target Oriented High Performance' work teams made up of 10 to 15 workers. This has replaced a system whereby orders were handed down from above through different, fragmented departments with one that is organised around the flow of production through the team approach.

The traditional, highly time-consuming system involved specialised demarcation of responsibilities for sales, stocks, production and distribution and following from this a high level of bureaucratic managerial control and top-heavy administration. In place of this have been put smaller, flexible work teams with wider responsibilities, allowing the barriers between administration and production to be abolished.

The results are impressive. The strategy 'has cut the time for making high-voltage direct current transmission equipment from three to two years. The time for supplying customers with standard switch gear has fallen from three to five weeks to three to five days from receipt of the order to delivery. Cycle times in ABB's components division have been reduced from 86 to 35 days.' (*Financial Times,* February 1993). Meanwhile, employees have a better working environment, greater job interest, and the facility to upgrade their skills constantly.

Your task is to read the business pages of a quality newspaper on a regular basis and watch out for similar examples.

The function of groups

1.10 From the organisation's standpoint, groups can be used for:

(a) performing tasks which require the collective skills of more than one person;

(b) testing and ratifying decisions made outside the group;

(c) consulting or negotiating, especially to resolve disputes within the organisation;

(d) creating ideas (acting as a 'think tank');

(e) exchanging ideas, collecting and transmitting information;

(f) co-ordinating the work of different individuals or other groups;

(g) motivating individuals to devote more energy and effort into achieving the organisation's goals.

1.11 From the individual's standpoint groups also perform some important functions.

(a) They satisfy social needs for friendship and belonging.

(b) They help individuals to develop images of themselves (a person may need to see himself as a member of the corporate planning department or of the works snooker team).

(c) They enable individuals to help each other in matters which are not necessarily connected with the organisation's purpose (people at work may organise a baby-sitting circle).

(d) They enable individuals to share the burdens of any responsibility they may have in their work.

1.12 Brayfield and Crockett suggested that an individual will always identify with some group or other.

(a) If he identifies with a *social* group, outside work, this will have no effect on his job performance.

(b) On the other hand, if he identifies with his *work-mates* and work group his performance will be affected. However, a congenial work group, creating high morale, can either raise or lower productivity, depending on the 'norm' adopted by the group. The group norm will probably be based on its collective idea of a fair day's work for a fair day's pay but such an attitude is a subjective one, so what is 'fair' may be a high or low standard of efficiency.

(c) An employee who identifies with his *company* may not be the most productive, through caring for quality rather than output level. It has been suggested that the employees who are most likely to win recognition and promotion within the company are those who are critical of its policies, and not those who identify with them.

2. THE FORMATION OF GROUPS

2.1 Groups are not static. They mature and develop. Four stages in this development were identified by Tuckman: forming, storming, norming and performing.

Forming

2.2 During the first stage the group is just coming together, and may still be seen as a collection of individuals. Each individual wishes to impress his personality on the group, while its purpose, composition, and organisation are being established. The individuals will be trying to find out about each other, and about the aims and norms of the group. There will at this stage probably be a wariness about introducing new ideas, and the established line will be the respected and stated one: members will not wish to appear radical or unacceptable to the group, even if their minds are full of ideas which would 'rock the boat'. Individuals will also be 'feeling out' each other's attitudes and abilities: no-one will want to appear less informed or skilled than the others. The objectives being pursued may as yet be unclear and a leader may not yet have emerged.

2.3 This settling down period is essential, but may be time wasting: the team as a unit will not be used to being autonomous, and will probably not be an efficient agent in the planning of its activities or the activities of others. It may resort to complex bureaucratic procedures to ensure that what it is doing is at least something which will not get its members into trouble.

Storming

2.4 The second stage is called *storming* because it frequently involves more or less open conflict between group members. There may be changes agreed in the original objectives, procedures and norms established for the group. If the group is developing successfully this may be a fruitful phase as more realistic targets are set and trust between the group members increases. The element of risk enters solutions to problems and options may be proposed which are more far-reaching than would have been possible earlier. Whilst the first stage of group development involved 'toeing the organisational line', the second stage brings out the identification of team members with causes: this may create disagreement, and there may also be political conflict over leadership.

Norming

2.5 The third stage is a period of settling down. There will be agreements about work sharing, individual requirements and expectations of output. Group procedures and customs will be defined and adherence to them secured. The enthusiasm and brain-storming of the second stage may be less apparent, but norms and procedures may evolve which enable methodical working to be introduced and maintained. This need not mean that initiative, creativity and the expression of ideas are discouraged, but that a reasonable hearing is given to everyone and 'consensus' sought.

Performing

2.6 Once the fourth stage has been reached the group sets to work to execute its task. Even at earlier stages some performance will have been achieved but the fourth stage marks the point where the difficulties of growth and development no longer hinder the group's objectives.

2.7 It would be misleading to suggest that these four stages always follow in a clearly-defined progression, or that the development of a group must be a slow and complicated process. Particularly where the task to be performed is urgent, or where group members are highly motivated, the fourth stage will be reached very quickly while the earlier stages will be hard to distinguish. Some groups, moreover, never progress beyond storming, because the differences are irreconcilable.

3. GROUP BEHAVIOUR

Group norms

3.1 A work group establishes 'norms' or acceptable levels and methods of behaviour, to which all members of the group are expected to conform. This group attitude will have a negative effect on an organisation if it sets unreasonably low production norms. Groups often apply unfair treatment or discrimination against others who break their 'rules'.

3.2 'Norms' are partly the product of 'rôles' and rôle expectations of how people in certain positions behave, as conceived by people in related positions. Rôle theory will be important in understanding relationships within groups, and the conflicts and ambiguities that may arise.

In a classic experiment by Sherif, participants were asked to look at a fixed point of light in a black box in a darkroom. Although the point of light is fixed, it so happens that in the darkness, it *appears* to move. Each participant was asked to say how far the light moved, and their individual estimates were recorded.

They were next put into a small group where each member of the group gave their own estimates to the others. From this interchange of opinions, individuals began to change their minds about how far the light had moved, and a group 'norm' estimate emerged.

When the groups were broken up, each individual was again asked to re-state his estimate; significantly, they retained the group norm estimate and rejected their previous individual estimate.

The experiment showed the effect of group psychology on establishing norms for the individual himself; even when, as in the case of the experiment, there is no factual basis for the group norm.

3.3 The general nature of group pressure is to require the individual to share in the group's own identity, and individuals may react to group norms and customs with:

(a) compliance - 'toeing the line' without real commitment;
(b) internalisation - full acceptance and identification; or
(c) counter-conformity - rejecting the group and/or its norms.

3.4 Pressure is strongest on the individual when:

(a) the issue is not clear-cut;
(b) he lacks support for his own attitude or behaviour; and
(c) he is exposed to other members of the group for a length of time.

Intelligent and independent individuals conform less readily. Those who do conform easily tend to be more conventional in their social values, less self-confident and lacking in spontaneity.

3.5 Norms may be reinforced in various ways by the group.

(a) *Identification*: the use of 'badges', symbols, perhaps special modes of speech, 'in-jokes' and so on - the marks of belonging, prestige and acceptance. There may even be 'initiation rites' which mark the boundaries of membership.

(b) *Sanctions* of various kinds. Deviant behaviour may be dealt with by ostracising or ignoring the member concerned ('sending him to Coventry'), by ridicule or reprimand, even by physical hostility. The threat of expulsion is the final sanction.

In other words the group's power to induce conformity depends on the degree to which the individual values his membership of the group and the rewards it may offer, or wishes to avoid the negative sanctions at its disposal.

The Hawthorne Studies

3.6 Elton Mayo was an academic based at Harvard University. His work sheds light on the importance of groups within an organisation. His most interesting findings emerged from the 'Hawthorne studies', so called because they were conducted at the Hawthorne plant of the Western Electric Company.

3.7 The experiments arose from an attempt by Western Electric to find out the effects of lighting standards on worker productivity. As a test, it moved a group of girls into a special room with variable lighting, and moved another group of girls into a room where the lighting was kept at normal standards. To the astonishment of the company management, productivity shot up in both rooms. When the lighting was then reduced in the first room, as a continuation of the test, not only did productivity continue to rise in the first room, but it also rose still further in the second room. Mayo was called in to investigate further.

3.8 Mayo brought to his researches a background of psychology. As a professor of industrial research at the Graduate School of Business, Harvard University, from 1927 to 1947, he had a great influence on the development of the social sciences and on other students of management and practising managers as well.

3.9 The Hawthorne studies consisted of five stages.

(a) *Stage one*: the company itself carried out an investigation into the effects of lighting on production and morale. Output increased in both the control (consistent lighting) and experimental (varied lighting) groups - even when the experimental group was subjected to very poor lighting conditions. Mayo and his colleagues were called in to identify the 'mystery' factor at work in the groups.

(b) *Stage two*: the Relay Assembly Test Room stage. Six women were separated from the others in a different room, where they were observed under changing working conditions. Checks were kept, with records not only of working conditions, lighting, heating and rest periods but also of the women's private lives. In most cases, the women were consulted in advance about the changes. Productivity rose - whatever the changes, good or bad! Mayo concluded that this was what was later called 'the Hawthorne effect': the response of the women appeared to be affected by their sense of being a group singled out for attention.

(c) *Stage three*. Stage two had suggested that employee attitudes and values were important. The company set up an interview programme designed to survey attitudes towards supervision, jobs and working conditions. The major conclusion was that relationships with people at work were important to employees.

(d) *Stage four*: the Bank Wiring Observation Room stage, conducted by Roethlisberger and Dickson. Fourteen men were separated from the department, and put in an observation room to work under more or less normal conditions. The group was seen to set its own rules and attitudes and standards of output. Its behaviour became oriented towards its own interests - manipulating the incentive scheme, restricting output - in a way that seemed beyond the supervisor's control. In other words, the group had developed into a powerful, self-protecting informal organisation.

(e) *Final stage*. Employee counselling was used to work through problems, and to improve relationships and personal adjustment at work.

3.10 The conclusions of the studies were that individual members must be seen as part of a group, and that *informal* groups exercise a powerful influence in the workplace: supervisors and managers need to take account of social needs if they wish to secure commitment to organisational goals.

> 'Management, by consultation with the girl workers, by clear explanation of the proposed experiments and the reasons for them, by accepting the workers' verdict in several instances, unwittingly scored a success in two most important human matters - the girls became a self-governing team, and a team that co-operated wholeheartedly with management'.
>
> *Elton Mayo*

3.11 Some of the most interesting results emerged during stage four, in the Banking Wiring Room. The results of the experiments were as follows.

(a) The group developed a keen sense of its own identity; however, it divided into two separate cliques, one of which felt that it did more difficult, 'higher status' work than the other.

(b) The group as a whole developed certain 'norms' with regard to both output and supervision.

   (i) With regard to output:

      (1) the group appeared to establish a standard amount of production which was 'fair' for the pay they received;

      (2) members of the group who produced above this normal level of output, or who shirked work and did less than the norm, were put under 'social pressure' by work-mates to get back into line.

   (ii) With regard to supervision, the group view was that supervisors should not be officious and take unpleasant advantage of their position of authority. One officious supervisor was put under 'social pressure' by the group, so that he asked for a transfer out of the group.

(c) The group did not follow company policy on some issues, with regard to work practices. In addition, daily reports of output were 'fiddled'; sometimes workers recorded more output than they actually produced, and on other days they recorded less than they produced. The effect was to report constant volumes of daily production, whereas actual daily volume, according to tiredness or morale on the day, fluctuated considerably.

(d) Individual production rates varied significantly, but not according to individual ability or intelligence. Members of the 'high status' clique produced more than members of the other clique; but both the highest-producing employees and lowest-producing employees were the 'social outcasts' from either group. The men were paid a productivity incentive, and the high-status group felt that the other group were shirking; this resulted in considerable ill-feeling, and an eventual decline in the productivity of the group as a whole.

The Hawthorne Studies in perspective

3.12 The Hawthorne studies were the first major attempt to undertake genuine social research, and to redirect attention to human factors at work. They are enduringly popular with managers, not least because they have an apparent simplicity and a straightforward, enthusiastically 'sold' message, without over-attention to academic rigour, cautious qualification etc. The Hawthorne human-relations approach - like Maslow, Herzberg, Theory X and Y and other well-known examples - is simple and memorable, and so seems more 'practical' to managers than more complex, difficult to operate theories like Expectancy Theory (described in a later chapter).

3.13 By modern standards of research, however, the methodology was 'less than rigorous in many respects' (G A Cole, *Personnel Management*). A balanced evaluation of their value admits that:

(a) the studies have had practical impact on organisation practice - beyond the Western Electric company where the research was originally carried out. Organisations like Marks and Spencer, IBM and the John Lewis Partnership (admittedly, exceptionally) have flourished by the application of an approach consciously geared to positive human relations - though also ensuring rigorous selection procedures and tight management control;

(b) Elton Mayo's human relations ideas were applied in the Western Electric Company - but didn't work. The company set up a programme of employee counselling which commenced in 1936 with a team of five counsellors for 600 employees, and grew to 55 counsellors for 21,000 employees in 1948. However, the scheme declined and died out in 1956. The programme was time-costly, as well as expensive; enthusiasm waned as the founders left. The conditions which made the initial experiments a success (the sense of status enjoyed by the girls in the Relay Assembly Test Room, because of their participation in the research) were no longer present when the experimental situation (the counselling service) was made available organisation-wide.

3.14 Buchanan and Huczynski address the question of whether a 'happy' group is also a productive one, as the Hawthorne studies suggested. 'The Hawthorne studies signalled the birth of the Human Relations School of Management ... It was not until some time had passed that people started to question this relationship between productivity and satisfaction. Perhaps it had been a fortuitous coincidence rather than some iron law? Sociologists who reviewed the findings and compared them with other data swung to the former explanation.'

Exercise 2

A quick test, just to review some of the points made so far.

(a) One of the attributes of a group is loyalty. How is this normally expressed?
(b) In what ways might an individual react to pressure to conform to group norms?
(c) How may norms be reinforced?
(d) 'Happy bees make more honey.' Is this a universal truth?

Solution

The answers can all be found in the preceding sections. Just in case you miss it, the answer to (d) is very definitely *No*!

Group cohesion and competition

3.15 In an experiment reported by Deutsch (1949), psychology students were given puzzles and human relation problems to work at in discussion groups. Some groups ('co-operative' ones) were told that the grade each individual got at the end of the course would depend on the performance of his group. Other groups ('competitive' ones) were told that each student would receive a grade according to his own contributions.

3.16 No significant differences were found between the two kinds of group in the amount of interest and involvement in the tasks, or in the amount of learning. But the co-operative groups, compared with the competitive ones, had greater productivity per unit time, better quality of product and discussion, greater co-ordination of effort and sub-division of activity, more diversity in amount of contribution per member, more attentiveness to fellow members and more friendliness during discussion.

3.17 Another experiment, conducted in 1949 by Sherif and Sherif, set out to investigate how groups are formed, and how relationships between groups are created. The 'guinea pigs' of the experiment were 24 boys of about 12 years old who were taken to a summer camp. After a few days, 'natural' affinities were discounted by breaking up friendships which had formed, and dividing 'the boys' into two 'formal' groups, the Bulldogs and the Red Devils.

3.18 It was found that when the groups were formed there was a noticeable switch of friendships. Boys whose previous 'best friends' were moved into the other group began to switch 'best friendships' to someone else who belonged to their group. The group identity thus had a significant effect on the attitudes of individual members.

3.19 The experimenters also tried to create friction between the groups; these efforts were so successful that by the end of the experiment there was such intense inter-group rivalry that subsequent attempts to re-unite the entire camp were insufficient to restore common goodwill. From this, and other research, it is argued that new members of a group quickly learn the norms and attitudes of the others, no matter whether these are 'positive' or 'negative', friendly or hostile. It is also suggested that inter-group competition may have a positive effect on group cohesion and performance.

3.20 Within each competing group:

(a) members close ranks, and submerge their differences; loyalty and conformity are demanded;

(b) the 'climate' changes from informal and sociable to work and task-oriented; individual needs are subordinated to achievement;

(c) leadership moves from democratic to autocratic, with the group's acceptance;

(d) the group tends to become more structured and organised.

3.21 Between competing groups:

(a) the opposing group begins to be perceived as 'the enemy';

(b) perception is distorted, presenting an idealised picture of 'us' and a negative stereotype of 'them';

(c) inter-group communication decreases - facilitating the perceptual distortion.

3.22 In a 'win-lose' situation, where competition is not perceived to result in benefits for both sides, the *winning* group will:

(a) retain its cohesion;
(b) relax into a complacent, playful state ('fat and happy');
(c) return to group maintenance and concern for members' needs;
(d) be confirmed in its group 'self-concept' with little re-evaluation.

3.23 The *losing* group will:

(a) deny defeat if possible, or place the blame on the arbitrator, or the system;

(b) lose its cohesion and splinter into conflict, as 'blame' is apportioned;

(c) be keyed-up, fighting mad ('lean and hungry');

(d) turn towards work-orientation to regroup, rather than members' needs or group maintenance;

(e) tend to learn by re-evaluating its perceptions of itself and the other group. It is more likely to become a cohesive and effective unit once the 'defeat' has been accepted.

3.24 Members of a group will act in unison if the group's existence or patterns of behaviour are threatened from outside. Cohesion is naturally assumed to be the result of positive factors such as communication, agreement and mutual trust - but in the face of a 'common enemy' (competition, crisis or emergency) cohesion and productivity benefit.

Cohesion and 'group think'

3.25 It is possible for groups to be *too* cohesive, too all-absorbing. Handy notes that 'ultra-cohesive groups can be dangerous because in the organisational context the group must serve the organisation, not itself.'

3.26 If a group is completely absorbed with its own maintenance, members and priorities, it can become dangerously blinkered to what is going on around it, and may confidently forge ahead in a completely wrong direction. I L Janis describes this as 'group think': 'the psychological drive for consensus at any cost, that suppresses dissent and appraisal of alternatives in cohesive decision-making groups.'

3.27 The cosy consensus of the group prevents consideration of alternatives, constructive criticism or conflict. Symptoms of 'group think' include:

(a) a sense of invulnerability - blindness to the risk involved in 'pet' strategies;
(b) rationalisations for inconsistent facts;
(c) moral blindness - 'might is right';
(d) a tendency to stereotype 'outsiders' and 'enemies';
(e) strong group pressure to quell dissent;
(f) self-censorship by members - not 'rocking the boat';
(g) perception of unanimity - filtering out divergent views;
(h) mutual support and solidarity to 'guard' the decision.

3.28 Different types of group and organisational structures can foster this phenomenon. Group cohesion may have the effect both of creating a 'self-serving' purpose to the group - whereby loyalty to the group agenda may outweigh (and be in conflict with) the interests of the organisation as a whole - and of strengthening the hold which group norms exercise over individual actions, to quell dissent. The psychological dissonance experienced by an individual whose views are in conflict with those of fellow group members may be acute: the individual, under pressure *either* to 'devalue' the group's importance in his mind, *or* to bring his views into line with it, is - in a cohesive group - more likely to participate in the 'group think' phenomenon of self-censorship and solidarity.

Highly political organisations and areas of the organisation structure - for example the top and centre - also tend to put pressure on individuals to 'avoid rocking the boat' - whether this be expressed as cultural control (through tradition and 'shared' values) or as more overt rules and sanctions.

3.29 Since by definition a group suffering from group think is highly resistant to criticism, recognition of failure and unpalatable information, it is not easy to break such a group out of its vicious circle. It must be encouraged to:

(a) actively encourage self-criticism;
(b) welcome outside ideas and evaluation; and
(c) respond positively to conflicting evidence.

The 'risky-shift' phenomenon

3.30 'Risky shift' is a phenomenon suggested by the discussion of 'group think' above. It is the term given to the tendency of individuals in groups to take greater risks than their individual, pre-discussion preferences. Stoner illustrated the phenomenon in experiments with groups of management students in 1961.

3.31 Explanations for the phenomenon include:

(a) the sharing - and therefore diffusion - of any sense of responsibility for the outcome of the decision;

(b) the reinforcement during discussion of cultural values associated with risk, like courage, boldness, strength. In some cultures - especially those fostered in the mid 1980s by contemporary ideas on 'excellence', innovation, entrepreneurship and so on - such values may be put forward as desirable, and 'risk aversion' derided.

3.32 There are, however, problems of validity with the 'risky shift' research. The experiments are conducted on artificial, short-lived, leaderless groups in artificial situations. The decisions they are called on to make are purely hypothetical, and have no practical effect: it is easy to 'take risks' if there is no real 'downside' involved. The research cannot be used to formulate predictions about group decision-making in real-life contexts. In addition, results have varied; the size of the shift (and even its direction) are by no means consistent, from hypothetical situation to situation, from individual to individual, or from study to study.

4. COMMUNICATING IN GROUPS

4.1 In a well-known 'laboratory' test (1951) H J Leavitt examined the effectiveness of four communication networks for written communication between members of a small group. The four network patterns were:

(a) the *circle*:

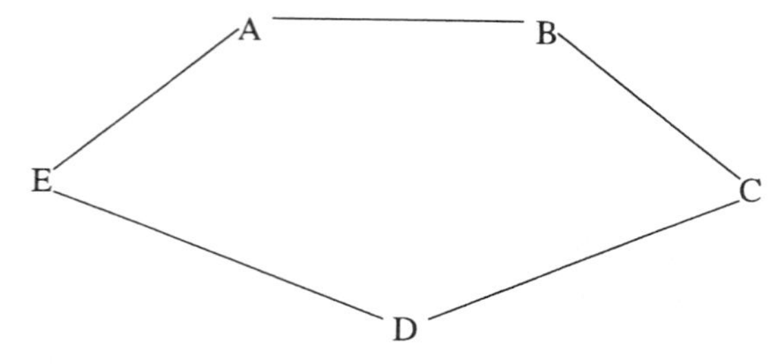

in which each member of the group could communicate with only two others in the group, as shown;

(b) the *chain*:

A - B - C - D - E

Similar to the circle, except that A and E cannot communicate with each other and are therefore at both ends of a communication chain;

(c) the *Y*:

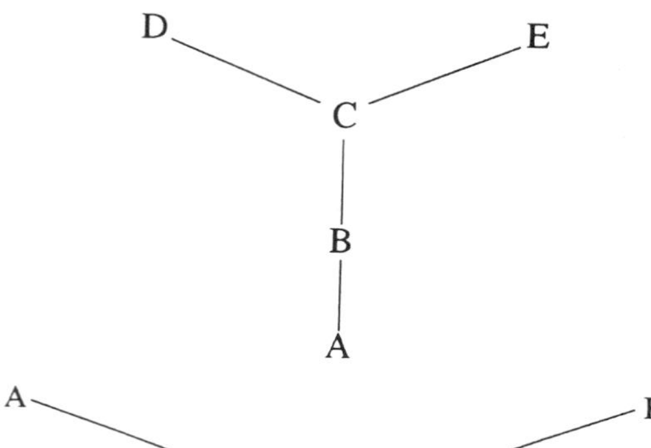

(d) the *wheel*:

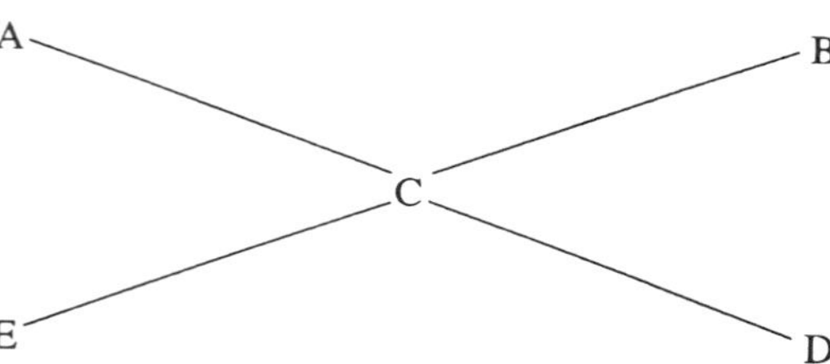

In both the 'Y' and the 'wheel' patterns, C occupies a more central position in the network.

4.2 In Leavitt's experiment, each member of a group of five people has to solve a problem and each has an essential piece of information. Only written communication, channelled according to one of the four patterns described above, was allowed.

The findings of the experiment were as follows:

(a) *speed of problem-solving*: the wheel was fastest, followed by the Y, the chain and finally the circle;

(b) *leadership*: in the circle, no member was seen as the leader, but in the other three types of groups, C was regarded as a leader (more so in the wheel and the Y);

(c) *job satisfaction*: the enjoyment of the job was greatest among members of a circle, followed by the chain, the Y and lastly the wheel (but see paragraph 4.5(c) below).

4.3 The progression (one way or the other) of Circle-Chain-Y-Wheel emerged in all these findings. Leavitt wrote 'the Circle, one extreme, is active, leaderless, unorganised, erratic and yet is enjoyed by its members. The Wheel at the other extreme is less active, has a distinct leader, is well and stably organised, is less erratic and yet is unsatisfying to most of its members'. He concluded that in organisations where there is minimal 'centrality' and 'peripherality' of individuals in a communication system, the organisation will be active, error-prone, leaderless, slow and enjoyed by its members; whereas in organisations where there is greater 'centrality' of some individuals and 'peripherality' of others, the organisation will be stable and efficient, consisting of leaders and the led (with low enjoyment among the members).

4.4 One significant communication pattern which was omitted from Leavitt's experiment was the *all-channel* communication system which might be practically employed in group working:

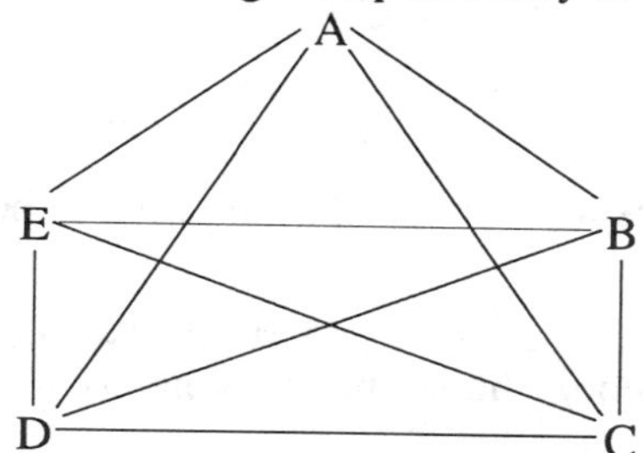

4.5 In a comparison of the all-channel system, the wheel and the circle, as methods of group communication and group decision making, the following results emerged.

(a) *Simple problem-solving*: the wheel system solves problems quickest, and the circle is the slowest, with the all-channel system in between.

(b) *Complex problem-solving*: the all-channel system, with its participatory style, and more open communication system, generally provides the best solutions to complex problems. The efficiency of the wheel depends on the ability of the leader, or central figure. In the circle, there is a lack of co-ordination and solutions to problems are poor.

(c) *Job satisfaction*: contrary to Leavitt's findings, it is now argued that job satisfaction in the circle is low, because of poor, or slow performance in decision-making and a lack of co-ordination and although job satisfaction in the wheel system is low for individuals away from the centre, the man at the centre has high satisfaction. The all-channel system provides fairly high job satisfaction to all group members (although the size of the group must not be so large that some individuals feel excluded from participation).

4.6 It must *not* be supposed that an all-channel system is best under all circumstances. It solves complex problems well, but slowly, and it tends to disintegrate under pressure (for example time pressure to get results) into a wheel system.

Meetings

4.7 Anthony Jay has argued that 'even the most apparently futile of meetings serves some wider corporate purpose as well as preserving the mental health of those who attend'. In other words the role of meetings is a wide-ranging one. Not all meetings are about 'getting things done', and 'getting things done' is only one of the purposes of having meetings at all.

4.8 The wider corporate purposes of meetings include the following.

(a) Ritual. If, as Minztberg suggests, a manager has a ceremonial role, it is quite possible that this is acted out in meeting. Often, meetings (between the managing directors of supplier and customer, say) might merely be the ritual ending of months of negotiation.

(b) Communication and personal contact. A meeting helps people to get to know one another. Establishing good personal relationships is important in organisational life.

(c) Meetings also enable colleagues to 'let off steam'. Arguably, there is sometimes a good purpose in having an argumentative session in which most grievances become aired. Disagreement might be seen as a useful way of generating new ideas.

(d) Even though meetings do not do anything, they may legitimise other managerial decisions. For example, the formal organisation structure may not be effective. Instead, a strong informal organisation may exist. A committee meeting may give authority to power.

(e) The fact, or at least the illusion, of participation in decisions may improve individual motivation.

(f) Meetings enable the various different interests in a decision to be represented as 'equals'.

(g) Meetings can be used to delay decisions. There are certain instances when this can be of benefit to the organisation or to managers.

(h) Meetings can also be used to put individuals on the spot. They can enforce commitment, or at least indicate where battle lines are drawn.

(i) Some meetings can be inspirational, if they are used to persuade, cajole or encourage a sense of values.

(j) Finally, bringing people together, underlines the fact that they belong to the same organisation, and in theory should be working to the same purpose. Perhaps this is a function of the ritual aspect outlined above.

5. EFFECTIVE AND INEFFECTIVE WORK GROUPS

Team building

5.1 Team building is the process of diagnosing problems in the functioning of a group and then devising exercises and training activities to help improve performance.

5.2 The management problem is how to create an effective, efficient work group. The criteria of group effectiveness are:

(a) fulfilment of task and organisational goals;
(b) satisfaction of group members.

5.3 Handy takes a contingency approach to the problem of group effectiveness, which, he argues, depends on:

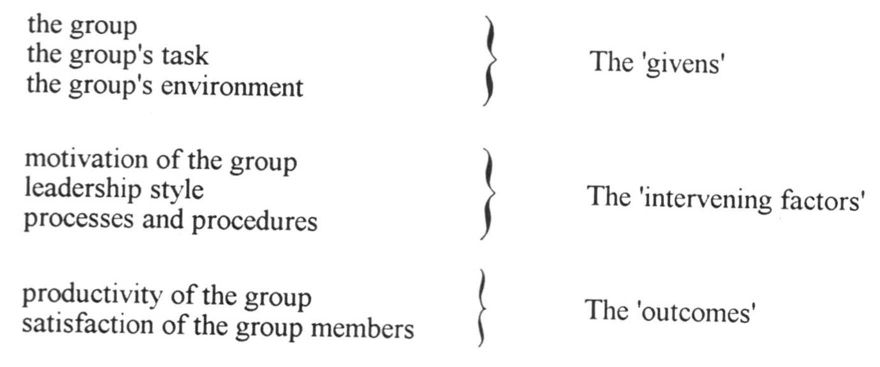

Exercise 3

Where have you met the term 'contingency approach' before in this text, and what does it mean? *Before* reading on, suggest what you think might be the gist of Handy's argument.

The givens

5.4 The personalities and characteristics of the individual members of the group, and the personal goals of these members, will help to determine the group's personality and goals. An individual is likely to be influenced more strongly by a small group than by a large group in which he may feel like a small fish in a large pond, and therefore unable to participate effectively in group decisions.

The group: team roles and Belbin

5.5 It has been suggested that the effectiveness of a work group depends on the blend of the individual skills and abilities of its members. For example, a project team might be most effective if it contains:

(a) a person of originality and ideas;

(b) a 'get-up-and-go' person with considerable energy, enthusiasm and drive;

(c) a quiet, logical thinker, who ponders carefully and criticises the ideas of others;

(d) a plodder, who is happy to do the humdrum routine work; and

(e) a conciliator, who is adept at negotiating compromises or a consensus of thought between other members of the group.

5.6 Belbin, in a study of business-game teams at Carnegie Institute of Technology in 1981, discovered that a differentiation of influence among team members (that is, agreement that some members were more influential than others) resulted in higher morale and better performance.

5.7 Belbin's picture of the most effective character-mix in a team (which many managers have found a useful guide to team working) involves eight necessary roles which should ideally be balanced and evenly 'spread' in the team:

(a) the *co-ordinator* - presides and co-ordinates; balanced, disciplined, good at working through others;

(b) the *shaper* - highly strung, dominant, extravert, passionate about the task itself, a spur to action;

(c) the *plant* - introverted, but intellectually dominant and imaginative; source of ideas and proposals but with disadvantages of introversion;

(d) the *monitor-evaluator* - analytically (rather than creatively) intelligent; dissects ideas, spots flaws; possibly aloof, tactless - but necessary;

(e) the *resource-investigator* - popular, sociable, extravert, relaxed; source of new contacts, but not an originator; needs to be made use of;

(f) the *implementer* - practical organiser, turning ideas into tasks, scheduling, planning and so on. Trustworthy and efficient - but not excited; not a leader, but an administrator;

(g) the *team worker* - most concerned with team maintenance - supportive, understanding, diplomatic; popular but uncompetitive - contribution noticed only in absence;

(h) the *finisher* - chivvies the team to meet deadlines, attend to details; urgency and follow-through important, though not always popular.

Belbin has also identified a ninth team-role, the *specialist*, who joins the group to offer expert advice when needed. Examples are legal advisers, PR consultants, finance specialists and the like.

Exercise 4

(a) Which of these categories do you think best describes your own character?

(b) Which do you think is most appropriate for a company secretary? Would you give the same answer whatever type of organisation was involved and whatever the range of responsibilities of the company secretary?

(c) How well does Belbin's picture describe a team of which you have been a member?

The task

5.8 The nature of the task, the second of Handy's 'givens', must have some bearing on how a group should be managed. If a job must be done urgently, it is often necessary to dictate how things should be done, rather than to encourage a participatory style of working. Jobs which are routine, unimportant and undemanding will be insufficient to motivate either individuals or the group as a whole. If individuals in the group want authoritarian leadership, they are also likely to want clearly defined group targets.

The environment

5.9 The group's environment relates to factors such as the physical surroundings at work and to inter-group relations. An open-plan office, in which the members of the group are closely situated, is more conducive to group cohesion than a situation in which individuals are partitioned into separate offices, or geographically distant from each other. Group attitudes will also be affected, as described previously, by relationships with other groups, which may be friendly, neutral or hostile.

Intervening factors: processes and procedures

5.10 Of Handy's 'intervening factors', motivation and leadership are discussed at length in separate chapters of this text. With regard to processes and procedures, research indicates that a group which tackles its work systematically will be more effective than a group which lives from hand to mouth, and muddles through - but this is often true of individuals as well.

5.11 In an ideal functioning group:

(a) each individual gets the support of the group, a sense of identity and belonging which encourages loyalty and hard work on the group's behalf;

(b) skills, information and ideas are 'pooled' or shared, so that the group's capabilities are greater than those of the individuals;

(c) new ideas can be tested, reactions taken into account, persuasive skills brought into play in group discussion for decision-making and problem-solving;

(d) each individual is encouraged to participate and contribute and thus becomes personally involved in and committed to the group's activities;

(e) goodwill, trust and respect can be built up between individuals, so that communication is encouraged and potential problems more easily overcome.

5.12 Unfortunately, group working is rarely such an undiluted success. There are certain constraints involved in working with others.

(a) Awareness of 'role' and group norms for behaviour, the desire to be acceptable to the group, may restrict individual personality and flair. This may perhaps create pressure or a sense of 'schizophrenia' for the individual concerned who can't 'be himself' in a group situation.

(b) Conflicting roles and relationships (where an individual is a member of more than one group) can cause difficulties in communicating effectively, especially if sub-groups or cliques are formed in conflict with other groupings.

(c) The effective functioning of the group is dependent upon each of its members, and will suffer if members dislike or distrust each other, or if one member is so dominant that others cannot participate, or so timid that the value of his ideas is lost, or so negative in attitude that constructive communication is rendered impossible.

(d) Rigid leadership and procedures may strangle initiative and creativity in individuals. On the other hand, differences of opinion and political conflicts of interest are always likely and if all policies and decisions are to be determined by consultation and agreement within the group, decisions may never be reached and action never taken.

The outcomes

5.13 High productivity may be achieved if work is so arranged that satisfaction of individuals' needs coincides with high output. Where teams are, for example, allowed to set their own improvement goals and methods and to measure their own progress towards those goals, it has been observed (by Peters and Waterman among others) that they regularly exceed their targets.

5.14 Individuals may bring to the group their own 'hidden agendas' for satisfaction: these are goals which may have nothing to do with the declared aims of the group such as protection of a sub-group, impressing the boss, inter-personal rivalry and so on. The danger is that the more cohesive the group, the more its own maintenance and satisfactions may take precedence over its task objectives, and the more collective power it has - even to sabotage organisational goals.

5.15 Like organisations, therefore, groups have 'cultures' which will contribute importantly to the effectiveness or otherwise of the group's operation and member satisfaction. Peters and Waterman (*In Search of Excellence*) outline the cultural attributes of successful *task force* teams. They should:

(a) be small - requiring the trust of those who are not involved;

(b) be of limited duration and working under the 'busy member theorem', in other words 'get off the damned task force and back to work';

(c) be voluntary - which ensures that the business is 'real';

(d) have an informal structure and documentation, in other words no bulky paperwork, and open communication;

(e) have swift follow-up, in other words be *action* oriented.

The characteristics of effective and ineffective work teams

5.16 If a manager is to try to improve the effectiveness of his work team he must be able to identify the different characteristics of an effective group and an ineffective group. Some pointers to group efficiency are quantifiable measures; others are more qualitative factors and are difficult to measure. A number of different factors should be considered, because a favourable showing in some aspects of work does not necessarily mean that a group is operating effectively. No one factor on its own is significant, but taken collectively the factors will present a picture of how well or badly the group is operating.

Quantifiable factors

Effective work group	*Ineffective work group*
(1) Low rate of labour turnover	(1) High rate of labour turnover
(2) Low accident rate	(2) High accident rate
(3) Low absenteeism	(3) High absenteeism
(4) High output and productivity	(4) Low output and productivity
(5) Good quality of output	(5) Poor quality of output
(6) Individual targets are achieved	(6) Individual targets are not achieved
(7) There are few stoppages and interruptions to work	(7) Much time is wasted owing to disruption of work flow
	(8) Time is lost owing to disagreements between superior and subordinates

Qualitative factors

Effective work group	*Ineffective work group*
(1) There is a high commitment to the achievement of targets and organisational goals	(1) There is no understanding of organisational goals or the role of the group
(2) There is a clear understanding of the group's work	(2) There is a low commitment to targets
(3) There is a clear understanding of the role of each person within the group	(3) There is confusion and uncertainty about the role of each person within the group
(4) There is trust between members and free and open communication	(4) There is mistrust between group members and suspicion of the group's leaders
(5) There is idea sharing	(5) There is little idea sharing
(6) The group is good at generating new ideas	(6) The group does not generate any good new ideas
(7) Group members try to help each other	(7) Group members make negative and hostile criticisms about each other's work
(8) There is group problem solving which gets to the root causes of the work problem	(8) Work problems are dealt with superficially, with attention paid to the symptoms but not the cause
(9) There is an active interest in work decisions	(9) Decisions about work are accepted passively
(10) Group members seek a united consensus of opinion	(10) Group members hold strongly opposed views
(11) The members of the group want to develop their abilities in their work	(11) Group members find work boring and do it reluctantly
(12) The group is sufficiently motivated to be able to carry on working in the absence of its leader	(12) The group needs its leader there to get work done

Chapter roundup

- A group has certain attributes that a crowd does not possess; groups have many important functions in organisations.
- You should be able to describe how groups form, and a variety of features of group behaviour such as 'group think'.
- Think hard about effective and ineffective groups to which you have belonged: the problems of turning the latter into the former form the basis of many a 'case-study' style examination question.
- As a summary of certain points, it is worth noting that for all its opportunities for exchange of ideas and knowledge, immediate feedback, 'brainstorming' and so on, the 'group' as a work unit is not necessarily superior to the individual in terms of performance in all situations.
    - Decision-making may be a cumbersome process where consensus has to be reached - and it has been shown (rather surprisingly) that teams take *riskier* decisions than the individuals comprising them.
    - Group norms may work to lower the standard rate of unit production. However, individuals need groups psychologically: isolation can produce stress and hostile behaviour, and can impair performance.
    - Group cohesion may provide a position of strength from which to behave in hostile or deviant (from the organisation's point of view) ways. On the other hand, teams can be committed to organisational objectives, and many provide a much stronger impetus for change than individuals.
    - Groups have been shown to produce fewer - though better evaluated - ideas than the individuals of the group working separately. However, a group *will* often produce a better solution to a problem than even its best individual, since 'missing pieces' can be added to his performance.

Test your knowledge

1. What is the function of groups, from the organisation's standpoint and from the individual's? (see paras 1.10, 1.11)
2. Outline Tuckman's model of group formation. (2.1-2.6)
3. What are norms? (3.1)
4. Evaluate the Hawthorne studies. (3.12-3.14)
5. Can group cohesion be counter-productive for the organisation? (3.25-3.32)
6. Which type of group communication is best for simple problem solving, and for complex problem solving? (4.4)
7. List six possible purposes of a meeting. (4.8)
8. Define team building. (5.1)
9. What character/ability mix is required for an effective group? (5.5, 5.6)
10. Identify twelve features of an ineffective work team. (5.15)

Now try question 5 at the end of the text

Chapter 6

LEADERSHIP AT WORK

This chapter covers the following topics.

1. Managers
2. Supervisors
3. Managers and supervisors as leaders
4. Theories of leadership
5. Entrepreneurs
6. The usefulness of leadership theories

Signpost

- 'Followership' is an important part of leadership and so this chapter should not be read in isolation from Chapter 5.
- Authority and power were discussed in Chapter 1. Organisation theories like the human relations approach and contingency theory were described in Chapter 2, and organisation culture in Chapter 3.
- Management development is a topic in Chapter 11.

1. MANAGERS

Functions of management

1.1 Henri Fayol made one of the earliest systematic attempts to define the functions of management.

(a) *Planning*. This involves selecting objectives, and the strategies, policies, programmes and procedures for achieving the objectives either for the organisation as a whole or for a part of it. Planning might be done exclusively by managers who will later be responsible for performance: however, advice on planning decisions might also be provided by 'staff management' who do not have 'line' authority for putting the plans into practice. Expert advice is nevertheless a part of the management planning function.

(b) *Organising*. This involves the establishment of a structure of tasks which need to be performed to achieve the goals of the organisation, grouping these tasks into jobs for an individual, creating groups of jobs within sections and departments, delegating authority to carry out the jobs, and providing systems of information and communication, and for the co-ordination of activities within the organisation.

(c) *Commanding*. This involves giving instructions to subordinates to carry out tasks over which the manager has authority for decisions and responsibility for performance.

(d) *Co-ordinating*. This is the task of harmonising the activities of individuals and groups within the organisation, which will inevitably have different ideas about what their own goals should be. Management must reconcile differences in approach, effort, interest and timing of these separate individuals and groups. This is best achieved by making the individuals and groups aware of how their work is contributing to the goals of the overall organisation.

(e) *Controlling*. This is the task of measuring and correcting the activities of individuals and groups, to ensure that their performance is in accordance with plans. Plans must be made, but they will not be achieved unless activities are monitored, and deviations from plan identified and corrected as soon as they become apparent.

1.2 Fayol's analysis of management functions is only one of several similar types of analysis. Other functions which might be identified, for example, are *staffing* (filling positions in the organisation with people), leading (unlike commanding, 'leading' is concerned with the interpersonal nature of management) and acting as the organisation's representative in dealing with other organisations (an ambassadorial or public relations role).

1.3 Peter Drucker added a further basic function of management in business: *economic performance*. In this respect, the business manager is different from the manager of any other type of organisation. Management of a business can only justify its existence and its authority by the economic results it produces, however significant the non-economic results which occur as well.

1.4 Drucker described the jobs of management within this basic function of economic performance as:

(a) *managing a business*. The purposes of the business are:

(i) to create a customer; and
(ii) innovation;

(b) *managing managers*. The requirements here are:

(i) management by objectives;
(ii) proper structure of managers' jobs;
(iii) creating the right spirit in the organisation;
(iv) making a provision for the managers of tomorrow;
(v) arriving at sound principles of organisation structure;

(c) *managing worker and work*.

Managerial roles: Mintzberg

1.5 Another way of looking at the manager's job is to observe what managers actually do, and from this to draw conclusions about what 'roles' they play or act out. This is known as the *managerial roles* approach.

1.6 *Henry Mintzberg* identified ten managerial roles, which may be taken on as appropriate to the personality of the manager and his subordinates and the nature of the task in hand.

Interpersonal roles

(a)	Figurehead	(performing ceremonial and social duties as the organisation's representative, for example at conferences)
(b)	Leader	(of people, uniting and inspiring the team to achieve objectives)
(c)	Liaison	(communication with people outside the manager's work group or the organisation)

Informational roles

(d)	Monitor	(receiving information about the organisation's performance and comparing it with objectives)
(e)	Disseminator	(passing on information, mainly to subordinates)
(f)	Spokesman	(transmitting information outside the unit or organisation, on behalf of the unit or organisation)

Decisional roles

(g)	Entrepreneur	(being a 'fixer' - mobilising resources to get things done and to seize opportunities)
(h)	Disturbance-handler	(rectifying mistakes and getting operations - and relationships - back on course)
(i)	Resource allocator	(distributing resources in the way that will most efficiently achieve defined objectives)
(j)	Negotiator	(bargaining - eg for required resources and influence)

1.7 The mix of roles varies from job to job and situation to situation: a manager will, as it were, put on the required 'hat' for his task. A manager will, however, wear some hats more than others: senior officials are more likely to be called upon to act as figureheads than sectional managers and supervisors, who will be more concerned with resource allocation and disturbance-handling.

1.8 In modern management theories, particular emphasis has been placed on leadership and entrepreneurship, at team level as well as organisational level: involving and committing employees to achieving goals, and focusing on creative action and resource mobilisation to get things done.

1.9 The importance of Mintzberg's theories is that they cast significant doubt on the view that real managers spend their mornings planning, and their afternoons controlling, and that they rely on sophisticated information systems. Managerial work is often interrupted, and managers are more interested in the grapevine than rows of figures.

Exercise 1

Here is an extract from Tom Peters *Liberation Management*, talking about working in a project team in a forward looking organisation.

'Self-managed teams. Scheduling. Budgeting. Quality control. Projects. These are the new worker's daily fare. But such activities have long been the almost exclusive province of the senior and middle ranks. It's hardly surprising, then, that many execs are asking: *Is the worker up to the challenge?* To understand why in 9 cases out of 10 ... the answer is a resounding "yes", just look at life *off* the job. "Managing" day-to-day affairs encompasses most big-business challenges.'

Peters goes on to give examples of this last point, including the following, which is relevant to the question of management.

'*Self-management*. Experts say the shift to self-managed teams takes years of training and preparation. Really? What's a family but a "self-managed team"? Complex, "flex-time scheduling algorithms" for a seven-person work team pale beside the logistics of a family with two working adults (holding three jobs between them) and a pair of teenagers.'

Do you agree that there is no difference between the management of the family unit and the management of a small team at work? Answer by reference to Fayol's and Mintzberg's views of the tasks and roles of management

Solution

Peters' idea may have some appeal (probably depending upon your own experience of family life!), and if you worked through each of Fayol's and Mintzberg's points in turn you probably found many analogies. However, consider the different relationships and emotional ties in either situation before subscribing to it completely. And what about the worker whose family role provides much needed relief from his or her role at work, or vice versa?

Being a manager: the views of Handy

1.10 Charles Handy suggested that a definition of a manager or a manager's role is likely to be so broad as to be fairly meaningless. His own analysis of being a manager was divided into three aspects.

(a) The manager as a general practitioner
(b) The managerial dilemmas
(c) The manager as a person

The manager as a general practitioner

1.11 A manager is the first recipient of an organisation's problems and he must:

(a) identify the symptoms in the situation (low productivity, high labour turnover, severe industrial relations problems and so on);

(b) diagnose the disease or cause of the trouble;

(c) decide how it might be dealt with - that is, develop a strategy for better health;

(d) start the treatment.

1.12 Typical strategies for health were listed as follows.

(a) *People*: changing people, either literally or figuratively:

(i) hiring and firing;
(ii) re-assignment;
(iii) training and education;
(iv) selective pay increases;
(v) counselling or admonition.

(b) *The work and the structure*:

(i) re-organisation of reporting relationships;
(ii) re-definition of the work task;
(iii) job enrichment;
(iv) re-definition of roles.

(c) *The systems and procedures*. In other words to amend or introduce:

(i) communication systems;
(ii) rewards systems (payment methods, salary guides);
(iii) information and reporting systems;
(iv) budgets or other decision-making systems (eg stock control, debtor control).

The managerial dilemmas

1.13 Managers are paid more than workers because they face constant dilemmas which they have to resolve. These dilemmas are as follows.

(a) *The dilemma of the cultures*. The cultures of organisations were described earlier in this text. It is management's task to decide which culture of organisation and management is required for his particular task. As a manager rises in seniority, he will find it necessary to behave in a culturally diverse manner to satisfy the requirements of his job and the expectations of his employees. In other words, managers must be prepared to show flexibility and good judgement in their choice of organisation culture.

> The manager 'must be flexible but consistent, culturally diverse but recognisably an individual with his own identity. Therein lies the dilemma. Those ....who relapse into a culturally predominant style will find themselves rightly restricted to that part of the organisation where their culture prevails. Middle layers of organisations are often overcrowded with culturally rigid managers who have failed to deal with this cultural dilemma'

(b) *The dilemma of time horizons*. This is the problem of responsibility for both the present and the future at the same time. Concentration on short-term success may be at the expense of the evolution and innovation required for survival and growth in the long term.

(c) *The trust-control dilemma*. This is the problem of balance between management's wish to control the work for which they are responsible, and the necessity to delegate work to subordinates, implying trust in them to do the work properly. The greater trust a manager places in subordinates, the less control he retains himself, which can be risky and stressful. Retaining control implies a lack of trust in subordinates. 'The managerial dilemma is always how to balance trust and control'.

(d) *The commando leader's dilemma*. In many organisations, junior managers show a strong preference for working in project teams, with a clear task or objective, working outside the normal bureaucratic structure of a large formal organisation and then disbanding. Unfortunately, there can be too many such 'commando groups' for the stability of the total organisation. A manager's dilemma is to decide how many entrepreneurial groups he should create to satisfy the needs of his subordinates and the demand for innovation, and how much bureaucratic organisation structure should be retained for efficiency, consistency and 'safety'.

The manager as a person

1.14 Management is developing into a 'semi-profession' and managers expect to be rewarded for their professional skills. The implications for individual managers are that 'increasingly it will come to be seen as the individual's responsibility to maintain, alter or boost his skills, to find the right market for his skills and to sell them to the appropriate buyer'. The manager must be regarded as an individual in his own right, with his own objectives: he does not exist solely within and for the benefit of the organisation.

1.15 Another consequence of this is that the 'traditional' view that an organisation should employ 'raw recruits' and nurture them into its management structure might in future no longer be accepted. 'There will be no obligation to continue to employ the individual when the benefits of his skills begin to be less than their costs'.

2. SUPERVISORS

What is a supervisor?

2.1 A supervisor is a person who is ranked below management level in an organisation, but who is responsible for directing other staff below him. One definition of a supervisor is 'a person selected by management to take charge of a group of people, or special task, to ensure that work is carried out satisfactorily. 'Supervisors exist in all functions of an organisation where numbers of junior employees or operatives work. There are supervisors for factory workers (ie foremen) and for clerical workers.

2.2 You will probably have your own ideas about what supervisors do, and you may well have identified for yourself already the following characteristics of supervision.

(a) A supervisor is a 'front line' manager, dealing directly with people who do the bread-and-butter work. Managers, in contrast, usually have supervisors or other managers as subordinates.

(b) A supervisor does not spend all his time 'managing' other people. Much of his time will be spent doing operational work himself.

(c) A supervisor is responsible for getting things done.

(d) Supervisory work is really management work, but at a lower level in the organisation than that of managers.

What do supervisors do?

2.3 Perhaps the best way of looking at a supervisor's job is to regard it as a junior management job. The tasks of supervision can then be listed under similar headings to the tasks of management. (You are certainly not expected to learn these lists, but glance over them to get a feel for the diversity of 'supervisorship'.)

2.4 *Planning*

(a) Planning work so as to meet work targets or schedules set by more senior management

(b) Alternatively, planning work so as to meet targets set by the supervisor himself

(c) Planning the work load for each employee or machine; making estimates of overtime required; making allowances for employees absent on holiday or with long-term illness

(d) Planning the total resources required by the section (machines, employees) to meet the total work-load

(e) Planning work methods and procedures

(f) Planning maintenance work

(g) Planning safety procedures

(h) Attending departmental planning meetings

(i) Preparing budgets for the section

(j) Planning staff training and staff development

(k) Planning the induction of new staff

(l) Planning improvements in the work environment (decorating the room, obtaining extra furniture, changing room layout)

2.5 *Organising and overseeing the work of others*

(a) Ordering materials from internal stores
(b) Ordering materials from external suppliers
(c) Ordering new equipment for the section
(d) Authorising spending by others on materials, sundry supplies or equipment
(e) Interviewing and selecting staff
(f) Authorising overtime
(g) Allocating work to staff/machines
(h) Allocating machines/equipment to staff
(i) Reorganising work (for example when urgent jobs come in)

(j) Establishing performance standards for staff/machines
(k) Organising transport
(l) Deciding job priorities
(m) Maintaining stock levels for the section
(n) General 'housekeeping' duties
(o) Maintaining liaison with more senior management

2.6 *Controlling: making sure the work is done properly*

(a) Keeping records of: materials requisitioned and purchased, and materials used; total time worked on the section, analysed by individual employees, time spent on each job, time spent idle

(b) Deciding when sub-standard work must be re-done

(c) Reporting work progress and attending progress control meetings

(d) Dealing with trade union representatives

(e) Dealing with personal problems of staff

(f) Disciplining staff (for late arrival at work and so on)

(g) Counselling staff

(h) Dismissing staff

(i) Ensuring that work procedures are followed

(j) Ensuring that the quality of work is sustained to the required levels

(k) Ensuring that safety standards are maintained

(l) Checking the progress of new staff/staff training

(m) Co-ordinating the work of the section with the work of other sections

(n) Ensuring that work targets are achieved, and explaining the cause to senior management of any failure to achieve these targets

2.7 *Motivating employees and dealing with others: discipline*

(a) Dealing with staff problems
(b) Dealing with people in other sections
(c) Reporting to a senior manager
(d) Dealing with customers
(e) Motivating staff to improve work performance
(f) Applying disciplinary measures to subordinates who act unreasonably or work badly
(g) Helping staff to understand the organisation's goals and targets
(h) Training staff, and identifying the need for more training

2.8 *Communicating*

(a) Telling employees about plans, targets and work schedules

(b) Telling managers about the work that has been done, and the attitudes of staff to the work and work conditions

(c) Calling and holding discussion meetings

(d) Filling in reports (for example absentee reports for the personnel department, job appraisal forms)

(e) Writing memos, notes and reports

(f) Passing information between employees and managers, and between sections (by word of mouth, or in writing)

(g) Collecting information and distributing it to the other persons interested in it. Keeping up-to-date with developments

2.9 *Doing*

(a) Doing operational work
(b) Standing in for a senior manager when he or she is on holiday or otherwise absent
(c) Repairing equipment
(d) Giving advice to others to help solve problems

Legal and trade union influences on supervision

2.10 Legislation in the UK such as the Health and Safety at Work Act has placed responsibility on the supervisor for safe methods of working at the work place. The supervisor as well as the company can be prosecuted for breaches of statutory safety regulations.

2.11 The legislation regarding protection of employment makes provision for employees to take cases of what they believe are 'unfair dismissal' to a tribunal where compensation can be awarded. In view of this, most companies do not allow supervisors to dismiss employees, but only to recommend dismissal, the final decision being taken by senior management.

2.12 The influence of shop stewards and workplace representatives, who are 'leaders' in their own right, has also had implications for supervision. Many companies have evolved and negotiated a 'disciplinary and grievance procedure' with the trade unions, which allows the shop steward to be present when an employee is reprimanded by a manager, or feels that he is being badly treated.

2.13 Because of these developments it is important that supervisors and managers are familiar with company rules and procedures and the general working of industrial relations within their particular company. Many of the more progressive companies arrange courses, discussions and seminars on these topics.

3. MANAGERS AND SUPERVISORS AS LEADERS

3.1 Leadership is the process of influencing others to work willingly towards an organisation's goals, and to the best of their capabilities. ' The essence of leadership is *followership*. In other words it is the willingness of people to follow that makes a person a leader' (Koontz, O'Donnell, Weihrich). Leadership would also be listed as *one* of the functions of management.

3.2 Leadership comes about in a number of different ways.

(a) A manager is appointed to a position of authority within the organisation. He relies mainly on the (legitimate) authority of that position. Leadership of his subordinates is a function of the position he holds (although a manager will not necessarily be a 'leader', if he lacks leadership qualities).

(b) Some leaders (for example in politics or in trade unions) might be elected.

(c) Other leaders might emerge by popular choice and through their personal drive and qualities. Unofficial spokesmen for groups of people are leaders of this style.

3.3 Leaders are *given* their roles by their putative followers; their 'authority' may technically be removed if their followers cease to acknowledge them. The *personal, physical or expert* power of leaders is therefore more important than position power alone.

3.4 Leaders are the creators and 'sellers' of culture in the organisation. 'The [leader] not only creates the rational and tangible aspects of organisations, such as structure and technology, but is also the creator of symbols, ideologies, language, beliefs, rituals and myths.' (*Pettigrew*)

3.5 If a manager had indifferent or poor leadership qualities his subordinates would still do their job, but they would do it ineffectually or perhaps in a confused manner. By providing leadership, a manager should be able to use the capabilities of subordinates to better effect: leadership is the 'influential increment over and above mechanical compliance with the routine directives of the organisation' (Katz and Kahn, *The Social Psychology of Organisations*). Managers therefore need - and should seek - to become leaders in situations where they require co-operation, not merely compliance by their immediate subordinates.

> 'Leadership over human beings is exercised when persons with certain motives and purposes mobilize, in competition or conflict with others, institutional, political, psychological and other resources so as to arouse, engage and satisfy the motives of followers.' *Gregor Burns*

3.6 Since leadership is concerned with influencing others, it is necessary to have some understanding about what motivates people to work. Motivation is the subject of an earlier chapter of this text, but it may be summarised briefly as the process which determines how much effort, energy and excitement a person is prepared to expend in his work. Koontz, O'Donnell and Weihrich formulate the principle that 'since people tend to follow those whom they see as a means of satisfying their own personal goals, the more managers understand what motivates their subordinates and how these motivations operate, and the more they reflect this understanding in carrying out their managerial actions, the more effective leaders they are likely to be.'

Exercise 2

Just above we use the terms 'personal', physical', 'expert' and 'position' power. What do these terms mean? Where did you first encounter them in this Study Text and in what context?

Solution

Use the index - and do this whenever you encounter terms that have been explained but whose meaning you have forgotten. Otherwise, how can you expect to understand what you are reading?

---

4. THEORIES OF LEADERSHIP

Trait theories

4.1 Early writers like Taylor believed that the capacity to 'make others do what you want them to do' was an inherent characteristic: you either had it, or you didn't: leaders were 'born, not made'. Studies on leadership concentrated on the personal *traits* of existing and past leadership figures.

One study by Ghiselli did show a significant correlation between leadership effectiveness and the personal traits of intelligence, initiative, self assurance and individuality. This is logical enough, and what you would expect. However, consider the following.

(a) The approach does not take account of the individuality of the *subordinates* and other factors in the *leadership situation*: 'A person does not become a leader by virtue of the possession of some combination of traits, but the pattern of personal characteristics of the leader must bear some relevant relationship to the characteristics, activities and goals of the followers.' (*Stodgill*)

(b) Jennings (1961) wrote that 'Research has produced such a variegated list of traits presumed to describe leadership, that for all practical purposes it describes nothing. Fifty years of study have failed to produce one personality trait or set of qualities that can be used to distinguish between leaders and non-leaders.'

(c) The 'great man' approach does not help organisations to make better managers or leaders - it merely attempts to help them recognise a leader when they see one.

(d) The full list of traits is so long that it appears to call for a man or woman of extraordinary, even superhuman, gifts to be a leader.

Though superficially attractive, trait approaches are now largely discredited.

The style approach

4.2 The conclusion was that if leadership is not an innate gift, a *style* of leadership appropriate to a given work situation could be learned and adopted.

4.3 *Charisma* is frequently identified as a leadership trait and it is certainly accepted as an ingredient of leadership (as opposed to management), since it enables a person to get work done through others without the exercise of pure positional authority: in other words it relates to *personal power*. It has traditionally been surrounded by a certain 'mystique', but House (*A Theory of Charismatic Leadership*, 1976) suggests that this power can be acquired and developed, through training in:

(a) role modelling - setting an example to followers;

(b) image creation - engaging in behaviour designed to create an impression of competence and success;

(c) confidence building - communicating confidence in and high expectations of followers;

(d) goal articulation - voicing goals laden with moral overtones;

(e) motive arousal - behaving in an inspiring manner.

4.4 Four different types or styles of leadership were identified by Huneryager and Heckman (1967).

(a) *Dictatorial style*: the manager forces subordinates to work by threatening punishment and penalties. The psychological contract between the subordinates and their organisation would be coercive.

Dictatorial leadership might be rare in commerce and industry, but it is not uncommon in the style of government in some countries of the world, nor in the style of parenthood in many families.

(b) *Autocratic style*: decision-making is centralised in the hands of the leader himself, who does not encourage participation by subordinates; indeed, subordinates' ideas might be actively discouraged and obedience to orders would be expected from them.

The autocratic style is common in many organisations, and you will perhaps be able to identify examples from your own experience. Doctors, matrons and sisters in hospitals tend to practise an autocratic style; managers/directors who own their company also tend to expect things to be done their way.

(c) *Democratic style*: decision-making is decentralised, and shared by subordinates in participative group action. To be truly democratic, the subordinate must be willing to participate.

(d) *Laissez-faire style*: subordinates are given little or no direction at all, and are allowed to establish their own objectives and make all their own decisions.

The leader of a research establishment might adopt a laissez-faire style, giving individual research workers freedom of choice to organise and conduct their research as they themselves want (within certain limits, such as budget spending limits).

The Ashridge studies

4.5 The research unit at Ashridge Management College carried out studies in UK industry in the 1960s and identified four styles.

(a) The autocratic or *tells* style. This is characterised by one-way communication between the manager and the subordinate, with the manager telling the subordinate what to do. The leader makes all the decisions and issues instructions, expecting them to be obeyed without question.

(b) The persuasive or *sells* style. The manager still makes all the decisions, but believes that subordinates need to be motivated to accept them before they will do what he wants them to. He therefore tries to explain his decisions in order to persuade them round to his point of view.

(c) The *consultative* style. This involves discussion between the manager and the subordinates involved in carrying out a decision, but the *manager retains the right to make the decision himself*. By conferring with his subordinates before making any decision, the manager will take account of their advice and feelings. Consultation is a form of limited participation in decision-making for subordinates, but there might be a tendency for a manager to appear to consult his subordinates when really he has made up his mind beforehand. Consultation will then be false and a facade for a 'sells' style of leadership whereby the manager hopes to win acceptance of his decisions by subordinates by pretending to listen to their advice.

(d) The democratic or *joins* style. This is an approach whereby the leader joins his group of subordinates to make a decision on the basis of consensus or agreement. It is the most democratic style of leadership identified by the research study. Subordinates with the greatest knowledge of a problem will have greater influence over the decision. The joins style is therefore most effective where all subordinates in the group have equal knowledge and can therefore contribute in equal measure to decisions.

4.6 The Ashridge studies made one or two very interesting findings with regard to leadership style and employee motivation. (You should compare these with the views of other writers described in the rest of this chapter.)

(a) There was a clear preference amongst the subordinates for the *consultative* style of leadership but managers were most commonly thought to be exercising the 'tells' or 'sells' style.

(b) The attitudes of subordinates towards their work varied according to the style of leadership they thought their boss exercised. The most favourable attitudes were found amongst those subordinates who perceived their boss to be exercising the *consultative style*.

(c) The least favourable attitudes were found amongst subordinates who were unable to perceive a consistent style of leadership in their boss. In other words, subordinates are unsettled by a boss who chops and changes between autocracy, persuasion, consultation and democracy. The conclusion from this finding is that *consistency* in leadership style is important.

4.7

Style	Characteristics	Strengths	Weaknesses
Tells (autocratic)	The manager makes all the decisions, and issues instructions which must be obeyed without question.	(1) Quick decisions can be made when speed is required. (2) It is the most efficient type of leadership for highly-programmed routine work.	(1) It does not encourage the sub-ordinates to give their opinions when these might be useful. (2) Communications between the manager and subordinate will be one-way and the manager will not know until afterwards whether the orders have been properly understood. (3) It does not encourage initiative and commitment from subordinates.
Sells (persuasive)	The manager still makes all the decisions, but believes that subordinates have to be motivated to accept them in order to carry them out properly.	(1) Employees are made aware of the reasons for decisions. (2) Selling decisions to staff might make them more committed. (3) Staff will have a better idea of what to do when unforeseen events arise in their work because the manager will have explained his intentions.	(1) Communications are still largely one-way. Subordinates might not accept the decisions. (2) It does not encourage initiative and commitment from subordinates.
Consults	The manager confers with subordinates and takes their views into account, but has the final say.	(1) Employees are involved in decisions before they are made. This encourages motivation through greater interest and involvement. (2) An agreed consensus of opinion can be reached and for some decisions consensus can be an advantage rather than a weak compromise. (3) Employees can contribute their knowledge and experience to help in solving more complex problems.	(1) It might take much longer to reach decisions. (2) Subordinates might be too inexperienced to formulate mature opinions and give practical advice. (3) Consultation can too easily turn into a facade concealing, basically, a sells style.
Joins (democratic)	Leader and followers make the decision on the basis of consensus.	(1) It can provide high motivation and commitment from employees. (2) It shares the other advantages of the consultative style (especially where subordinates have expert power).	(1) The authority of the manager might be undermined. (2) Decision-making might become a very long process, and clear decisions might be difficult to reach. (3) Subordinates might lack enough experience.

4.8 It is important to get the consultative and 'joins' styles in perspective. A leader cannot try to forget that he is the 'boss' by being friendly and informal with subordinates, or by consulting them before making any decision. Douglas McGregor (*Leadership and Motivation*) wrote about his own experiences as a college president that : 'It took a couple of years, but I finally began to realise that a leader cannot avoid the exercise of authority any more than he can avoid responsibility for what happens in the organisation.' A leader can try to avoid acting dictatorially, and he can try to act like 'one of the boys', but he must accept all the consequences of being a leader. McGregor wrote that 'since no important decision ever pleases everyone in the organisation, he must also absorb the displeasures, and sometimes severe hostility, of those who would have taken a different course'.

Task and people: Blake's grid

4.9 By emphasising style of leadership and the importance of human relations, it is all too easy to forget that a manager is primarily responsible for ensuring that tasks are done efficiently and effectively.

Robert R Blake and Jane S Mouton attempted to address the balance of management thinking, with their *management grid* (1964) based on two aspects of managerial behaviour, namely:

(a) concern for production, ie the 'task'; and
(b) concern for people.

4.10 The results of their work were published under the heading of 'Ohio State Leadership Studies', but are now commonly referred to as *Blake's grid*.

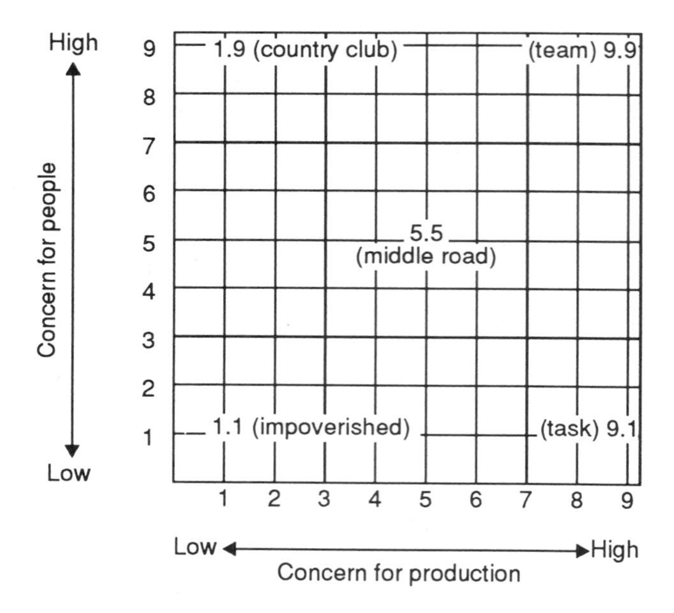

4.11 The extreme cases shown on the grid are defined by Blake as being:

(a) 1.1 *impoverished*: manager is lazy, showing little effort or concern for staff or work targets;

(b) 1.9 *country club*: manager is attentive to staff needs and has developed satisfying relationships. However, little attention is paid to achieving results;

(c) 9.1 *task management*: almost total concentration on achieving results. People's needs are virtually ignored and conditions of work are so arranged that people cannot interfere to any significant extent;

(d) 5.5 *middle of the road or the dampened pendulum*: adequate performance through balancing the necessity to get out work while maintaining morale of people at a satisfactory level;

(e) 9.9 *team*: high performance manager who achieves high work accomplishment through 'leading' committed people who identify themselves with the organisational aims.

4.12 The conclusion is that the most efficient managers combine concern for the task with concern for people.

4.13 It is worth being clear in your own mind about the possible usefulness of Blake's grid. Its primary value is obtained from the appraisal of a manager's performance, either by the manager himself or by his superiors. For example, a manager rated 3.8 has further to go in showing concern for the task itself than for developing the work of his subordinates.

However the grid is rather 'two-dimensional' compared to:

(a) contingency approaches which bring in other variables; and

(b) W J Reddin's 3-dimensional version of the grid, which recognises that whether a manager's concern for task and concern for people is high or low, he may still be effective or ineffective, depending on the circumstances. Blake assumes that a 1.1 manager is ineffective - but he may be effective if he is a rule-follower or bureaucrat in a bureaucratic environment.

Reddin's point is most important in limiting the effectiveness of the grid as a tool for management development. There is a tendency for Blake's framework to look as if concern for people and concern for task are ends in themselves, whereas the crucial consideration in leadership behaviour is the propensity to produce results (whether in terms of group satisfaction or task achievement), ie 'effectiveness'. A 9.9 manager on Blake's scale is, according to Blake, by definition effective, but Reddin points out that a 9.9 can still be ineffective, if he is a 'compromiser': his *concern* for human relations and the task will not necessarily translate into positive action.

Tannenbaum and Schmidt

4.14 Tannenbaum and Schmidt built on several contemporary theories to the effect that managers adopt either a task-centred or an employee-centred approach, an authoritarian or democratic approach. On these polarised dimensions Tannenbaum and Schmidt constructed a continuum, somewhere along which will be shown a manager's style in given circumstances. They include elements of contingency thinking in recognising that managers make *choices* in selecting and exercising a leadership style, affected by 'forces' including the manager himself, the subordinates and the situation.

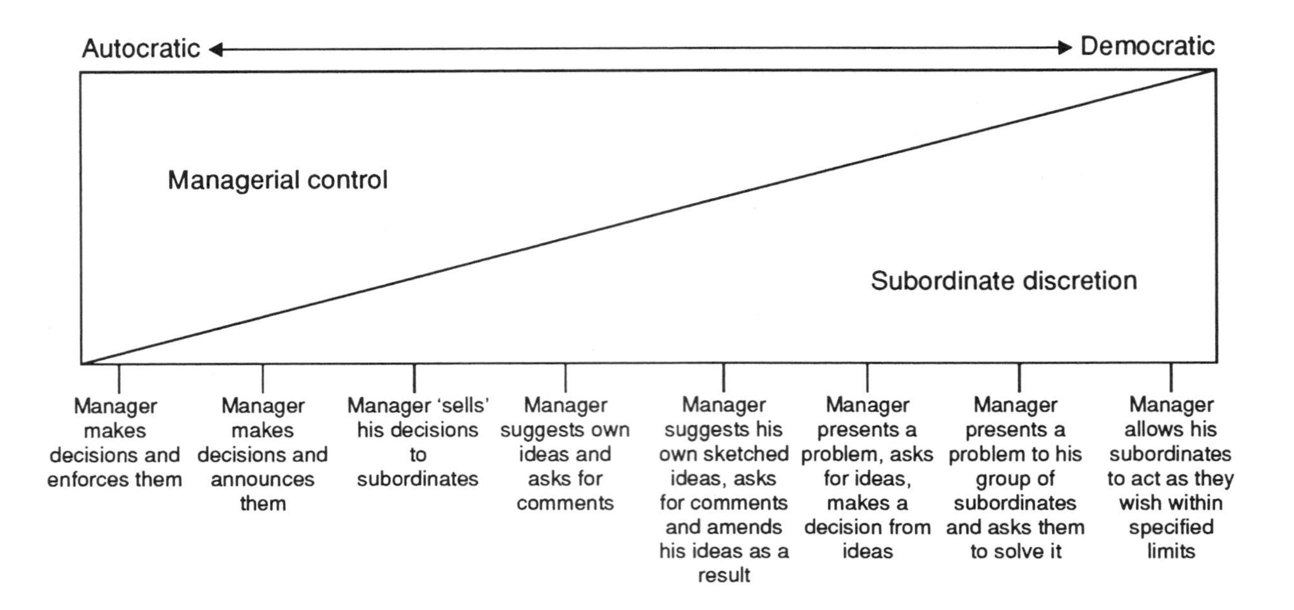

4.15 The continuum is a flexible model which addresses a range of situations, although only on one dimension - the extent to which the manager retains and exercises control. The theory has the advantage of getting managers to recognise that management style is a flexible thing, and has been used to this end in management development programmes.

Rensis Likert

4.16 Rensis Likert's model is another style theory, a further variation on the authoritarian/ democratic polarisation and the combined concern for task and people. He distinguished four systems of management.

(a) System 1: (*Exploitative-authoritative*): use of hierarchical position power, no teamwork, poor communication, use of sanctions for discipline. Mediocre productivity.

(b) System 2: (*Benevolent-authoritative*): similar to 1, but with some leeway for consultation and decentralisation. Productivity fair to good - but not 'happy'.

(c) System 3: (*Consultative authoritative*): good communication, some teamwork, consultation in goal-setting and decision making, positive reinforcement used. Good productivity.

(d) System 4: (*Participative group management*): Likert's 'ideal' system. Participation, good communication, growth needs catered for, commitment secured. Excellent productivity.

4.17 Effective management, for Likert, was a combination of:

(a) expecting high levels of performance;
(b) being employee-centred, taking time to get to know workers and develop trust;
(c) not practising close supervision;
(d) operating a participative style of management as a natural style.

4.18 He emphasised that *all* four features must be present for a manager to be truly effective. For example, if a manager is employee-centred, if he delegates and is participative, then he will have a happy working environment but he will not produce a high performance unless he also establishes high standards. A manager's concern for people must be matched by his concern for achieving results.

This linking of the human relations approach with scientific management targets will provide the recipe for effective performance. It is important to remember that management techniques such as time and motion study, financial controls etc are used by high producing managers 'at least as completely as by the low producing managers, but in quite different ways.' The different application is caused by a better understanding of the motivations of human behaviour.

Contingency or 'Situational' theories

John Adair: action-centred leadership

4.19 J Adair's Action-centred, or 'Functional' Leadership model (1973) is part of the contingency school of thought. Like other contingency thinkers, Adair saw the leadership process in a context made up of three main variables, all of which are interrelated and must be examined in the light of the whole situation.

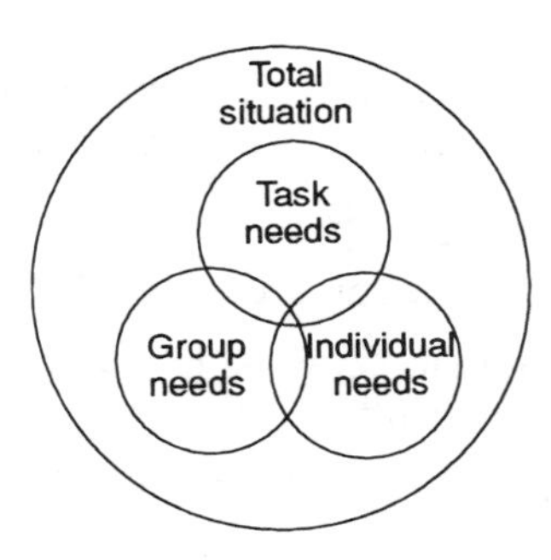

4.20 The total situation dictates the relative priority that must be given to each of the three sets of needs. Effective leadership is identifying and acting on that priority to create a balance between the needs. Meeting of the various needs can be expressed as specific management roles.

Task roles	*Group maintenance roles*	*Individual maintenance roles*
Initiating	Encouraging	Goal-setting
Information-seeking	Peace-keeping	Feedback
Diagnosing	Clarifying	Recognition
Opinion-seeking	Standard-seeking	Counselling
Evaluating		Training
Decision-making		

4.21 Around this framework, Adair developed a scheme of leadership training based on precept and practice in each of eight leadership activities as applied to task, team and individual.

(a) Defining the task
(b) Planning
(c) Briefing
(d) Controlling
(e) Evaluating
(f) Motivating
(g) Organising
(h) Setting an example

4.22 Adair argued that the common perception of leadership as 'decision-making' was inadequate to describe the range of action required by the complex situation in which the manager finds himself. This model is therefore more practical.

F E Fiedler's contingency theory

4.23 Perhaps the leading advocate of contingency theory is F E Fiedler. He studied the relationship between style of leadership and the effectiveness of the work group. Two styles of leader were identified.

(a) *Psychologically distant managers* (PDMs) who maintain distance from their subordinates by:

(i) formalising the roles and relationships between themselves and their superiors and subordinates;

(ii) being withdrawn and reserved in their inter-personal relationships within the organisation (despite having good inter-personal skills);

(iii) preferring formal consultation methods rather than seeking opinions of their staff informally.

PDMs judge subordinates on the basis of performance, and are primarily task-oriented: Fiedler found that leaders of the most effective work groups tend to be PDMs.

(b) *psychologically close managers* (PCMs) who:

(i) do not seek to formalise roles and relationships with superiors and subordinates;

(ii) are more concerned to maintain good human relationships at work than to ensure that tasks are carried out efficiently;

(iii) prefer informal contacts to regular formal staff meetings.

4.24 Fiedler went on to develop his contingency theory in *A Theory of Leadership Effectiveness*. He suggested that the effectiveness of a work group depended on the situation, made up of three particular variables:

(a) the relationship between the leader and his group - ie the amount of trust and respect they have for him;

(b) the extent to which the task is defined and structured;

(c) the power of the leader in relation to the group (his authority, and power to reward and punish).

4.25 A situation is 'favourable' to the leader when:

(a) the leader is liked and trusted by the group;
(b) the tasks of the group are clearly defined;
(c) the power of the leader to reward and punish with organisation backing is high.

4.26 Fiedler concluded that:

(a) a structured (or psychologically distant) style works best when the situation is either very favourable, or very unfavourable to the leader;

(b) a supportive (or psychologically close) style works best when the situation is moderately favourable to the leader.

'Group performance will be contingent upon the appropriate matching of leadership styles and the degree of favourableness of the group situation for the leader, that is, the degree to which the situation provides the leader with influence over his group members.' (*Fiedler*)

Charles Handy: the leadership 'environment'

4.27 Handy has also suggested a contingency approach to leadership. The factors in any situation which contribute to a leader's choice of style (on a scale from 'tight' to 'loose' control) and effectiveness are:

(a) the leader himself - his personality, character and preferred style of operating;

(b) the subordinates - their individual and collective personalities, and their preference for a style of leadership;

(c) the task - the objectives of the job, the technology of the job, methods of working;

(d) the environment.

4.28 The *environment* of leadership is important.

(a) *The position of 'power' held by the leader in the organisation and the relationship of the leader and his group.* A person with great power has a bigger capacity to set his own style of leadership, select his own subordinates and re-define the task of his work group.

(b) *Organisational 'norms' and the structure and technology of the organisation.* No manager can flout convention and act in a manner which is contrary to the customs and standards of the organisation. If the organisation has a history of autocratic leadership, it will be difficult to introduce a new style. If the formal organisation is highly centralised, there will be limits to how far a task can be re-structured by an individual manager.

(c) *The variety of tasks and the variety of subordinates.* In many groups, tasks vary from routine and simple to complex 'one-off' problem-solving. Managing such work is complicated by this variety. Similarly, the individuals in a work group might be widely different. One member of the group might seek participation and greater responsibility, whereas another might want to be told what to do. Furthermore, where labour turnover is frequent, the individuals who act as leaders or subordinates are constantly changing; such change is unsettling because the leadership style will have to be altered to suit the new situation.

4.29 The 'environment' can be improved for leaders within an organisation if top management act to ensure that:

(a) leaders are given a clear role and 'power';

(b) organisational 'norms' can be broken - the culture is responsive and adaptive;

(c) the organisational structure is not rigid and inflexible;

(d) subordinates in a work group are all of the same quality or type;

(e) labour turnover is reduced, especially by keeping managers in their job for a reasonably lengthy period of time.

Exercise 3

Grouped below are a number of key words used frequently in the preceding section. Beside each word write the name or names of the theorist(s) who uses the term, or a very similar one. (One has been done for you.) Are all of the theorists using the terms in the same way? Do you think our grouping of terms is appropriate?

Persuasion -
Benevolent-authoritative -

Task centred -
Variety of tasks - *Fiedler, Handy*
Organisational norms -
PDMs -

Consultative -
Participation -

Dictatorial -
Power -
Exploitative-authoritative -

Autocratic -

Consistency -

Employee-centred -
PCMs -
Group/individual maintenance -

Democratic -

Laissez-faire -

5. ENTREPRENEURS

5.1 A good deal of research has been done into the entrepreneurial personality. In 1964 Collins and Moore *(The enterprising man)* drew up a psychoanalytic model of entrepreneurship. They found that entrepreneurs had a young adulthood characterised by:

(a) disorientation, goal-lessness and testing;
(b) non-conformity, rebelliousness;
(c) enjoyment of setbacks (martyrdom, masochism);
(d) a high need for control;
(e) suspiciousness;
(f) fear of being victimised;
(g) scanning the environment.

5.2 In adulthood these characteristics create an organisation that is:

(a) authoritarian;

(b) centralised;

(c) lacking trust and delegation;

(d) a work environment of high dependency and power that is a function of centrality or closeness to the entrepreneur;

(e) unresolved regarding succession.

(Collins and Moore, as summarised in Barbara J Bird, *Entrepreneurial Behaviour* (1989))

5.3 Other writers have supported and developed these views. A famous article by Manfred F R Kets de Vries in the *Harvard Business Review* (November/December 1985) is revealing even in its title: 'The Dark Side of Entrepreneurship'. For example, Kets de Vries writes of the striking ambivalence of entrepreneurs when an issue of control arises: 'They seem to fear that their grandiose desires will get out of control and place them ultimately at the mercy of others.' (Robert Maxwell, perhaps, springs to mind.) Others use the word 'obsessive' or 'fixated' to describe the entrepreneurial drive. Much academic research has observed that entrepreneurs are abnormally (and sometimes excessively) competitive people. Some have gone so far as to suggest that entrepreneurs are psychopaths. However Drucker (*Innovation and Entrepreneurship*, 1985) has little patience with attempts to pigeon-hole the entrepreneur: 'People of all kinds and temperaments apparently do equally well'.

5.4 The question 'what is an entrepreneur' is particularly relevant to the recent history of the UK and the desire of the Thatcher government to create an enterprise culture.

Exercise 4

In *The Naked Entrepreneur* (1990) David Robinson characterises an entrepreneur as follows.

'Stripped of the self-seeking promotion and the adulation of the gossip columns we find a workaholic dedicated to the pursuit of wealth and power. Complex in his makeup, often careless of personal relationships, he has a view of the world which is very different from the rest of us. He is in charge of his destiny, embraces extraordinary risks but in many respects acts with caution and circumspection. His enthusiasm for his career gains momentum rather than loses it as material success is achieved. Winning the competitive "game" is more important than the financial stakes which become mere counters on the board. Megalomania waits in the wings while the world both envies and dislikes continued success. Seldom loveable but always exciting to work with, and for, entrepreneurs are mainsprings in an enterprise culture. Long may they remain so.'

If this is an accurate picture, how do you think the organisations created by such people in the 1980s will fare in the 'caring 90s'?

6. THE USEFULNESS OF LEADERSHIP THEORIES

6.1 Few of the concepts discussed above attempt to distinguish 'leadership' behaviour from 'management' behaviour, other than peripherally in the form of 'motivation' or 'setting an example'. It is arguable, therefore, that although they may encourage better management - through their application in management assessment and development - they are incapable of making better *leaders*, other than by adding insight into management functions for those who are leaders already.

6.2 It is also doubtful to what extent study of a concept, even where it is built into a training scheme, can in fact change a given individual's behaviour. Individual managers/leaders will not change their values in response to a theory, especially where it is one of many - often conflicting - frameworks. Even if *willingness* to change management values and style exists, conditions in the organisation may not allow it.

6.3 The trouble with theories is that many of them do not take relevant organisational conditions into account, and for a manager to model his own behaviour on a formula which is successful in theory or in a completely different situation will not necessarily be helpful. In addition, it may be that:

(a) the manager's personality (or 'acting' ability) is simply not flexible enough to utilise leadership theories by attempting to change styles to suit a situation. A manager may not be able to be participative in some circumstances and authoritative in others where his personality and personal goals are incompatible with that style;

(b) the demands of the task, technology, organisation culture and other managers constrains the manager in the range of 'styles' and leadership behaviours open to him. If the manager's own boss believes in and practises an authoritarian style, and the group members are incompetent and require close supervision, no amount of theorising on the desirability of participative management will be effective;

(c) consistency is important to subordinates. If a manager practises a contingency approach to leadership, subordinates may simply perceive him to be 'fickle', or may suffer insecurity and distrust the 'changeful' manager.

'There is therefore no simple recipe which the individual manager can use to decide which style to adopt to be most effective. Management style probably can be changed, but only if management values can be changed ... It is not enough to present managers with research findings and try to convince them with logical argument that change is necessary.'
(Buchanan and Huczynski, *Organisational Behaviour*)

6.4 Moreover, it must be reiterated that 'the essence of leadership is followership' (Koontz et al). It is follower response that will ultimately decide the competence of leadership - and follower response is not necessarily subject to leadership theory, however well supported with research findings from other situations.

Chapter roundup

- There are conflicting interpretations as to the roles of the manager.
  - Fayol suggests planning and controlling.
  - Mintzberg on the other hand emphasises a variety of interpersonal, informational and decisional roles.
- Being a *leader* is not the same as being a manager. A leader's authority can be less formal than that of a manager, whose power derives from his or her position.
- Some believe that leadership is an inherited quality.
- Others talk of a 'leadership' style which can be taught to an extent. Leadership styles can be: dictatorial, autocratic, democratic, laissez-faire, or tells, sells, consults, joins.
- The contingency approach argues that the leadership style adopted depends on the task, the environment and subordinates.
- Many writers have attempted to identify the essential characteristics of entrepreneurs.

Test your knowledge

1 Outline Fayol's management functions. (see para 1.1)
2 What is Mintzberg's assessment of the nature of managerial work? (1.6)
3 Identify four characteristics of supervision. (2.2)
4 Distinguish between leadership and management. (3.1- 3.6)
5 What are the flaws in trait theories of leadership? (4.1)
6 Describe eight leadership styles. (4.4, 4.5)
7 What two aspects of management behaviour are measured on Blake's grid? (4.9)
8 List Adair's three types of role and eight leadership activities. (4.21)
9 What is a PDM and who used this term? (4.23)
10 Describe the leadership environment. (4.28)
11 What is the essence of leadership, according to Koontz et al? (6.4)

Now try question 6 at the end of the text

PART C

PEOPLE AT WORK

Chapter 7

HUMAN RESOURCE MANAGEMENT

This chapter covers the following topics.

Signpost

- This chapter gives the background to the 'human resource' aspect of the *Organisation and the Human Resource* paper
- The distinction between line and staff functions is developed from Chapter 3. Quality issues were also first discussed in Chapter 3.

1. PERSONNEL MANAGEMENT OR HUMAN RESOURCE MANAGEMENT?

The personnel function

1.1 Organisations are made up of people, and at the very least it is commonly recognised that *someone* in every organisation will need to be responsible for the many matters which arise in connection with the recruitment, selection, training, motivation and payment of staff, and compliance with the various laws relating to employment.

1.2 The personnel 'function' *may* consist of a specialist department or departments under the control of a personnel manager (or director, where the function is represented on the Board) - but may, in a smaller organisation, be the owner, manager, company secretary or other officers undertaking the necessary tasks.

1.3 Indeed, it has been suggested that a separate personnel function need not exist at all: *all* line managers need to achieve results through the efforts of other people, and must therefore take an interest in personnel policy and practice as the basis of both employee relations and task performance.

1.4 The Institute of Personnel Management (IPM) has said [italics ours] that:

> 'Personnel Management is that *part of management* concerned with people at work and with their relationships within an enterprise. It applies not only to industry and commerce but to all fields of employment. Personnel management aims to achieve both efficiency and justice, neither of which can be pursued effectively without the other.
>
> It seeks to bring together and develop into an effective organisation the men and women who make up an enterprise, enabling each to make his or her own best contribution to its success both as an individual and as a member of a working group. It seeks to provide fair terms and conditions of employment, and satisfying work for those employed.... Personnel management must also be concerned with the human and social implication of change in internal organisation and methods of working, and of economic and social changes in the community.'

Note that the emphasis is both on people/relationships/justice etc *and* on the development of effective organisation, and contribution to success.

Personnel management: the 'odd job' view

1.5 The traditional view of personnel management has been essentially task-, activity- or technique-based. Dr Dale Yoder of the Graduate School of Business, Stamford University, defines the personnel management function as follows.

(a) Setting general and specific management policy for employment relationships and establishing and maintaining a suitable organisation for leadership and co-operation.

(b) Collective bargaining.

(c) Staffing and organisation: finding, getting and holding prescribed types and numbers of workers.

(d) Aiding the self-development of employees at all levels, providing opportunities for personal development and growth as well as requisite skills and experience.

(e) Incentivating: developing and maintaining the motivation in work.

(f) Reviewing and auditing manpower and management in the organisation.

(g) Industrial relations research, carrying out studies designed to explain employment behaviour, and thereby improve manpower management.

1.6 In 1968, Crichton (*Personnel Management in Context*) complained that personnel management was often a matter of 'collecting together such odd jobs from management as they are prepared to give up.'

Other writers shared this view, notably Peter Drucker, who - while recognising the importance of human resources in the organisation - saw the personnel function of the time as 'a collection of incidental techniques without much internal cohesion'. According to Drucker, the personnel manager saw his role as 'partly a file clerk's job, partly a housekeeping job, partly a social worker's job and partly 'fire fighting' to head off union trouble or to settle it.' (*The Practice of Management*, 1955).

1.7 A more imaginative set of job titles for personnel managers - albeit just as depressing, insofar as they reflect a purely administrative and reactive role for Personnel - was suggested by Nick Georgiades (*Personnel Management*, February 1990).

(a) 'The administrative handmaiden' - writing job descriptions, visiting the sick, and so on.

(b) 'The policeman' - ensuring that both management and staff obey the rules and do not abuse the job evaluation scheme, and keeping a watchful eye on absenteeism, sickness and punctuality.

(c) 'The toilet flusher' - administering 'downsizing' policies (cutting staff numbers).

(d) 'The sanitary engineer' - ensuring that there is an awareness of the unsanitary psychological conditions under which many people work.

1.8 The status and contribution of the personnel function is still often limited by the image of 'fire-fighting', an essentially reactive and defensive role. The personnel manager is judged according to his effectiveness in avoiding or settling industrial disputes, preventing accidents and ill-health (and their associated costs), filling vacancies and so on.

> 'People may be regarded as a vital resource - at least plenty of lip service is paid to this concept by company chairmen in their annual statements - but many managers find it difficult to appreciate where the personnel department fits in, except in the simplest terms as a procurement and fire-fighting function.' Armstrong, *A Handbook of Personnel Management Practice*

This is a vicious circle. As long as personnel policy and practice is divorced from the strategy of the business, and fails to be proactive and constructive, it will be perceived by line management to have little to do with the 'real' world of business management, or the 'bottom line' (profitability): personnel specialists therefore command scant respect as business managers, and their activities continue to be limited to areas of little strategic impact.

1.9 The situation has, however, been changing, and a new approach has emerged which has been labelled *Human Resource Management*.

> 'The real requirement is proactive and constructive rather than defensive and reactive. To discharge their true role, personnel managers must anticipate the needs of the organisation in the short and the long term. They must develop the policies to produce solutions to anticipated problems resulting from the external and internal environment, whilst influencing and creating the attitudes amongst employees needed for the enterprise's survival and success.'
>
> Livy, *Corporate Personnel Management*

Human Resource Management (HRM)

1.10 'HRM' is a recently fashionable term, which some personnel specialists have seized upon and applied to themselves (whether or not they are in fact doing anything more than staff administration or manpower planning) in the interests of their status and self-esteem. HRM is not yet being carried out by personnel functions in the majority of organisations, and to some extent, therefore, widespread use of the term is unhelpful. It is rather a case of the 'Emperor's New Clothes': no-one wants to admit that the term has little meaning for him.

1.11 However, a precise and positive interpretation of HRM centres on the following notions.

(a) Personnel management has been changing in various ways in recent years. Many of its activities have become more complex and sophisticated and, particularly with the accelerating pace of change in the business environment, less narrowly concerned with areas previously thought of as personnel's sole preserve (hiring and firing, training, industrial relations and manpower planning). The personnel function has become centrally concerned with *issues of broader relevance to the business and its objectives*, such as change management, the introduction of technology, and the implications of falling birthrates and skill shortages for the resourcing of the business.

(b) Personnel management can and should be integrated with the *strategic* planning of the business, that is, with management at the broadest and highest level. The objectives of the personnel function can and should be directly related to achieving the organisation's goals for growth, competitive gain and improvement of 'bottom line' performance.

(c) Personnel managers can and should be businessmen or women - even entrepreneurs.

1.12 In a survey of 20 top personnel directors (reported in *Personnel Management*, October 1989) Michael Armstrong identifies the key issues of HRM, with implications for the personnel function of the future.

(a) HRM implies a shift of emphasis in personnel management from the peripheral 'staff' role of the past to mainstream business management. Armstrong suggests that 'twenty years ago people in equivalent positions would have been more likely to talk about [personnel activities] as if these were techniques or areas of knowledge which had intrinsic value and did not need to be considered in terms of fit with business strategies or impact on business results.'

Now, Barry Curnow (a former IPM President), suggests that: 'We've moved through periods when money has been in short supply and when technology has been in short supply. Now it's the people who are in short supply. So personnel directors are better placed than ever before to make a real difference - a bottom-line difference. The scarce resource, which is the people resource, is the one that makes the impact at the margin, that makes one firm competitive over another.'

(b) A definition of entrepreneurship is the 'shifting of economic resources out of the area of lower and into an area of higher productivity and greater yield' (JB Say), and this is essentially what HRM embodies, in terms of finding, obtaining and developing - getting the best out of - the human resources of the business.

(c) This implies a close match between corporate objectives and the objectives of the human resource function. *All* business planning should recognise that the ultimate source of 'value' is people, and should appreciate the human resource implications and potential constraints associated with any long-term strategies evolved.

(d) The integration of personnel and overall corporate objectives firmly establishes personnel as an *enabling* function, creating a framework and culture within which effective contributions can be made.

(e) This 'enabling' role brings us back to the term 'HRM'. A major part of personnel's relatively new-found concern for performance management is the re-orientation towards *resourcing* in its broadest sense. Personnel's strategic contribution to a business is the definition of relationships between business requirements and organisational and human requirements: the human resourcing of the business is how this works in practice, and includes not only the obtaining of an increasingly scarce resource (people) but the maximisation of their contribution through development, reward, organisational culture, succession planning and so on.

1.13 S Tyson and A Fell (*Evaluating the Personnel Function*) suggest four major *roles* for personnel/human resource management which illustrate the shift in emphasis from the 'odd job' to the 'strategic' viewpoint.

(a) To represent the organisation's central value system (or 'culture').

(b) To maintain the boundaries of the organisation (its identity and the flow of people 'in' and 'out' of it).

(c) To provide stability and continuity [through planned succession, flexibility and so on].

(d) To adapt the organisation to change.

1.14 Don Beattie, personnel director of STC, has defined the job of personnel directors as follows.

> 'To know what business they are in, to know where it is going, and to ensure that the input to get there is available from a human resourcing and organisational capability point of view.'

Exercise 1

In June 1993 *Personnel management* magazine (sub-titled 'The magazine for human resource professionals') published an article entitled 'The mystery of the missing human resource manager', reporting on a survey whose results suggested amongst other things, that:

(a) only 17% of establishments in the UK employing more than 25 people have a personnel specialist;

(b) only 44 out of the 2,061 respondents used the 'HRM' title;

(c) however, there was some evidence to support the possibility that HRM is found at head offices, 'a strategic phenomenon at the strategic apex of the organisation'.

The article continues:

'Perhaps predictably, the title does not necessarily reflect practice. IBM, for example, has retained the personnel title despite being frequently held up in the past as an exemplar of human resource management. Even in the United States, where it has been estimated that over half of the top 50 corporations now use the human resource title, IBM has kept the personnel label. One explanation is that it only makes sense to change if it provides some competitive advantage. Maybe establishments in the UK have taken that message on board. Certainly, we can refute the 'old wine in new bottles' argument that some of us promulgated. There may be some good wine around - some good human resource policy and practice - but it is still marketed in the old bottles.'

Do you think the *name* of the function matters?

Solution

It clearly matters to some of the people who perform the function, for their own sense of self-esteem. Adoption of the title may be a sign of conflict and power struggles within the organisation. An important consideration is whether the humans whom the function manages view the idea of themselves as 'resources' positively or negatively.

2. PERSONNEL'S RELATIONSHIP WITH LINE MANAGEMENT

Ambiguities in the role of personnel

2.1 Personnel managers can, therefore, operate in different areas and at different levels - from crisis handling and routine administration on the one hand, to policy-making and innovation on the other.

Charles Handy (*Understanding Organisations*) suggested that the effectiveness of the 'traditional' personnel department is reduced by ambiguity about the role of the department, or conflict between different roles of the department. Personnel managers are expected to act in a variety of ways.

(a) As line managers - for example, in negotiating and implementing industrial agreements with trade unions; perhaps also enforcing safety procedures

(b) As advisers - for example, in welfare work, or in helping with problems of human attitudes and behaviour in various departments

(c) As a service department - for example, in recruitment, training and education

(d) As 'auditors' - for example, in checking on the effectiveness of appraisal schemes, or the training structure, or perhaps in reporting on the effectiveness of the management of the organisation (ie carrying out management audit)

(e) As co-ordinators and planners - for example, in manpower planning

2.2 Handy suggested that this ambiguity leads to inefficiency and loss of authority for the personnel function, which attempts to bolster its influence by expansion, and by proliferation of rules and procedures as a form of control over line management. In other words, the personnel department develops into a bureaucratic institution, regardless of the type of organisation it serves. One solution, put forward by Handy, was to split the personnel function, within the organisation structure, into separate 'executive', 'advisory', 'service', 'auditing' and 'planning and co-ordinating' functions.

2.3 Karen Legge suggested a *matrix* structure with the key functions of personnel (such as manpower planning, selection, training and welfare) on one axis of the grid, set against key *process* activities (such as policy-making, innovation, the establishing of routines and crisis handling) on the other. Key personnel roles can then be allocated in a more rational manner.

Exercise 2

What is the role of the personnel function in your organisation? What areas of activity and what kinds of decision-making is it involved in? (You might find it helpful to draw them up in a grid such as that described in Paragraph 2.3 above.) Have you noticed (or been involved in) any recent changes in that role?

If you can, get hold of the departmental/individual job description of the personnel department/officer in your organisation, or ask some of the people concerned how they see their job. Does that tally with the way the rest of the organisation sees it?

2.4 A further ambiguity may exist for the personnel function in determining whose interests to serve: who is its 'client'?

According to the IPM's Code of Professional Practice in Personnel Management, the personnel manager has three principal areas of responsibility. [Italics ours.]

(a) 'A personnel manager's primary responsibility is to his *employer*.'

(b) He will 'resolve the conflict which must sometimes exist between his position as a member of the management team and his special relationship with the *workforce* in general and with individual *employees*.'

(c) He will 'use his best endeavours to enhance the standing .... of his *profession*', in dealings with other bodies.

2.5 Personnel managers may have to occupy the middle ground between employee and employer on some occasions. In a sense both employers/management and employees are their clients. Yet the personnel manager is not in any formal sense the 'representative' of the workforce, and is paid to be part of the organisation's management team, as both representative and adviser. The 'diplomatic' role of personnel may thus pose a dilemma of dual allegiance - particularly where there is conflict in the relationship between the personnel function and other (line) members of the management team, as we will now go on to discuss.

Line and staff relationships

2.6 There are two ways of looking at the distinction between line and staff management.

(a) Line and staff can be used to describe *functions* in the organisation. Line management consists of those managers directly involved in achieving the objectives of an organisation (usually production and sales managers, in a manufacturing company). Every other manager is staff (including accounting, research and development and personnel).

(b) Line and staff can be used to denote *relationships* of authority. A line manager is any manager who has been given (delegated) authority over a subordinate down the chain of command. Thus in personnel (a 'staff' department), the personnel director will have line authority over the managers in charge of recruitment and training. In other words, the 'line' represents the chain of command through the management hierarchy. Staff authority, on the other hand, resides not in official position, but influence - by means of special knowledge or expertise, resources or other desirable qualities. The legal expert would thus have staff authority in the personnel department - whatever his official 'rank'.

2.7 Staff functions and relationships exist in many organisations where there is a need for specialisation of management: accountants, personnel administrators, economists, data processing experts and so on. Where this expertise is 'syphoned off' into a separate department, the problem naturally arises as to whether the experts:

(a) exist to *advise* line managers, who may accept or reject the advice given; or

(b) can step in to *direct* the line managers in what to do - in other words, to assume authority themselves. This is usually defined as 'functional' authority, based on the recognition of expertise in areas which are vital to the achievement of line objectives: a personnel manager might have such authority over industrial relations decisions throughout the organisation, for example.

2.8 The ambiguities of line/staff relationships create a potential for conflict, where:

(a) staff managers are held in low regard as overheads who contribute nothing to the bottom line but have considerable nuisance power in imposing rules and policies;

(b) staff managers *are* in fact divorced from the business objectives of line management, but impose rules and procedures in the interests of their profession or speciality, or to enhance their power;

(c) staff managers undermine - intentionally or otherwise - the authority of line managers;

(d) staff managers report over the heads of line managers, as advisers to senior management;

(e) staff managers are not held accountable for the results of their advice - or *are* held accountable for the results of their advice, without having the authority to get it properly implemented.

2.9 As we saw earlier, the solutions to these problems are easily stated, but not easy to implement in practice.

(a) Authority must be clearly defined, and distinctions between line authority and staff advice clearly set out (for example, in job descriptions).

(b) Senior management must encourage line managers to make positive efforts to discuss work problems with staff advisers, and to be prepared to accept their advice. The use of experts should become an organisational way of life, part of the culture.

(c) Staff managers must be fully informed about the aspects of the business in which, as functional experts, they will become involved. They should then be less likely to offer impractical advice.

(d) When staff advisers are used to plan and implement changes in the running of the business, they must be kept involved during the implementation, monitoring and review of the project. Staff managers must be prepared to accept responsibility for their failures and this is only really possible if they advise during the implementation and monitoring stages.

Personnel as a staff function

2.10 Whatever the size of the organisation, the finding and keeping of high-quality staff will necessitate the existence of the personnel function. In a small organisation, this may be a non-specialist function, the responsibility of any appropriate manager or administrator.

In larger organisations, more time is likely to be taken up with organisational and personnel problems. It may then be felt that a specialist member of staff is required to take on full-time responsibility for:

(a) advising management on personnel matters; and
(b) implementing well-defined policies.

2.11 This is not a situation in which management simply 'hives off' its responsibility for 'people management': it is an acknowledgement that line managers need to have available (and listen to) a specialist because:

(a) there are areas developing in personnel management in which expertise is increasingly required; and

(b) there is a need for consistency in the design and application of personnel management practice.

> 'Heads of departments basically have technical responsibilities; the personnel specialist's job is to explain to them reasons for social behaviour which may not be obvious. Why, for example, are decisions that are technically and economically sound sometimes met by seemingly irrational resistance by subordinates? What emotional factors are at play when this sort of thing happens? Offering advice to clarify such situations is a task that demands an expert who has been trained to perceive the ways in which individuals and groups interact within an organisation, and who can convey his understanding in practical terms to departmental heads who are too pre-occupied with their technical activities to have time to consider the social system of the workplace as a whole.'
>
> M W Cuming (*The Theory and Practice of Personnel Management*)

2.12 Several factors have contributed to the perceived need for a specialised personnel function.

(a) The need to comply with increasing regulation and legislation means that expert attention will have to be given to such matters as recruitment and selection (to avoid racial or sexual discrimination), termination of the employment contract (especially, for example, in circumstances involving trade union membership, or maternity), and health and safety at work.

(b) High staffing costs, and the changing workforce (as technology becomes widespread, skill requirements change, women enter or re-enter the market, temporary and home working increases and so on) also require that recruitment and staffing policies be designed and implemented by individuals with current knowledge of the labour market.

(c) There is continuing pressure for social responsibility towards employees, with information to be provided, schemes for worker participation and rulings on working conditions - from the European Commission among others.

(d) Behavioural sciences (chiefly psychology, sociology and social psychology) have become highly developed in organisational contexts, and are gaining currency as important elements in managers' mental equipment. Up-to-date research on such matters as motivation, stress, resistance to change, industrial fatigue and response to leadership style should be monitored.

(e) Trade unions and their officials, Industrial Tribunals, the Advisory, Conciliation and Arbitration Service (ACAS) and others have a continuing role in industrial relations: familiarity with legislation, rulings and practice, liaison and information will be required.

> 'Managers, if one listens to the psychologists, will have to have insights into all kinds of people. They will have to be in command of all kinds of psychological techniques. They will have to understand an infinity of individual personality structures, individual psychological needs, and individual psychological problems... But most managers find it hard enough to know all they need to know about their own immediate area of expertise, be it heat-treating or cost accounting or scheduling.'
> Drucker: *Management*

2.13 In an article in *The Administrator* (May 1985) entitled 'The Personnel Function - A shared responsibility', Laurie Mullins writes: 'It is the job of the personnel manager to provide specialist knowledge and services to line managers, and to support them in the performance of their jobs. It is *not* the job of the personnel manager to manage people, other than direct subordinates. The personnel manager has no direct control over other staff except where a specific responsibility is delegated directly by top management, for example, if nominated as safety officer under the Health and Safety at Work Act 1974. The personnel manager has executive authority for such delegated responsibility and for the management of the personnel department and its staff... In all other respects the personnel manager's relationship with other managers, supervisors and staff of the organisation is indirect, that is an advisory or 'functional' relationship.'

2.14 Mullins goes on to explore how this role works itself out in practice. He suggests that the line managers have authority over staff in their own departments, and therefore that they retain immediate responsibility for personnel management in the context of their own areas of work, ie operational aspects of personnel management, such as:

(a) the organisation and allocation of work;
(b) minor disciplinary matters and staff grievances;
(c) standards of performance;
(d) on-the-job training and induction;
(e) communication; and
(f) safe working practices and maintenance of a tidy, healthy work environment.

2.15 Line managers 'have both the right and the duty to be concerned with the effective operation of their own department, including the management and well-being of their staff.' After all, it is the line managers who are directly affected by the symptoms of poor personnel management: lateness, low morale, hostility, absenteeism, incapacity to perform to standard or whatever.

2.16 Personnel managers will, however, act as advisers and arbitrators where necessary, even in this 'front-line' territory, where legislation and other areas of expertise are involved.

The personnel manager sees the overall effect of personnel policies and practices in the organisation. At organisational level, his role is that of specialist adviser, and interpreter/ executor of personnel policy - in consultation and co-operation with line managers.

2.17 At departmental level, the personnel manager is still likely to be concerned mainly with the broader effects of personnel policy on staff, such as:

(a) manpower planning and problems of turnover;

(b) recruitment and selection systems; training systems;

(c) consultations and negotiations with trade union representatives and outside bodies, like ACAS, professional associations, wages councils;

(d) compliance with employment legislation; and

(e) maintenance of records.

The authority of the personnel manager

2.18 In effect, the personnel manager's authority depends on several means of *influence*.

According to Michael Armstrong (*A Handbook of Personnel Management Practice*): 'What in effect the personnel manager says is that "this is the personnel policy of the company, ignore it at your peril". He can seldom forbid anyone to do anything, except where it contravenes a law or negotiated procedure, but he can refuse to authorise something - say, a pay increase - if it is in his power to do so. And he can refer a matter to higher authority (the joint superior of the two managers concerned) and request that the action be delayed until a ruling is made.'

2.19 The personnel specialist may lack 'positional' or 'legal power' in the organisation, in relation to line departments, but may possess:

(a) 'resource' power - through control over manpower and pay; and
(b) 'expert' power - where his specialist knowledge is recognised.

2.20 In addition, he can 'influence' (direct and modify the behaviour of others) through:

(a) establishing rules and procedures;
(b) bargaining and negotiation;
(c) persuasion; and
(d) 'ecology' or environmental control.

'The design of work, the work, the structure of reward and control systems, the structure of the organisation, the management of groups and the control of conflict are all ways of managing the environment in order to influence behaviour. Let us never forget that although the environment is all around us, it is not unalterable, that to change it is to influence people, that ecology is potent, the more so because it is often unnoticed.' (*Handy*)

2.21 The establishment of policy and procedure is a particularly potent source of influence for the personnel function, although, as we saw in paragraph 2.2, it can have adverse effects. Formal policies are helpful in terms of consistency, standardisation throughout the organisation, impartiality and clarity. But as Armstrong (*Handbook of Personnel Management Practice*) notes, they 'may be inflexible, or platitudinous, or both, and therefore useless or dangerous.' If the policy is couched in purely abstract terms, it may lack credibility: if it is fixed, it may become a stranglehold on the organisation's flexibility and power in negotiation (if a pay policy specifies a certain salary level, for example). Fixed policies may also inhibit initiative and creative problem-solving, and may not allow junior managers, in particular, to develop if their activities are closely proscribed.

Exercise 3

Think about your job. Is it 'line' or 'staff' in the nature of its authority? What kinds of authority or power do you have personally in your job (whether or not formal authority is involved)? *How* do you influence other people, when you need them to do something for you at work?

Does the personnel function in your organisation have power, and if so what kind? What tactics does the personnel department use to 'get its way' in the organisation?

2.22 Whatever the range of the personnel function's formal responsibilities, it is important to remember that it operates largely by *consent*, and the *perception of its authority* by line management. Personnel managers frequently suffer from lack of credibility related to:

(a) lack of definition of the personnel function and its contribution to business objectives;

(b) the perceived idealistic and/or theoretical basis of people-oriented activity;

(c) the perceived ambiguity in personnel's 'clientele', whereby its loyalty to the management position may be seen as suspect;

(d) the traditionally administrative role of personnel, lacking the creativity and pro-activity expected of a management function.

> 'Sometimes it seems as though a great many very talented personnel specialists are wasting an awful lot of time. They carefully watch developments in the industrial relations, political and labour market environments; they develop sensible, well thought-out personnel policies that would make their company one of the most progressive and highly respected of employers. And then they see their efforts continually frustrated and subverted by a management team that seems determined to ignore most of what the personnel department does.'
>
> Brewster and Richbell, *Getting Managers to Implement Personnel Policies*

3. HR VISION: MANAGING A QUALITY WORKFORCE

'Customer care' and the personnel function

3.1 Values to do with the *customer* ('the customer is king, 'know your customer') and his needs have been a major feature of the perception of quality, and the achievement of success, in the 1980s and 1990s. Strategically, there has been a move away from a technical or production orientation, to a marketing orientation: the objectives of a business can best be met by identifying the market, its needs and wants, and by fulfilling those needs and wants in the most efficient and effective way possible.

3.2 'Customer care' initiatives have been a feature of this orientation, since it has been recognised that:

(a) customers have an ever-widening choice of products and services available to them; and

(b) service (including sales service, delivery, after-sales service and, in general, communication and contact between the customer and the organisation) is of major importance in winning and retaining customers.

3.3 One consequence of a customer orientation for the *personnel* function is that it must identify with the organisation's objectives of winning and retaining *external customers*, in order to achieve business success. So, for example, if the organisation establishes a service improvement or customer care policy, the personnel department's objectives will be to staff the organisation (and particularly those units which deal directly with customers) with people who have a strong service ethic, good communication skills and so on - whether through recruitment and transfer, training or retraining.

3.4 One conspicuous example might be in banks, which have hitherto been regarded as non-responsive bureaucratic organisations. They are currently making particular efforts to establish a customer-friendly culture at branch level. Staff are being trained in complaint handling, face-to-face transactions ('personal banking'), explaining bank procedures and technology and so on. Quality/service surveys are being used to find out what customers think of the service they get from staff.

Internal customers

3.5 An equally important concept for the personnel function, however, is that of the *internal customer*.

3.6 Market forces can be a useful way of compelling departments or functions in an organisation to reappraise their performance, and their relationship to each other. Personnel - like any other unit - may focus on its activity for its own sake, as if it had no objective, no purpose outside the department. It may take for granted its relationship to other units, having a 'take it or leave it' attitude to the service it provides, being complacent about quality because it appears to have an effective monopoly on that service or task ('if we don't do it - it doesn't get done'). The concept of the internal customer aims to change all that.

3.7 As the term suggests, the internal customer concept implies the following.

(a) Any unit of the organisation whose task contributes to the task of other units (whether as part of a process, or in a 'staff' or 'service' relationship) can be regarded as a supplier of services like any other supplier used by the organisation. The receiving units are thus *customers* of that unit.

(b) The concept of *customer choice* operates within the organisation as well as in the external market environment. If an internal service unit fails to provide the right service at the right time and cost, it cannot expect customer loyalty: it is in *competition* with other internal and external providers of the service. Although there are logistical and control advantages to retaining the provision of services within the organisation, there is no room for complacency.

(c) The service unit's objective thus becomes the efficient and effective *identification of* and *satisfaction of customer needs* - as much within the organisation as outside it. This has the effect of integrating the objectives of service and customer units throughout the organisation. (It also makes units look at the *costs* of providing their services and what *added value* they are able to offer.)

3.8 The internal customers of personnel include:

(a) line managers, who expect the right quality and quantity of labour resources to meet their own objectives;

(b) senior management and shareholders, who expect the strategic objectives of the organisation to be met through human resource management; and

(c) employees, who expect their contract of employment to be fulfilled, and to have their interests preserved (insofar as they do not conflict with those of the other internal customers).

3.9 Personnel must find out what the needs and wants of these customers are (yet another reason why personnel managers need to be closely involved in objective-setting at all levels of the organisation). These needs and wants must then be satisfied as far and as efficiently as possible. The personnel department, in its various activities, may be in competition with:

(a) external service providers (say, training companies, recruitment consultancies, industrial relations services like ACAS, or computer bureaux for record-keeping and modelling); and/or

(b) internal service providers - notably, line managers who might wish to decentralise many personnel functions and perform them themselves.

Exercise 4

Peter, who works in the word processing department, types a letter drafted by Sarah, the human resource manager.

Who is the 'internal customer' in this transaction?

Solution

Arguably, if you identified only *one* of the parties involved you have missed the point. Sarah supplies Peter with work to do; but she is wasting Peter's time if she provides a bad draft that has to be altered extensively. Peter supplies Sarah with the finished product and he will waste Sarah's time if it is badly typed and needs extensive correction. Thus they are *both* customers.

Total quality management (TQM) and the personnel function

3.10 Total quality management (TQM) is a management technique derived from Japanese companies and focusing on the belief that total quality is essential to survival in a global market. One of the basic principles of TQM is that the cost of preventing mistakes is less than the cost of correcting them once they occur. The aim should therefore be to *get things right first time*, consistently. (This will be a familiar idea to any personnel manager who has recruited the wrong person - or too many people - and then had to dismiss them or make them redundant....) The other main principle is that organisations should strive to *improve continuously* - make things even better next time.

3.11 Another aspect of TQM for many organisations seems to be the development of close supportive relationships with suppliers of services, goods and materials - along the lines of traditional Japanese businesses. This ensures quality all along the line from raw materials to after-sales service and delivery.

3.12 This, too, can affect the personnel function. *Personnel Management* (September 1992) reported a number of companies to be training and developing suppliers - and even customers! ICL, Rank Xerox and Texas Instruments were developing training links with suppliers, and also programmes training customers to use products - with spin-off benefits in the feedback thus obtained on customer needs and abilities. The article noted that this approach is essential in new technology fields, but that 'personnel managers ... in other industries should be helping colleagues to think along similar lines', especially since the British Standard for Quality Assurance Systems (BS5750) includes the registered organisation's obligation to ensure adequate staff training by its *suppliers* - not just within the organisation itself.

3.13 Stephen Connock (*HR Vision: Managing a Quality Workforce*) suggests that there is a renewed focus on the need for a 'productive, trained, flexible and innovative workforce', which he sums up as 'a *quality* workforce'. He suggests that human resource managers should use vision and strategies to create and maintain this quality workforce in support of business objectives: this is the source of perceived added value for the HR function.

3.14 Connock lists what he considers the dimensions of quality for employers.

(a) A customer services orientation.
(b) Taking personal responsibility for quality output.
(c) Well trained and developed staff to meet quality requirements.
(d) Employee involvement in all aspects of quality.
(e) Maintaining quality standards.
(f) Communication and recognition programmes which reinforce quality.
(g) Searching for continuous improvement.
(h) Knowledge of and identification with quality from staff at all levels.

Example: quality at Mars' Four Square division

3.15 'Four Square (suppliers of Klix drinks systems, among others) have strong quality principles, communicated from Mars.

(a) The concept of the 'internal customer' is highly developed and considerable time has been devoted to the 'Putting customers first' campaign, involving both internal and external customers. Each associate in the company has to have a company service objective included in his/her annual standards of performance.

(b) There is a seven-step programme - including communication and training - to develop all suppliers to be suppliers of excellence.

(c) On the customer side, drinks system distributors' staff are not only given technical training in servicing, fault finding and learning but also a lot of sales and customer-care training.'

Personnel Management September 1992

3.16 If the role of the personnel department in organisations is going to move away from the maintenance or fire-fighting sphere, and away from pure routine, into more creative and pro-active areas, criteria such as the above should be taken seriously - rather than purely quantitative measurements like speed of action, cost savings, or attendance records.

As the examiner noted in his report to a June 1986 question on 'excellence' in the personnel department: 'If only more personnel departments set these standards for themselves, then conceivably the function would increase its status and help to implement the kind of cultural change that so many organisations desperately need.'

Chapter roundup

- 'People may be regarded as a vital resource ... but many managers find it difficult to appreciate where the personnel department fits in, except in the simplest terms as a procurement and fire-fighting function.' (Armstrong)
- Personnel is a 'staff' rather than a 'line' function, and so must exploit to the full its resource, expert and ecological power and influence, rather than the rational/legal power possessed by line managers.
- This view is changing, with the concept of human resource management, or HRM, which basically views the personnel specialists' job as a strategic one: 'To know what business they are in, to know where it is going, and to ensure that the input to get there is available from a human resourcing and organisational capability point of view.' (Beattie)

Test your knowledge

1 List seven traditional functions of personnel management. (see para 1.5)

2 What is 'human resource management'? (1.10 - 1.12)

3 List four strategic roles for personnel/human resource management. (1.13)

4 Distinguish between 'line' and 'staff' management. (2.6)

5 What (a) enhances and (b) undermines the authority of the personnel specialist? (2.18 - 2.22)

6 What is the 'internal customer concept'? Who is the customer of the personnel function? (3.17, 3.18)

7 What are Connock's eight 'dimensions of quality'? (3.14)

Now try question 7 at the end of the text

Chapter 8

MANPOWER PLANNING

This chapter covers the following topics.

1. The need for manpower planning
2. The process of manpower planning
3. Labour turnover
4. Management succession and promotion
5. Manpower planning in the 1990s: labour flexibility

Signpost

- This chapter is largely concerned with the *internal* job market; the external market is discussed in Chapter 9. Skill development is the concern of Chapters 11 and 12. Shedding labour is the topic of Chapter 15.
- Labour flexibility is also a topic in Chapter 3; hours of work were discussed in Chapter 4.

1. THE NEED FOR MANPOWER PLANNING

1.1 'Manpower' is a general term for people at work, or 'labour', or 'human resources': people are one of the resources of an organisation which management must control. They are a relatively difficult resource to control because they are not only subject to environmental factors which cause fluctuations in supply and demand, but also to the complexities of human behaviour at individual and group level.

Thus, for example, a *shortage* of a particular skill which affects the performance of an organisation may be caused by: long-term declines in education and training, or in population (nationally or in the local area); the immediate effects of a competitor entering the market or area and employing some of the pool of skilled labour; increases in demand for the product or service for which the skill is required; or the relocation, resignation or demotivation of key skilled people - for all sorts of personal and circumstantial reasons.

1.2 It is arguable that as control over the labour resource becomes more *difficult* - because of increasing uncertainty and rate of change - it also becomes more *necessary*, because the risks of 'getting it wrong' are correspondingly greater.

Manpower planning is a form of *risk management*. It involves realistically appraising the present and anticipating the future (as far as possible) in order to get the *right people* into the *right jobs* at the *right time*.

1.3 In simple terms, manpower planning is a form of supply and demand management, aiming to minimise the risk of either surplus (and therefore inefficiency) or shortage (and therefore ineffectiveness) of relevant kinds of labour. The process may be broadly outlined as follows.

(a) Forecast demand for each grade and/or skill of employee.

(b) Forecast supply of each grade and/or skill of employee, both within and outside the organisation.

(c) Plan to remove any discrepancy between demand and supply. If there is a shortage of labour, for example, you would need to reduce demand (say, through improved productivity), or improve supply (through training and retention of current staff, or recruitment from outside, for example).

Proactive or reactive?

1.4 So far, we have assumed that manpower planning is about anticipating and preventing problems (as far as possible) before they occur: a proactive and forward-looking process. However, it may seem to you (perhaps on the basis of your past experience in an organisation) that this is impractical - and not really necessary - for most small- to medium-sized concerns. If you are short of staff, you hire some, or train or promote one of your existing employees. If your activities, on the other hand, decline and you have superfluous staff, you make some redundancies, or 'downsize' your staff.

1.5 Things are not, in fact, so simple. An attempt to look beyond the present and short-term future, and to prepare for contingencies, is increasingly important.

(a) Jobs often require experience and skills which cannot easily be bought in the marketplace, and the more complex the organisation, the more difficult it will be to supply or replace highly specialised staff quickly. It takes time to train and develop technical or specialist personnel (say, an airline pilot or computer programmer) so there will be a lead time to fill any vacancy: the need will have to be anticipated in time to initiate the required development programmes.

(b) Employment protection legislation and general expectations of 'social responsibility' in organisations make staff shedding a slower and more costly process. Attempted redundancies in the coal mining industry in October 1992 indicated the extent to which the general public - recession-hardened as it may be - still responds to job losses with a sense of outrage.

(c) Rapid technological change is leading to a requirement for manpower which is both more highly skilled and more adaptable, in response to replacement of human labour by technology in routine areas of activity, and changes in work practices. Labour flexibility, in particular, is a major issue, and means that the career and retraining *potential* of staff are at least as important as their actual qualifications and skills - and need to be assessed in advance of requirements. (In fact, 'trainability' as a major criteria for selection is one of the most popular innovations of the HRM era of personnel management.)

(d) Recessionary pressures highlight problems of productivity and manpower costs, and the need to make efficient use of labour. Meanwhile, with contraction rather than expansion the order of the day, the organisation will need to rethink its policies on job security, promotion and other means of retaining staff. Attempts to 'prune' staff levels can have adverse effects on the ratio of indirect to direct labour, on the age distribution of the workforce (through early retirement policies) and so on.

(e) The UK continues to suffer from specific skill shortages (with wide local variations), despite high unemployment levels. This is exaggerated by demographic trends, discussed later. Skill shortages also create a tendency to greater career mobility, which complicates the manpower planners' assumptions about labour wastage rates (the number of people leaving the organisation for any reason).

(f) The amount and variety of markets, competition and labour is being increased by political and economic moves such as the unification of Germany, ethnic separatist movements in the Balkans, the opening of Eastern Europe and the completion of the European market in 1992.

(g) Computer technology has made available techniques which facilitate the monitoring and planning of manpower over fairly long time spans: manipulation of manpower statistics, trend analysis, 'modelling' and so on.

> 'Manpower planning has maintained its imperatives for several reasons: (i) a growing awareness of the need to look into the future, (ii) a desire to exercise control over as many variables as possible which influence business success or failure, (iii) the development of techniques which make such planning possible.
> Bryan Livy, *Corporate Personnel Management*

2. THE PROCESS OF MANPOWER PLANNING

2.1 The complexity of the supply and demand equation with regard to the vital human resource of an organisation necessitates an integrated, systematic approach each of the following.

(a) Forecasting *manpower requirements*, by grades and skills, to meet the long- and short-term needs of the organisation (the manpower strategy). The forecast will be based on the objectives of the organisation (and any likely changes), present (and anticipated future) manpower utilisation - including productivity - and any external factors which will influence demand for labour (such as new technology, market expansion or recession, or employee demands for longer holidays).

(b) Forecasting the *manpower supply* available to meet these needs, taking into account labour turnover, the potential production capacity of the existing labour force, and the accessible pool of labour in the environment.

(c) Acquiring or shedding manpower where necessary and controlling the rate and direction of its flow through the organisation (through conditions of employment, pay, job enrichment, career development, redeployment, employee relations, termination of employment and other succession and retention strategies).

(d) Developing required skills and abilities within the organisation, to enhance the workforce's capacity, and providing the conditions and resources necessary to enhance or maintain productivity.

> 'Manpower planning is an interdisciplinary activity. It requires the combined technical skills of statisticians, economists and behavioural scientists together with the practical knowledge of managers and planners... At the level of the firm, manpower planning deals with problems of recruitment, wastage, retention, promotion and transfer of people within the firm and in relation to its environment.'
> *(Bartholemew)*

2.2 The *demand* for labour must be forecast by considering:

(a) the *objectives* of the organisation, and the long and short-term plans in operation to achieve those objectives. Where plans are changed, the effect of the changes must be estimated: proposed expansion, contraction or diversification of the organisation's activities will obviously affect the demand for labour in general or particular skills, and may be estimated by market research, competitive analysis, trends in technological advances and so on (although sudden changes in market conditions complicate the process: the effect of the collapse of the Soviet Union on defence spending, for example);

(b) *manpower utilisation* - how much labour will be required, given the expected productivity or work rate of different types of employee and the expected volume of business activity. Note that productivity will depend on capital expenditure, technology, work organisation, employee motivation and skills, negotiated productivity deals and a number of other factors;

(c) the *cost* of labour - including overtime, training, benefits and so on - and therefore what financial constraints there are on the organisation's manpower levels;

(d) *environmental factors* and trends in technology and markets that will require organisational change, because of threats or opportunities. The recession in the early 1990s has created conditions in which expectations of labour demand in the short term are low: downsizing of staffs and delayering of organisation structures are the current trend.

Exercise 1

Is the need for manpower in your own organisation growing, shrinking or simply 'moving' into new skill areas? What, if anything, is the organisation doing about this?

2.3 The available *supply* of labour will be forecast by considering:

(a) wastage (turnover through resignations and retirements), promotions and transfers, absenteeism and other staff movements. This will require information on:

- (i) the age structure of staff (forthcoming retirement or family start-up);
- (ii) labour turnover for a comparable period;
- (iii) the promotion potential and ambitions of staff;

(b) the production level and potential of the existing work force, and its structure (age distribution, grades, location, sex, skills, hours of work and rates of pay);

(c) the potential supply of new labour with the relevant skills from the environment - the external labour market. This is the area expected to be hardest hit by the so-called 'demographic time bomb'.

UK demographic trends

2.4 The 'demographic timebomb' or (less sensationally) 'demographic downturn' has been a major HR issue for some years, and although it has lost some of its importance with recessionary falls in the demand for labour, it still has knock-on effects, and may re-emerge as a major problem once it is no longer masked by the recession.

Demography is the study of statistics to do with births, deaths, population distribution and so on. The changes facing the UK in the 1990s appear to be the following.

(a) Falling birthrates following the post-war 'baby boom' mean that the number of young people (16-19 year olds) entering the labour market has been falling since the mid 1980s and is expected to fall to a minimum in 1994.

(b) At the same time, improved health and improved opportunities for women will create an expanding adult workforce, including significantly higher numbers of women and older workers (say, over 45).

2.5 This means that:

(a) employers who previously recruited large numbers of school leavers will need to alter their strategies;

(b) the need to employ more women will require consideration of equal pay and opportunities, discrimination and sexual harassment in the workplace, and facilities for workers with children;

(c) the need to employ older workers may mean higher entry grades of pay, consideration of retraining and working conditions.

Exercise 2

If you are not based in the UK, what are the key demographic issues facing your own country?

Closing the gap between demand and supply

2.6 Shortages or surpluses of labour which emerge in the process of formulating the position survey may be dealt with in various ways.

(a) A *deficiency* of labour may be met through:

(i) internal transfers and promotions, training and development;
(ii) external recruitment;
(iii) the extension of temporary contracts, or the contracts of those about to retire;
(iv) reducing labour turnover, by reviewing possible causes, including pay and conditions;
(v) the use of freelance/temporary/agency staff;
(vi) encouraging overtime working;
(vii) productivity bargaining; or
(viii) automation (increasing productivity, and/or encouraging the elimination of jobs).

(b) A *surplus* of labour may be met by:

(i) running down manning levels by natural wastage (or 'accelerated wastage' - encouraging labour turnover by reducing job satisfaction, pay or other incentives to stay);

(ii) restricting or 'freezing' recruitment;

(iii) redundancies (voluntary and/or compulsory);

(iv) early retirement;

(v) a tougher stance on discipline, enabling more dismissals;

(vi) part-time working or job sharing;

(vii) eliminating overtime; or

(viii) redeployment of staff to areas of labour shortage. This may necessitate diversification by the organisation, to find new work for the labour force.

2.7 Manpower strategy thus requires the integration of policies for:

(a) pay and conditions of employment;
(b) promotion;
(c) recruitment;
(d) training;
(e) industrial relations.

Tactical plans can then be made, within this integrated framework, for pay and productivity bargaining; management and career development; organisation and job specifications, recruitment, downsizing and so on.

The manpower plan

2.8 The manpower plan is prepared on the basis of the analysis of manpower requirements, and the implications for productivity and costs. The plan may consist of various elements, according to the circumstances.

(a) *The recruitment plan:* numbers and types of people, and when required; the recruitment programme.

(b) *The training plan:* numbers of trainees required and/or existing staff needing training; training programme.

(c) *The re-development plan:* programmes for transferring or retraining employees.

(d) *The productivity plan:* programmes for improving productivity, or reducing manpower costs; setting productivity targets.

(e) *The redundancy plan:* where and when redundancies are to occur; policies for selection and declaration of redundancies; re-development, re-training or re-location of redundant employees; policy on redundancy payments, union consultation and so on.

(f) *The retention plan:* actions to reduce avoidable labour wastage.

The plan should include budgets, targets and standards. It should allocate responsibilities for implementation and control (reporting, monitoring achievement against plan).

Exercise 3

Compare the degree to which effective manpower planning is possible and desirable within any two of the following types of organisation:

(a) a company designing, manufacturing and selling personal computers;
(b) a large local authority;
(c) an international airline, or publicly owned international airline.

(ICSA June 1986)

Notes

This question is basically about the practical difficulties of implementing manpower planning in different situations, as well as the particular manpower requirements of certain types of organisation. A general framework for an answer can be drawn from Paragraph 1.5 of this chapter. In addition, you might note that:

(a) the computer company is in an extremely volatile and changing market, which will present planning difficulties. It is also a highly skilled business, however, with long training times and skill shortages - so despite the difficulties, manpower planning will be important, especially since the company is likely to have a flexible structure;

(b) when the question was set, the local authority was a more stable (even ponderous) bureaucratic structure, and there was little volatility in its immediate environment. Reductions in funding and contracting out of services will make manpower planning more difficult in the 1990s;

(c) the international airline is a business which is potentially volatile (transatlantic price wars, political/terrorist action, closely protected airspace, crashes affecting popularity etc) and which is critically dependent on scarce employees with long training cycles.

How 'scientific' is manpower planning?

2.9 Manpower planning is regarded as a scientific, statistical exercise, but it is important to remember that:

(a) statistics are not the only element of the planning process, and are subject to interpretation and managerial judgements (about future growth, say, or potential for innovation) that are largely qualitative and even highly speculative;

(b) trends in statistics are the product of social processes, which are *not* readily quantifiable or predictable: staff leave for various social reasons in (unpredictable) individual cases, to get married, relocate or whatever. The growth of the temporary and freelance workforce is a social trend, as are the buying patterns which dictate demand for goods and services.

2.10 Forecasting is not an exact science. Few exponents of even the most sophisticated techniques would claim that they are wholly accurate, although:

(a) the element of guesswork has been substantially reduced by the use of computer and other models to test various assumptions, and to indicate trends;

(b) the general principles can still be applied to indicate problems, and stimulate control action.

The uncertainty of the future is the main problem for manpower planning of any long-range nature. This is not to say that the exercise is in itself futile (indeed it will be even more necessary, in order to assess and control the *risk* of manpower resourcing problems in the future) but a measure of flexibility should be built into the plan, so that it can be adapted to suit likely or even unforeseen contingencies. Above all, it should not be seen and communicated as an inflexible plan, as if it were based on certainty.

2.11 *Statistical methods* are varied in their approach and degree of sophistication. Computerisation has greatly enhanced the speed, ease and accuracy with which they can be applied. Simple extrapolation, regression analysis and sensitivity analysis can be used to create a more accurate model of the future than simple subjective estimates. Even so, there are a number of assumptions involved.

Moreover, the results are limited in value: they are *quantitative* - for example, numbers of staff required - where *qualitative* information may be required as well: the effects of change, re-staffing, or management style on the culture of the organisation and individual/group behaviour and so on.

2.12 *Work study methods* aim to set standards of man-hours per unit of output in order to achieve maximum productivity. Where end-products are measurable, work-study techniques can offer a reasonably accurate forecast of manpower requirements, for direct workers at least. In service sectors and 'knowledge work', however, end products and output may not be easily subject to standard-setting: the number of telephone calls, interviews, customers served, or ideas generated is likely to fluctuate widely with the flow of business and the nature of particular transactions.

2.13 *Managerial estimates* form the simplest and cheapest method of assessment: as such, they may be the most appropriate - and are the most common - method for small organisations. At the best of times, however, this method has the disadvantage of a high degree of subjectivity, and although this can, to an extent, be controlled (by requiring managers to support their estimates with reasons and to reconcile their estimates with those of senior management), it is a source of potential risk.

'Clearly, the more precise the information available, the greater the probability that manpower plans will be accurate. But, in practice, they are subject to many imponderable factors, some completely outside an organisation's control ... international trade, general technological advances, population movements, the human acceptance of or resistance to change, and the quality of leadership and its impact on morale. The environment, then, is uncertain, and so are the people whose activities are being planned. Manpower plans must therefore be accepted as being continuous, under constant review, and ever-changing. Since they concern people, they must also be negotiable.' Cuming, *Personnel Management*

The manpower audit

2.14 Many large organisations have instituted manpower planning systems to meet their staffing requirements. It is important to ensure that such systems work as they are intended to do and that the plans they incorporate are properly implemented. This is the process of *manpower audit*.

2.15 Regular headcount summaries should be produced to ensure that the manpower trend is in line with forecasts. The summaries should show breakdowns of actual manpower against flexed budgets for the organisation as a whole and for divisions and departments within it.

2.16 Another common feature of the manpower audit is checking through a batch of personnel records to identify that each change (promotion, transfer, redundancy, recruitment, etc) has been properly approved.

This process may uncover:

(a) inadequate authorisation of particular types of change. For example, it may be common to transfer employees within the same department without proper approval or reference to the overall manpower plan;

(b) unauthorised or unnecessary use of agency or temporary personnel.

2.17 Actual manning levels should be checked against manning standards. There are various ways in which actual manning may become out of line with agreed standards.

(a) If the standards were originally agreed on the basis that manning levels would gradually decrease through natural wastage, it is important to ensure that such wastage is allowed to happen. It is a natural tendency for managers to seek replacements for any staff losses, even those which have been budgeted for.

(b) The standards themselves may be (or have become) inappropriate. The manpower plan is constantly evolving and it is necessary to review and revise manpower standards as a regular process.

2.18 A further step is the audit of manpower utilisation. Organisations need to review how well they use their employees in the skill categories concerned. It may be that this process will uncover a need for some fundamental change (such as a complete restructuring of work practices). This might become evident, for example, from instances of employees leaving because they feel under-employed. Apart from the satisfaction and development of employees, it should be clear that under-utilisation of a skill category is an inefficient use of the organisation's resources. Paying a shorthand typist a shorthand typist's wage to perform copy-typing duties is one obvious example of wastage.

The labour market

2.19 Do not forget that labour supply is potentially located both inside and outside the organisation. The market for labour, within which organisations compete for skills and experience and individuals compete for jobs, is likewise both internal and external.

The external labour market will be covered in the following chapter on recruitment.

Three aspects of the *internal* labour market, which we will discuss here, are:

(a) labour turnover, or wastage;
(b) management succession and promotion; and
(c) labour flexibility.

3. LABOUR TURNOVER

3.1 There are different ways of measuring labour turnover. Most simply, actual gross numbers of people leaving may provide a basis for identifying recruitment numbers - but the statistic does not say anything about whether or not these people *need* replacing. To measure labour turnover in a more systematic and useful way, an index may be used, allowing managers to make comparisons between, for example, different organisations in the same industry, or different departments in the same organisation: the *significance* of the wastage figures then emerges.

For example:

(a) *Crude labour turnover rate* (the BIM Index, British Institute of Management, 1949)

Here we express turnover as a percentage of the number of people employed.

$$\frac{\text{Number of leavers in a period}}{\text{Average number of people employed in the period}} \times 100 = \%\ \textit{turnover}$$

This is normally quoted as an annual rate and may be used to measure turnover per organisation, department or group of employees. The disadvantage of this index is that it does not indicate the length of service of leavers, which makes it impossible to identify long term employees and therefore the size of the *stable* workforce. (In fact, most wastage occurs among young people and those in the early stages of their employment in an organisation: stability tends to increase with length of service.)

(b) *Labour stability*

Here we try to eliminate short-term employees from our analysis, thus obtaining a better picture of the significant movements in the workforce.

$$\frac{\text{Number of employees with one or more years' service}}{\text{Number of employees employed at the beginning of the year}} \times 100\% = \%\ \textit{stability}$$

Particularly in times of rapid expansion, organisations should keep an eye on stability, as a meaningful measure.

Suppose a company has 20 employees at the beginning of 19X2, and 100 at the end of the year. Disliking the expansion, 18 of the original experienced labour force resign. At the end of 19X2, the company works out that it has:

BIM Index: $\frac{\text{18 leavers}}{\text{60 (average) employees}} \times 100 = 30\%$ turnover

This is not uncommon, and would cause no undue worries. However:

Stability index: $\frac{\text{2 year-servers}}{20} \times 100 = 10\%$ stability

Only 10% of the labour force is stable (and therefore offering the benefits of experience and acclimatisation to the work and culture of the organisation)! A crude turnover rate has disguised the significance of what has happened

(c) The labour stability index ignores new starts during the year but does not consider actual length of service, which may be added to the measurement via *length of service analysis*, or *survival rate analysis*, whereby the organisation calculates the proportion of employees who are engaged within a certain period who are still with the firm after various periods of time. There may be a survival rate of 70% after two years, for example, but only 50% in year three: the distribution of losses might be plotted on a survival curve to indicate trends.

Advantages and disadvantages of labour turnover

3.2 It would be wrong to think of labour turnover as purely disadvantageous to every organisation. Potential advantages include:

(a) opportunities to inject 'new blood' into the organisation: new people bringing new ideas and outlooks, new skills and experience in different situations. Absence of labour turnover would simply create an increasingly aged workforce;

(b) the creation of opportunities for promotion and succession. If there were no labour turnover, junior staff and management would face a long career without development, waiting for someone higher up the ladder to retire, with no incentives or encouragement to ambition;

(c) the ability to cope with labour surpluses, in some grades of job, without having to make redundancies. Natural wastage can save industrial relations problems in this way.

3.3 However, disadvantages of labour turnover include the following.

(a) It breaks the continuity of operations, culture and career development. It is generally recognised by employers that continuity offers stability and predictability, which is beneficial to efficiency. When people leave, there is bound to be a hiatus while a replacement is found and - through induction, training and experience - brought 'on line' to the same level of accustomed expertise of the previous job-holder.

(b) It may be perceived by other employees as a symptom of job dissatisfaction, causing the problem to escalate. High labour turnover can foster a culture low in morale and loyalty.

(c) The costs of turnover can be high, including:

(i) *replacement costs:* the cost of recruiting, selecting and training replacements; loss of output or efficiency during this process; possible wastage, spoilage and inefficiencies because of inexperience in new staff; and also

(ii) *preventative costs:* the cost of retaining staff, through pay, benefits and welfare provisions, maintaining working conditions or whatever.

3.4 It is common to hear that turnover is bad when it is high - but this cannot be assessed in isolation. What is an acceptable rate of turnover and what is excessive? There is no fixed percentage rate of turnover which is the borderline between acceptable and unacceptable. Labour turnover rates will be a signal that something is possibly wrong when:

(a) they are higher than the turnover rates in another similar department of the organisation. For example, if the labour turnover rate is higher at branch A than at branches B, C and D in the same area, something might be wrong at branch A;

(b) they are higher than they were in previous years or months. In other words, the situation might be deteriorating;

(c) the costs of labour turnover are estimated and are considered too high - although they will be relative to the costs of *preventing* high turnover by offering employees rewards, facilities and services that will keep them in the organisation.

Otherwise, the organisation may live with high rates because they are the norm for a particular industry or job, because the organisation culture accepts constant turnover, or because the cost of keeping employees is greater than the cost of replacing them.

Causes and control of labour turnover

3.5 A systematic investigation into the causes of undesirable turnover will have to be made, using:

(a) information given in *exit interviews* with leaving staff, which should be the first step after an employee announces his intention to leave. It must be recognised, however, that the reasons given for leaving may not be complete, true, or those that would be most useful to the organisation: a person may say that they are 'going to a better job', while the real *reason* for the move is dissatisfaction with the level of interest in the current job. The interviewer should be trained in interview techniques, and should be perceived to be 'safe' to talk to and objective in his appraisal of the situation (and therefore probably *not* the supervisor against whom the resigning employee has a complaint or the manager who is going to write a reference);

(b) information gleaned from interviews with leavers, in their homes, shortly after they have gone. This is an occasionally-used practice, intended to encourage greater objectivity and frankness, but one which requires tact and diplomacy if it is not to be resented by the subject;

(c) attitude surveys, to gauge the general climate of the organisation, and the response of the workforce as a whole to working conditions, management style and so on. Such surveys are notoriously unreliable, however;

(d) information gathered on the number of (interrelated) variables which can be *assumed* to correlate with labour turnover. Some of these are listed below.

3.6 Labour turnover might be influenced by any of the following factors:

(a) *the economic climate and the state of the jobs market.* When unemployment is high and jobs are hard to find, labour turnover will be much lower;

(b) *the age structure and length of service of the work force.* An ageing workforce will have many people approaching retirement. However, it has been found in most companies that labour turnover is highest amongst:

(i) young people, especially unmarried people with no family responsibilities;
(ii) people who have been in the employment of the company for only a short time.

The employment life cycle usually shows a decision-point shortly after joining, when things are still new and perhaps difficult: this is called the 'first induction crisis'. There is then a period of mutual accommodation and adjustment between employer and employee (called the 'differential transit' period): in the settling of areas of conflict, there may be further turnover. A second (less significant) induction crisis occurs as both parties come to terms with the new status quo. Finally, the period of 'settled connection' begins, and the likelihood of leaving is much less;

(c) *the organisation climate or culture, and its style of leadership.* An organisation might be formal and bureaucratic, where employees are expected to work according to prescribed rules and procedures. Other organisations are more flexible, and allow more scope for individual expression and creativity. Individuals will prefer - and stay with - one system or the other;

(d) *pay and conditions of employment.* If these are not good enough, people will leave to find better terms elsewhere, or will use this as a catalyst to express their discontent in other areas;

(e) *physical working conditions.* If working conditions are uncomfortable, unclean, unsafe, or noisy, say, people will be more inclined to leave;

(f) *career prospects and training.* If the chances of reaching a senior position before a certain age are low, an ambitious employee is likely to consider leaving to find a job where promotion is likely to come more quickly. The same may be true where an employee wants training for a qualification or skill development, and opportunities are limited in his current job.

3.7 Some reasons for leaving will be genuine and largely unavoidable, or unforeseeable, such as:

(a) illness or accident, although transfer to lighter duties, excusing the employee from shiftwork or other accommodations might be possible;

(b) a move from the locality for domestic reasons, transport or housing difficulties;

(c) marriage or pregnancy. Many women still give up working when their family situation changes;

(d) retirement; and

(e) career change.

3.8 However, where factors such as those discussed in Paragraph 3.6 above are identified as sources of employee dissatisfaction and departure, problems can be addressed, if it is considered worthwhile to do so. Selection systems may be revised to ensure that future recruits are more compatible with the culture and leadership of the organisation, and with the demands of their jobs (since some people will be able to handle monotony, pressure, responsibility or lack of it better than others).

4. MANAGEMENT SUCCESSION AND PROMOTION

4.1 Promotion and succession policies are a vital part of the manpower plan, as a form of risk management associated with the internal supply of labour. The planned development of staff (not just skills training, but experience and growth in responsibility) is essential to ensure the *continuity* of performance in the organisation. This is particularly so for *management* planning. The departure of a senior manager with no planned or 'groomed' successor could leave a gap in the organisation structure: the lead time for training and developing a suitable replacement may be very long.

4.2 Promotion is useful from the firm's point of view, in establishing a management succession, filling more senior positions with proven, experienced and loyal employees. It is also one of the main forms of reward the organisation can offer to its employees, especially where, in the pursuit of equity, employees are paid a rate for the job rather than for their performance in the job: pay ceases to be a prime incentive. In order to be a motivator, promotion must be seen to be available, and fair. It can also cause political and structural problems in the organisation if it is not carefully planned.

4.3 A coherent policy for promotion is needed. This may vary to include provisions such as the following.

(a) All promotions, as far as possible, and all things being equal, are to be made from within the firm. This is particularly important with reference to senior positions if junior ranks are not to be discouraged and de-motivated. Although the organisation will from time to time require new blood if it is not to stagnate, it will be an encouragement to staff to see that promotion is open to them, and that the best jobs do not always go to outsiders.

(b) Merit and ability (systematically appraised) should be the principal basis of promotion, rather than seniority (age or years of service). Loyalty and experience will obviously be considered but should not be the sole criterion. Younger staff may grow impatient if they feel that they are simply waiting to grow old before their prospects improve. Promotion on ability is more likely to have a motivating effect. Management will have to demonstrate to staff and unions, however, that their system of appraisal and merit rating is fair and fairly applied if the bases for promotion are to be trusted and accepted.

(c) Vacancies should be advertised and open to all employees.

(d) There should be full opportunities for all employees to be promoted to the highest grades.

(e) Personnel and appraisal records should be kept and up-dated regularly.

(f) Training should be offered to encourage and develop employees of ability and ambition in advance of promotion.

(g) Scales of pay, areas of responsibility, duties and privileges of each post and so on should be clearly communicated so that employees know what promotion means - in other words, what they are being promoted *to*.

4.4 The decision of whether to promote from within or fill a position from outside will hinge on many factors. If there is simply no-one available on the current staff with the expertise or ability required (say, if the organisation is venturing into new areas of activity, or changing its methods by computerisation), the recruitment manager will obviously have to seek qualified people outside. If there is time, a person of particular potential in the organisation could be trained in the necessary skills, but that will require an analysis of the costs as compared to the possible (and probably less quantifiable) benefits.

4.5 Where the organisation has the choice, it should consider the following points.

(a) Management will be familiar with an internal promotee: there will be detailed appraisal information available from employee records. The outside recruit will to a greater extent be an unknown quantity - and the organisation will be taking a greater risk of unacceptable personality or performance emerging later.

(b) A promotee has already worked within the organisation and will be familiar with its:

(i) culture, or philosophy; informal rules and norms as well as stated policy;
(ii) politics; power-structures and relationships;
(iii) systems and procedures;
(iv) objectives;
(v) other personnel (who will likewise be familiar with him).

(c) Promotion of insiders is visible proof of the organisation's willingness to develop people's careers. This may well have an encouraging and motivating effect. Outsiders may well invite resentment.

(d) Internal advertisement of vacancies contributes to the implementation of equal opportunities policies. Many women are employed in secretarial and clerical jobs from which promotion is unlikely - and relatively few are in higher-graded roles. Internal advertising could become a route for opening up opportunities for women at junior levels.

(e) On the other hand, an organisation must retain its ability to adapt, grow and change, and this may well require new blood, wider views, fresh ideas. Insiders may be too socialised into the prevailing culture to see faults or be willing to 'upset the applecart' where necessary for the organisation's health.

4.6 A comprehensive *promotion programme*, as part of the overall manpower plan (for getting the right people into jobs at the right time) will include:

(a) establishment of the relative significance of jobs, by analysis, description and classification, so that the line and consequence of promotion are made clear;

(b) establishment of methods of assessment of staff and their potential for fulfilling the requirements of more senior positions;

(c) planning in advance for training where necessary to enhance potential and develop specific skills;

(d) policy with regard to internal promotion or external recruitment and training.

4.7 The personnel manager will have to use the utmost sensitivity in applying any policies formulated. It is often difficult to persuade departmental managers to agree to the promotion of a subordinate out of the department, especially if he has been selected as having particular ability: the department will be losing an able member, and will have to find, induct and train a replacement. Moreover, if the manager's resistance were made known, there would be a motivational problem to contend with. The personnel manager will have to be able to back his recommendation with sound policies for providing and training a replacement with as little loss of the department's efficiency as possible.

5. MANPOWER PLANNING IN THE 1990s: LABOUR FLEXIBILITY

5.1 Flexibility is an area of current interest in human resource management, as economic pressures require more efficient use of manpower. It involves the development of *versatility* in the labour resource, and the management of that versatility through the complex human relations problems surrounding the application of flexibility in many of its forms.

(a) For the organisation, it offers a cost effective, efficient way of utilising the labour resource. Under competitive pressure, technological innovation, and a variety of other changes, organisations need a flexible, 'lean' workforce for efficiency, control and predictability: the stability of the organisation in a volatile environment depends on its ability to adapt swiftly to meet changes, without incurring cost penalties or suffering waste. If employee flexibility can be achieved with the co-operation of the employees and their representatives, there may be an end to demarcation disputes, costly redundancy packages and other consequences of apparently rational organisation design.

(b) From the point of view of the employee, the erosion of rigid specialisation, the micro-division of labour and the inflexible working week can also offer: a higher quality of working life; an accommodation with non-work interests and demands; greater job satisfaction, through variety of work; and perhaps job security and material benefits, since a versatile, mobile, flexible employee is likely to be more attractive to employers and have a higher value in the current labour market climate.

5.2 Organisational and cultural mechanisms for developing flexibility include the following.

(a) *The erosion of demarcations between job areas*. Workers can be encouraged and organised to work across the boundaries of their job or craft. This has, historically, been difficult to achieve because craft and occupational groups have supported demarcation in order to protect jobs and maintain skills, standards and differentials. With the need for adaptability, however, rigid job descriptions and specialisation have gone out of fashion and versatility is much prized on the labour market.

(b) *Flexibility in the deployment of the workforce in terms of man hours*. With the shrinking demand for some categories of labour, ideas about full employment, full-time employment and 'one man, one job' have had to be revised. There are various ways in which individuals can be given a flexible job - in terms of working hours - and organisations can avoid overmanning and idle time.

(i) Fixed-term contracts, job sharing and splitting, or annual hours contracts are ways of countering various problems of seasonal or task-oriented fluctuations in the demand for labour, avoiding redundancies and lay-offs.

(ii) The employment of non-permanent, non-career labour was a major growth sector of the employment scene in the 1980s and the trend is likely to continue into the 1990s. Part-time work, casual labour, temporary working (the 'temping' boom, not only in secretarial work but in accounting, nursing and teaching, for example) and consultancy are popular options, both for the workers and for the organisations who benefit from their services without long-term contractual obligations.

(iii) Flexitime is another area in which conventional rules and boundaries are increasingly bent or broken: the '9 to 5' is no longer the most effective scheme of working in large city areas where commuting is a problem, and in a workforce where parents and home makers are having to reconcile the requirements of family, household and work.

These options are discussed in more detail below.

(c) *Cultural flexibility*. All of the above methods will require cultural change - or will involve cultural upheaval - in the organisation and in society as a whole. Ideas about specialism, about career, about the relationship between work, leisure and self-identity are threatened.

Quite apart from the initial change of attitude required to smooth the way for these methods, there is an important role for personnel in the development and maintenance of an organisation culture which accepts and applauds flexibility generally: this will be the key to task- and customer-orientation, adaptability to environmental changes, innovation and entrepreneurship which - according to many current theories - is the basis of survival in the modern business environment.

5.3 The personnel function has a role to play in designing flexible systems for recruitment and deployment of labour (permanent and non-permanent, full-time and part-time) within employment legislation. It must organise and administer these systems. It must be responsible for the communication, consultation and negotiation that will be required not only to avoid industrial relations problems but to sell the ideas and values that will become a flexible organisation culture. There may be welfare considerations, counselling and so on as a side effect of major change. So the issue of employee flexibility will impact on much of the personnel function's activity in the organisation.

Exercise 4

How flexible is the workforce in your organisation? Is it versatile, multi-skilled and happy to work unorthodox hours or without a strict job description? What does the personnel department in your organisation do to encourage, or to discourage, flexibility of labour? (Think about job descriptions, training and development, working hours etc.)

Flexible working methods

5.4 G A Cole *(Personnel Management: Theory and Practice)* sums up the pressures on managerial decisions about the size and nature of the workforce as follows.

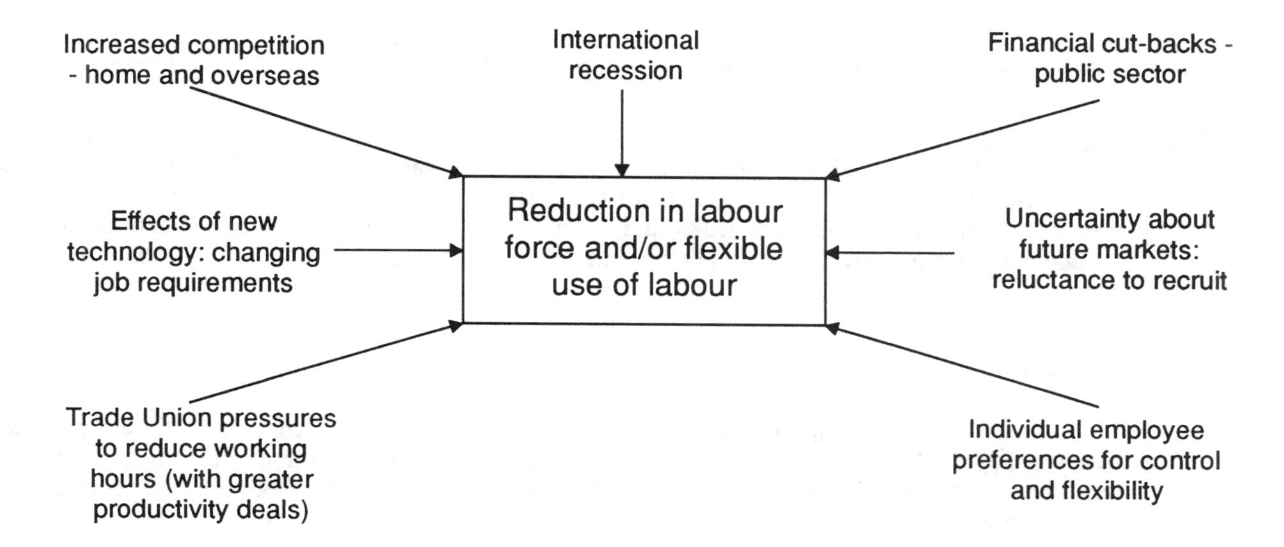

5.5 Torn between the need to downsize (or at least not expand) their workforce and the desire to offer a measure of security of employment, many organisations build themselves a 'buffer' by increasing the proportion of non-permanent labour in their workforce. Categories which generally supplement the permanent (full- or part-time) workforce include trainees, student or agency 'temps' (temporary workers), and subcontract workers.

Alternatively, *job-sharing* may be used: an existing job is split in two, so that two people can share it, paid pro-rata (much like ordinary part-time working).

5.6 Atkinson, in a 1984 paper for the Institute of Manpower Studies *(Flexible Manning: the way ahead)*, suggested that the workforce be divided into:

(a) a *core* group: permanent, functionally flexible (and trained and rewarded accordingly), and representing the lowest number of employees required by work activity at any time throughout the year; and

(b) a *peripheral* group, which can be taken on at any season when staff requirements rise above the minimal. The periphery may consist of:

(i) *top-up* staff, recruited as short-contract, casual or part-time staff; or

(ii) *permanent* employees in areas where there is traditionally a high level of mobility and therefore wastage/turnover: clerical, secretarial and so on.

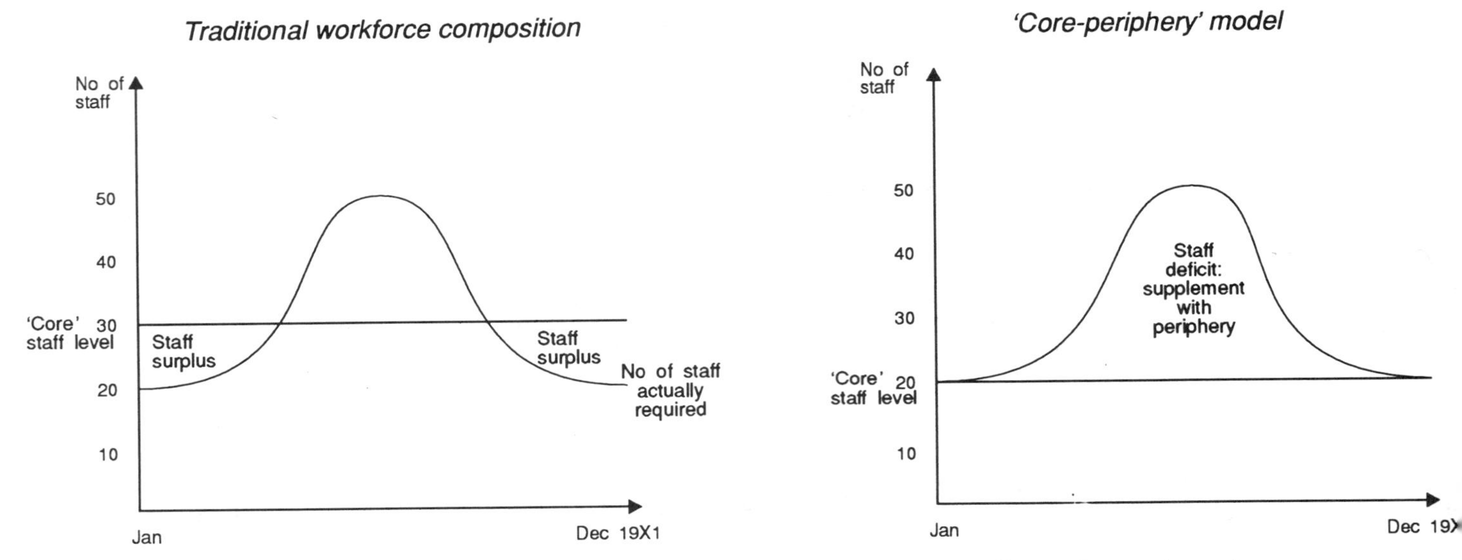

5.7 The question, touched on above, of *rewarding* functionally versatile employees is an important one. A *single status* policy, or the provision of common salary and benefit systems, facilitates redeployment of staff. Breaking down artificial status and reward barriers not only helps to secure greater transferability of skills and labour, but may have advantages of morale and the broadening of staff loyalty and commitment to embrace organisational, rather than sectional, goals.

Example: ICI

5.8 In *Personnel Management* (June 1991) it was reported that 'Manual workers at ICI are expected to agree a deal which would give them a 14 per cent increase and a 36-hour week in return for the introduction of flexible working practices ... The agreement will bring about the biggest change in working practices at ICI since 1969 and has been under discussion for two years.

ICI believes it will improve the competitiveness of the company against its international rivals in Germany, Japan and the US ... The unions recognised that practices would have to change and that the deal gave them some control as well as securing benefits for their members.'

Example: Calor Gas

5.9 In *Personnel Management Plus* (September 1992) it was reported that:

'Calor Gas has replaced more than a third of its hourly paid workers with temporary staff on lower rates of pay, after terminating the contracts of the entire manual workforce.

Two-thirds of the group have been re-employed on new contracts, including 500 drivers, and 100 skilled workers and supervisors, whose pay levels have been maintained. But they are now working more flexibly under a new agreement with the Transport and General Workers Union.

The temporary workers... have replaced full-time operatives who cleaned and filled gas bottles at nine depots around the country.

John Harris, operations and personnel director, said the company needed to cut costs, increase productivity and change its culture.

The market was seasonal and Calor Gas had always used temporary workers in the winter. The difference now was that there was a core of temps, many of whom he expected would work for the company on a long-term basis, as well as a fluctuating periphery.

The company had also replaced its single national pay rate with varying regional rates for the contract workers. "We have fewer people, paid less and they are very much more productive," said Harris.'

Flexible working hours

5.10 There are various techniques for flexibility of working hours, including the following.

(a) *Annual hours contracts*, compensating for seasonal patterns of demand by agreeing yearly (rather than weekly) hours of work: some or all of these hours may be committed to a rota schedule, while some may be held in reserve for unforeseen fluctuations. During the intensive work period, longer working days, or more intensive shifts may be used - without incurring 'overtime' costs, since the extra hours are compensated for by leisure time (from 'shift-free' days to long blocks of time) in the slack period.

(b) *Overtime working* (above standard rates of pay for hours worked in excess of standard). This is highly flexible, but can be expensive, and encourage slower working during standard hours.

(c) *Shift work*. This was described and discussed in Part A.

(d) *Flexi-time*. This was also described in Part A.

(e) *Part-time working*.

(f) *Homeworking/networking/teleworking*.

We will look at the last two of these, in a little more detail.

Part-time working

5.11 The number of part-time employees as a proportion of all employees in the UK grew from 19.1% in 1978 to 24.2% in December 1989. Of these, 80% are women. Stephen Connock *(HR Vision)* gives seven main reasons for the increase in part-time working.

(a) Employers can match working hours to operational requirements better.

(b) The personal circumstances of key staff can be accommodated through part-time working. This will be particularly relevant to women returners.

(c) The productivity of part-timers is generally higher than that of full-timers (hardly surprising, since work is undertaken in more concentrated time periods).

(d) The absence levels of part-timers are generally lower: domestic requirements can more easily be fitted into the free periods in the part-time schedule.

(e) A pool of trained employees is available for switching to full-time work, or extending working time temporarily.

(f) Difficulties in recruiting full-time staff have prompted organisations to recruit and train part-time staff. Women returners are more likely to be attracted to an organisation if the hours of work are suitable, which will generally mean working part-time.

(g) Part-time working can cut overtime costs, since it makes it possible to avoid paying premium rates.

5.12 There has for some years been concern about equal treatment for part-time workers in regard to terms and conditions - especially since, with a higher percentage of part-time workers being women, less favourable terms can be construed as indirect discrimination. (We discuss some of the EC measures in this area a bit later in this chapter.)

Teleworking and homeworking

5.13 Off-site or out-of-office working may be available to employers, with the availability of telecommunications and computer networking. We touched on this topic in an earlier chapter. 'Telecommuting' (*The Administrator*, August 1992) 'describes the process of working from home, or from a satellite office close to home, with the aid of computers, facsimile machines, modems or other forms of telecommunication equipment'. It offers:

(a) savings on overheads, particularly premises costs, in view of rising rents in many of the major cities in the UK;

(b) the opportunity to bring into employment skilled and experienced people for whom traditional working practices have hitherto been impracticable: single parents, mothers, the handicapped, carers and so on. (The greatest impact of this may be increased opportunities for the handicapped, according to the article in *The Administrator*);

(c) elimination of the need to commute - with consequent reductions in traffic congestion, fuel consumption, travel costs, pollution etc;

(d) potential reduction in stress, since there is less conflict with non-work goals and needs, and a more congenial environment (with no commuting to undergo).

5.14 In addition to 'telecommuters', home workers may be:

(a) traditional *outworkers*, such as home typists or wordprocessors and envelope-fillers (for mailshots), writers and editors, tele-canvassers and market-researchers. These may be employed by the firm or sub-contracted (freelance);

(b) *itinerants* such as salesmen, who do not have a permanent presence in the office, and use their home (and even car) for working on the move. These people may be employed by the firm, or self-employed; or

(c) those in personal services, like ironing and mending of clothes, out-of-salon hairdressing, music teaching and so on.

5.15 The Henley Centre for Forecasting predicted (as reported in the *Financial Times*, October 1989) that more than four million people will be working at home in the UK by 1995, in a full- or part-time capacity. However, there are certain problems associated with homeworking for both management and individuals, namely:

(a) *control*. There needs to be a shift of emphasis from managing the work process and methods - since supervision on a day-to-day basis is impossible - towards monitoring and controlling work output or results. Hours of work will need to be monitored;

(b) *communication*. Out-stationed workers may feel isolated from the culture and work-flow of the organisation. Communication for planning, control and co-ordination - as well as morale, motivation and team-building - will need to be constantly pursued;

(c) *co-ordination*. Employing outworkers increases the project management aspects of a task. Integrated plans, communication and control will be required to ensure that the outworkers' output fits into the overall operation in which they are involved;

(d) *employee appraisal and development*. Performance assessment, analysis of training needs, training and career planning must be given attention, given the inaccessibility of the worker most of the time. The basis of assessment may need to be adjusted, given that working conditions will be different and certain skills and attributes (such as teamworking) irrelevant;

(e) *risks associated with 'networking'* (a method by which remote computer terminals can be linked to a central computer at the office). Computer viruses, industrial espionage or 'hacking' and hardware/software incompatibility remain technical problems to be resolved;

(f) *health and safety*. Housing is not always well adapted to work: there may be problems with electrical wiring, fire precautions and other matters which can be more closely controlled at a central office.

Examples: working at home

5.16 Rank Xerox launched a successful networking scheme as long ago as 1981. They closed down a central London office costing them £300,000 a year, 'fired' the staff and then re-engaged them on networking contracts: each became a *separate company* working from home, linked direct to HQ, and guaranteed income if they supplied work on time.

The advantages seemed to be that:

(a) as 'self-employed' workers, the networkers developed the discipline and motivation to work conscientiously;

(b) the more they did, the more they earned, and the firm encouraged them to use their spare time to take on contracts outside Rank Xerox itself;

(c) networkers travelled in to Head Office only one day per week;

(d) some banded together to form multiple units, or shared office premises near their homes, thus overcoming any sense of isolation.

5.17 *The Administrator* (August 1992) cites further examples.

(a) At computer manufacturer DEC, 700 of the 7,000 UK staff have computers at home, and almost 100 work at home full time.

(b) The London Borough of Enfield has more than 50 teleworkers operating from home, with a terminal linked to the central computer. 'Supervisors, who also work from home, collect and deliver work about three times a week: more complex matters are dealt with at the council's headquarters. Homeworkers receive the same pay, terms and conditions as equivalent staff in the office, so trade unions, often wary of homeworking, have been supportive.'

UK and EC measures on flexible working

5.18 Three draft EC directives currently cover the employment of part-time and temporary staff, as part of the action programme for implementing the Social Charter.

(a) The first directive deals with equal treatment of part-time, temporary and full-time permanent staff. The introduction welcomes the positive effects on flexibility and job creation of the increase in part-time and temporary employment, but notes that the treatment of such employees is not always on a par with others, in terms of employment and working conditions.

(b) The second directive deals with the legal rights of part-time and temporary workers. It welcomes flexible work patterns, but warns that the different treatment of these contracts across the member states creates dangers of distorted competition due to differences in indirect wage costs.

(c) The third directive deals with the health and safety protection of temporary workers. It says that 'in general temporary workers are more exposed to the risk of accidents at work and occupational diseases than other workers'. In particular, night-work and shift-work are causes for concern, and the establishment of minimum rest periods is considered essential.

5.19 Temporary work is more regulated in most other member states of the EC than it is in the UK. Many EC employers have a legal obligation to justify temporary work, provide a contract, limit the duration of the contract and provide comparable or similar pay and conditions. Fixed-term contracts automatically become permanent in most states, if they over-run.

The proposed directives would mean some changes to UK legislation.

(a) All employment rights currently available to those working 16 hours or more would have to be extended, with the same conditions for those working eight hours or more. These include, for example, an itemised pay statement and a statement of terms and conditions after 13 weeks; a minimum period of notice and sickness benefit after one month; unfair dismissal protection, redundancy, maternity leave and right to return after 2 years.

(b) Occupational benefits such as access to pension schemes would have to be given to all those working more than eight hours; occupational sick pay and holidays would also have to be given on an equal *pro-rata* basis. Recent EC case law suggests that employers are already at risk on sex discrimination grounds if part-timers are not given the same benefits as full-time employees where the majority of part-timers are female.

(c) Most of the regulations on temporary staff would be new. However, apart from the requirement for a more detailed contract, most of the health and safety provisions for temporary staff would be covered by existing legislation.

5.20 The UK government is concerned that the proposals will add to costs, make it harder to compete internationally and, in the long term, damage job creation.

Exercise 5

It has been argued that employee flexibility will be a key issue for personnel management in the 1990s. How far do you agree? *(ICSA)*

Solution

Have a look at Paragraph 5.1, for reasons why flexibility will become an issue for organisations - and 5.2-5.3 for the personnel manager's role. Bear in mind, also, that it will be an 'issue' for other reasons, too, such as the legal, health and safety, and reward implications: Paragraphs 5.17, 5.19, 5.7 and 5.15 may give you some ideas.

Chapter roundup

- In the following chapters we go on to explore recruitment and selection, training, and termination of employment in much more detail. The contents of this chapter may be summarised diagrammatically: see the next page.

Test your knowledge

1 Outline a systematic approach to manpower planning (see para 2.1)

2 List three types of method used in manpower forecasting. Indicate to what degree each is a 'scientific' method. (2.9 - 2.13)

3 When is labour turnover to be regarded as a bad thing? (3.3, 3.4)

4 List the factors that might contribute to your leaving the organisation where you work. What (if anything) could or should your employers do to keep you? (3.6 - 3.8 for ideas, but think about your situation).

5 List four reasons why an organisation might wish to promote an existing employee rather than recruit someone from outside to fill the post. What might make this a less desirable option from the firm's point of view? (4.5)

6 What three broad 'types' of flexibility may be developed by an organisation? (5.2) List some examples of each. (5.2, 5.6, 5.10)

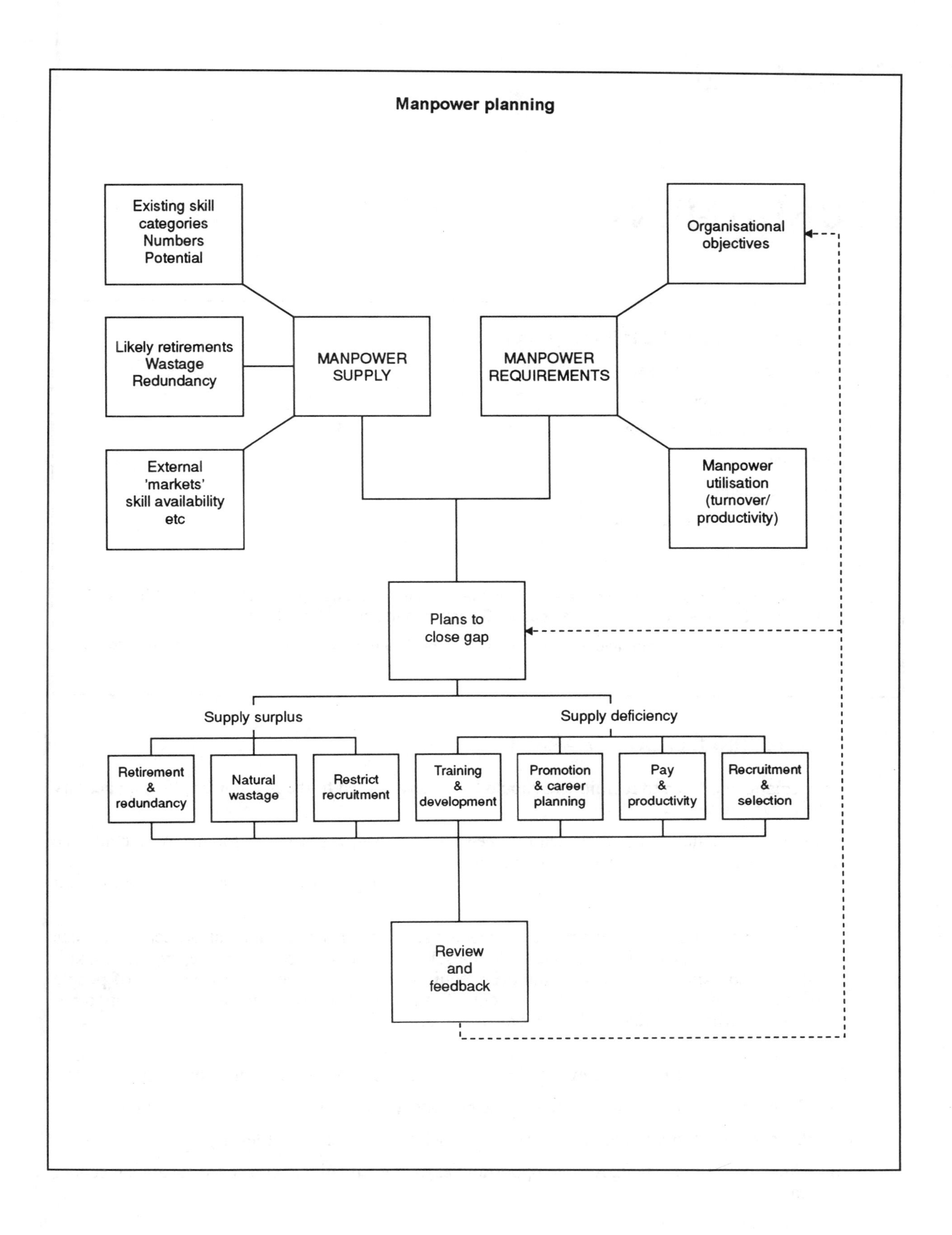

Now try question 8 at the end of the text

Chapter 9

RECRUITMENT

This chapter covers the following topics.

Signpost

- The recruitment plan is the first element in the overall manpower plan described in Chapter 8. Recruitment is often confused with selection. Selection is the subject of Chapter 10.
- Material on effective communication generally is to be found in Chapter 16. This is relevant to job advertisements.

1. THE EXTERNAL LABOUR MARKET

1.1 A systematic approach to recruitment should be closely related to the manpower planning activity of the organisation.

'A recruiter operates at the interface between an organisation's demand for manpower and the supply of people as may be available on the open market.'

Livy, *Corporate Personnel Management*

1.2 The labour market is variously defined by economists, but for the recruitment officer, it consists of the group of potential employees, internal, local or otherwise, with the types of skill, knowledge and experience that the employer requires at a given time. It thus consists of people within the organisation, people who are out of work at that time, and also people in other organisations who may wish to change jobs or employers.

1.3 The size and composition of the *external* market for labour depends on factors such as:

(a) Government policies (on matters such as benefits, taxation or employment protection);

(b) the level of activity in the economy (with associated unemployment levels);

(c) education/training standards and opportunities, and occupational choice trends among school leavers;

(d) wage and salary levels;

(e) competition between organisations for particular skills;

(f) trade union attitudes and the expectations of existing and potential employees (with regard to wages, the role of technology, the adoption of flexible working methods and so on);

(g) the extent to which new technology replaces human skills in a given area;

(h) population changes affecting the number of people in an area, the age or sex distribution of the population and other demographic factors.

1.4 The labour market has changed dramatically in the last couple of decades. Writers on manpower planning in the 1970s suggested that a 'seller's market' had been established, as technology increased the skills and therefore scarcity value of employees in certain jobs, and the scale of state benefits blunted the fear of unemployment: the initiative seemed to be with the employee, or with organised groups of employees. Economic recession and the more general application of technology, among other factors, has changed that situation: a 'buyer's market' for labour now gives employers considerable power, with a large pool of available labour created by redundancy and unemployment.

1.5 This shift in the market changed the purpose and practice of recruitment. In times of low unemployment, employers have to compete to attract desirable categories of labour (and may also have to downgrade their selection requirements if the competition is too 'stiff'). In times of high unemployment, and therefore plentiful supply, 'the problem is not so much of attracting candidates, but in deciding how best to select them' (Cole, *Personnel Management: Theory and Practice*). In times of low demand for labour, however, socially responsible employers may have the additional policy of using existing staff (internal recruitment) rather than recruiting from outside, in order to downsize staff levels through natural wastage and redeployment.

1.6 Even in conditions of high overall employment, particular skill shortages still exist and may indeed be more acute because of recessionary pressures on education and training. Engineers and software designers, among other specialist and highly trained groups, are the target of fierce competition among employers, forcing a revaluation of recruitment and retention policies.

1.7 Market intelligence will clearly have to be gathered, as part of the manpower planning process, in order to monitor fluctuations in the local, regional and national labour supply for the variety of skills the organisation requires.

2. APPROACHES TO RECRUITMENT AND SELECTION

2.1 The overall aim of the recruitment process in an organisation is to obtain the quantity and quality of employees required to fulfil the objectives of the organisation.

2.2 This process can be broken down into three main stages.

(a) Defining requirements, including the preparation of *job descriptions, job specifications and personnel specifications.*

(b) Attracting potential employees, including the evaluation and use of various methods of reaching sources of applicants. Remember that these sources may be inside as well as outside the organisation.

(c) Selecting the appropriate people for the job or the appropriate job for the people.

2.3 Note that there is a distinction between recruitment and selection.

(a) *Recruitment* is the part of the process concerned with finding the applicants: it is a positive action by management, going into the labour market (internal and external), communicating opportunities and information, generating interest.

(b) *Selection* is the part of the employee resourcing process which involves choosing between applicants for jobs: it is largely a 'negative' process, eliminating unsuitable applicants.

2.4 There is also a distinction between:

(a) the *selection approach* to recruitment ('fitting the person to the job'). This is based on the prime objective of fulfilling the organisation's precise manpower needs. It therefore involves close definition of those needs, the choice or design of techniques for selection - matching people to the organisation's requirements - and evaluation of recruitment on the basis of successful matching; and

(b) the *classification approach* to recruitment ('fitting the job to the person'). This focuses more on the needs and aspirations of the worker, and attempts to identify or even design a job whose tasks and physical environment will fit the limits, capacities and wants of an individual or 'type' of individual. It will focus on: techniques aimed at analysing those limits, capacities and needs; definitions of existing jobs in terms of the personal attributes required of the job holder (such as personnel specifications); ways in which jobs can be redesigned, or ergonomics applied, in order to make the task structure and working environment most suitable to the worker or 'type' of worker available.

2.5 Although we discuss selection in detail, as a systematic approach suitable to most organisations with fairly defined goals and structures, you should be aware that the classification approach may be equally valid in some circumstances. In a highly innovative market, technological environment or organisational culture, for example, rigid job descriptions would not be suitable: in order to 'thrive on chaos' (Tom Peters), organisations should be able to look at the skills and attributes of the people they employ, and gifted outsiders, and ask: 'What needs doing that this person would do best?' In a relatively informal environment, where all-round knowledge/skills/experience are highly valued and external labour resources scarce (say, in management consultancy), this approach would give much-needed flexibility. The additional effort to create a congenial job and environment may also be considered a valuable non-financial benefit which would help to retain staff.

2.6 In this text, we concentrate on a systematic approach to recruitment based on the selection model, which will involve the following stages.

(a) Detailed *manpower planning* defining what resources the organisation needs to meet its objectives.

(b) *Job analysis*, so that for any given job there is:

(i) a *job description*: a statement of the component tasks, duties, objectives and standards;

(ii) a *job specification*: a specification of the skills, knowledge and qualities required to perform the job; and

(iii) a *personnel specification*: a reworking of the job specification in terms of the kind of person needed to perform the job.

(c) An identification of vacancies, from the requirements of the manpower plan or by a *job requisition* from a department, branch or office which has a vacancy.

(d) Evaluation of the *sources of labour*, which again should be forecast and in the manpower plan. Internal and external sources, and media for reaching both, will be considered.

(e) Preparation and publication of *information*, which will:

   (i) attract the attention and interest of potentially suitable candidates;

   (ii) give a favourable (but accurate) impression of the job and the organisation; and

   (iii) equip those interested to make an attractive and relevant application (how and to whom to apply, desired skills, qualifications and so on).

(f) *Processing applications* and assessing candidates.

(g) *Notifying applicants* of the results of the selection process.

2.7 The personnel department will handle much of this activity, but if there is no such department, there are still strong arguments for centralising the process to some extent, in order to take advantage of experience and to economise on effort.

(a) The overall priorities and requirements of the organisation will be more clearly recognised and met, rather than the objectives of sub-systems such as individual departments.

(b) There will be a focal reference point for communication, queries and applications from outside the organisation.

(c) Communication with the environment will also be more standardised, and will be more likely to reinforce the overall corporate image of the organisation.

(d) Potential can be spotted in individuals and utilised in the optimum conditions - not necessarily in the post for which the individual has applied, if he or she might be better suited to another vacant post. (There is an element of the classification approach - see Paragraph 2.4 above - in this practice.)

(e) The volume of administration and the need for specialist knowledge (notably of changing legal and industrial relations requirements) may suggest a specialist function.

(f) Standardisation and central control should be applied to, for example, equal opportunities, pay provisions and performance standards, where fairness must be seen to operate.

Recruitment policy and procedure

2.8 Detailed procedures for recruitment should only be devised and implemented within the context of a coherent *policy*, or code of conduct.

2.9 A typical recruitment policy might deal with:

(a) internal advertisement of vacancies;
(b) efficient and courteous processing of applications;
(c) fair and accurate provision of information to potential recruits; and
(d) selection of candidates on the basis of qualification, without discrimination on any grounds.

2.10 The Institute of Personnel Management has itself issued a Recruitment Code.

> **The IPM Recruitment Code**
>
> 1 Job advertisements should state clearly the form of reply desired, in particular whether this should be a formal application form or by curriculum vitae. Preferences should also be stated if handwritten replies are required.
>
> 2 An acknowledgement of reply should be made promptly to each applicant by the employing organisation or its agent. If it is likely to take some time before acknowledgements are made, this should be made clear in the advertisement.
>
> 3 Applicants should be informed of the progress of the selection procedures, what there will be (eg group selection, aptitude tests, etc), the steps and time involved and the policy regarding expenses.
>
> 4 Detailed personal information (eg religion, medical history, place of birth, family background, etc) should not be called for unless it is relevant to the selection process.
>
> 5 Before applying for references, potential employers must secure permission of the applicant.
>
> 6 Applications must be treated as confidential.
>
> 7 The code also recommends certain courtesies and obligations on the part of the applicants.
>
> *Institute of Personnel Management 1985*

2.11 Detailed *procedures* should be devised in order to make recruitment activity systematic and consistent throughout the organisation (especially where it is decentralised in the hands of line managers). Apart from the manpower resourcing requirements which need to be effectively and efficiently met, there is a *marketing* aspect to recruitment, as one 'interface' between the organisation and the outside world: applicants who feel they have been unfairly treated, or recruits who leave because they feel they have been misled, do not enhance the organisation's reputation in the labour market or the world at large.

2.12 Basic procedures involved in recruitment are as follows.

(a) Obtain approval or authorisation for engagement (in accordance with the manpower budget).

(b) Prepare, or update and confirm job description and specification, as appropriate to the job requisition received from the departmental head.

(c) Select media of advertisement or other notification of the vacancy.

(d) Prepare advertising copy, and place advertisement.

(e) Screen replies, at the end of a specified period for application.

(f) Shortlist candidates for initial interview.

(g) Advise applicants accordingly.

(h) Draw up a programme for the selection process which follows.

Exercise 1

Find out, if you do not already know, what are the recruitment and selection procedures in your organisation, and who is responsible for each stage. The procedures manual should set this out, or you may need to ask someone in the personnel department.

Get hold of and examine some of the documentation your organisation uses. We show specimens in this chapter, but practice and terminology varies, so your own 'house style' will be invaluable. Try to find the job description for your job; the personnel specification for your job (if any); the application form(s) of the organisation; and an interview assessment form (if any).

2.13 We will now go on to discuss some of these procedures in more detail.

3. JOB ANALYSIS AND DESCRIPTION

Job analysis (or job appraisal)

3.1 According to the British Standards Institution, job analysis is 'the determination of the essential characteristics of a job', the process of examining a job to identify its component parts and the circumstances in which it is performed. Analysis may be carried out by observation, for routine or repetitive jobs. Irregular jobs with a lot of invisible work (planning, man management, creative thinking and so on) will require interviews and discussions with superiors and with the people concerned.

3.2 The product of the analysis is usually a *job specification* - a detailed statement of the activities (mental and physical) involved in the job, and other relevant factors in the social and physical environment.

3.3 Job analysis, and the job specification resulting from it, may be used by managers:

(a) in recruitment and selection - for a detailed description of the vacant job;

(b) for appraisal - to assess how well an employee has fulfilled the requirements of the job;

(c) in devising training programmes - to assess the knowledge and skills necessary in a job;

(d) in establishing rates of pay - this will be discussed later in connection with job evaluation;

(e) in eliminating risks - identifying hazards in the job;

(f) in re-organisation of the organisational structure - by reappraising the purpose and necessity of jobs and their relationship to each other.

3.4 Information which should be elicited from a job appraisal is both task-oriented information, and also worker-oriented information, including:

(a) *initial requirements of the employee*: aptitudes, qualifications, experience, training required;

(b) *duties and responsibilities of the job*: physical aspects; mental effort; routine or requiring initiative; difficult and/or disagreeable features; consequences of failure; responsibilities for staff, materials, equipment or cash etc;

(c) *environment and conditions of the job*: physical surroundings, with notable features such as temperature or noise; hazards; remuneration; other conditions such as hours, shifts, benefits, holidays; career prospects; provision of employee services - canteens, protective clothing etc;

(d) *social factors of the job*: size of the department; teamwork or isolation; sort of people dealt with - senior management, the public; amount of supervision; job status.

3.5 Opportunities for analyses occur when jobs fall vacant, when salaries are reviewed, or when targets are being set, and the personnel department should take advantage of such opportunities to review and revise existing job specifications.

3.6 The fact that a job analysis is being carried out may cause some concern among employees: they may fear that standards will be raised, rates cut, or that the job may be found to be redundant or require rationalisation. The job analyst will need to gain their confidence by:

(a) communicating: explaining the process, methods and purpose of the appraisal;

(b) being thorough and competent in carrying out the analysis;

(c) respecting the work flow of the department, which should not be disrupted; and

(d) giving feedback on the results of the appraisal, and the achievement of its objectives. If staff are asked to co-operate in developing a framework for office training - and then never hear anything more about it, they are unlikely to be responsive on a later occasion.

3.7 It is worth noting that there is some confusion over the terminology associated with job analysis. The following definitions were agreed between the Department of Employment and the industrial training boards.

(a) Job Analysis. The process of examining a 'job' to identify the component parts and the circumstances in which it is performed.

(b) Job Description. A broad statement of the purpose, scope, duties and responsibilities of a particular 'job'.

(c) Personnel Specification. An interpretation of the 'job specification' in terms of the kind of person suitable for the job.

Job description

3.8 A job description is a broad description of a job or position at a given time (since jobs are dynamic, subject to change and variation). 'It is a written statement of those facts which are important regarding the duties, responsibilities, and their organisational and operational interrelationships.' (Livy, *Corporate Personnel Management*)

3.9 In recruitment, a job description can be used:

(a) to decide what skills (technical, human, conceptual, design or whatever) and qualifications are required of the job holder. When formulating recruitment advertisements, and interviewing an applicant for the job, the interviewer can use the resulting job specification to match the candidate against the job;

(b) to ensure that the job:

(i) will be a full time job for the holder and will not under-utilise his capacity by not giving him enough to do;

(ii) provides a sufficient challenge to the job holder - job content may be a factor in the motivation of individuals;

(c) to determine a rate of pay which is fair for the job, if this has not already been decided by some other means.

3.10 Job descriptions can also be used in other areas of personnel management.

(a) For job evaluation (used in establishing wage rates).

(i) A standard format for analysing jobs makes it easier for evaluators to compare jobs.

(ii) Job descriptions focus attention on the job, not the job holder, which is important in the job evaluation process.

(iii) Job descriptions offer opportunities for the job holder and his manager to discuss any differences of opinion about what the job involves, allowing fairer and more accurate evaluation.

(b) In induction and training, to help new employees to understand the scope and functions of their jobs and to help managers to identify training needs of job holders on an ongoing basis.

(c) To pinpoint weaknesses in the organisation structure (such as overlapping areas of authority, where two or more managers are responsible for the same area of work; areas of work where no manager appears to accept responsibility; areas of work where authority appears to be too centralised or decentralised).

(d) To provide information for work study (or Organisation & Methods).

3.11 Townsend (*Up the Organisation*) suggested that job descriptions are only suited for jobs where the work is largely repetitive and therefore performed by low-grade employees: once the element of judgement comes into a job description it becomes a straitjacket. Management jobs are likely to be constantly changing as external influences impact upon them, so a job description is constantly out-of-date. Many of the difficulties that arise with job descriptions have arisen because they encourage demarcation disputes, where people adhere strictly to the contents of the job description, rather than responding flexibly to task or organisational requirements: this in turn leads to costly overmanning practices.

3.12 Where job descriptions are used, it should be remembered that:

(a) a job description is like a photograph, an image 'frozen' at one point in time;
(b) a job description needs constant and negotiated revision;
(c) a job description rigidly adhered to can work against flexibility.

The contents of a job description

3.13 A job description should be clear and to the point, and so ought not to be lengthy. Typically, a job description would show the following.

(a) *Job title* and department and job code number. The person to whom the job holder is responsible. Possibly, the grading of the job.

(b) *Job summary* - showing in a few paragraphs the major functions and tools, machinery and special equipment used. Possibly also a small organisation chart.

(c) *Job content* - list of the sequence of operations that constitute the job, noting main levels of difficulty. In the case of management work there should be a list of the main duties of the job, indicating frequency of performance - typically between 5 and 15 main duties should be listed. This includes the degree of initiative involved, and the nature of responsibility (for other people, machinery and/or other resources).

(d) The extent (and limits) of the jobholder's authority and responsibility.

(e) Statement showing relation of job to other closely associated jobs, including superior and subordinate positions and liaison required with other departments.

(f) Working hours, basis of pay and benefits, and conditions of employment, including location, special pressures, social isolation, physical conditions, or health hazards.

(g) Opportunities for training, transfer and promotion.

(h) Possibly, also, objectives and expected results, which will be compared against actual performance during employee appraisal - although this may be done as a separate exercise, as part of the appraisal process.

(i) The names and positions of the people/person who has

(i) prepared the job description;
(ii) agreed the job description;

(j) Date of preparation.

3.14 Two examples of job descriptions are shown below.

JOB DESCRIPTION

1 *Job title:* Baking Furnace Labourer.

2 *Department:* 'B' Baking.

3 *Date:* 20 November 19X0.

4 *Prepared by:* H Crust, baking furnace manager.

5 *Responsible to:* baking furnace chargehand.

6 *Age range:* 20-40.

7 *Supervises work of:* N/A

8 *Has regular co-operative contact with:* Slinger/Crane driver.

9 *Main duties/responsibilities*: Stacking formed electrodes in furnace, packing for stability. Subsequently unloads baked electrodes and prepares furnace for next load.

10 *Working conditions*: stacking is heavy work and requires some manipulation of 100lb (45kg) electrodes. Unloading is hot (35° - 40°C) and very dusty.

11 *Employment conditions*:

Wages £2.60 ph + group bonus (average earnings £158.50 pw)
Hours: Continuous rotating three-shift working days, 6 days on, 2 days off. NB must remain on shift until relieved.
Trade Union: National Union of Bread Bakers, optional.

MIDWEST BANK PLC

1 *Job title*: Clerk (Grade 2)

2 *Branch*: All branches and administrative offices

3 *Job summary*: To provide clerical support to activities within the bank

4 *Job content*: Typical duties will include:

(a) cashier's duties;
(b) processing of branch clearing;
(c) processing of standing orders;
(d) support to branch management.

5 *Reporting structure*:

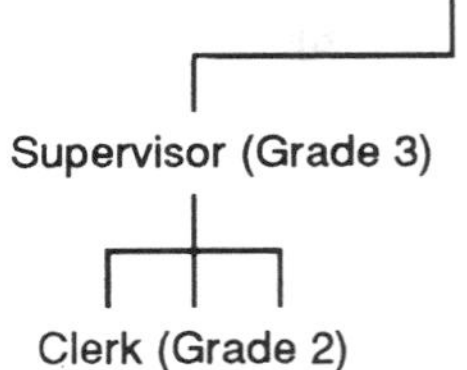

6 *Experience/Education*: experience not required, minimum 3 GCSEs or equivalent.

7 *Training to be provided*: initial on-the-job training plus regular formal courses and training

8 *Hours*: 38 hours per week

9 *Objectives and appraisal*: Annual appraisal in line with objectives above.

10 *Salary*: refer to separate standard salary structure.

Job description prepared by: Head office personnel department *Date:* March 19X9

Exercise 2

Without looking at the real thing, to start with, draw up a job description for your own job and for the job of a personnel officer in your organisation. Now look at the official job descriptions. Are they true, detailed and up-to-date, compared with the actual jobs as you saw them? If not, what does this tell you about (a) job descriptions and (b) perceptions of the personnel function?

4. PERSONNEL SPECIFICATION

4.1 Once the job has been clearly defined in terms of its organisational content and operational interrelationships, the organisation can decide what kind of person is needed to fill it effectively.

4.2 A personnel specification identifies the type of person the organisation should be trying to recruit: their character, aptitudes, educational or other qualifications, aspirations in their career and other attributes. Research has been carried out into what a personnel specification ought to assess. Two popular designs of specification are those of Professor Alec Rodger and J Munro Fraser.

4.3 Professor Alec Rodger was a pioneer of the systematic approach to recruitment and selection in Britain. He suggested that:

'If matching *[ie of demands of the job and the person who is to perform it]* is to be done satisfactorily, the requirements of an occupation (or job) must be described in the same terms as the aptitudes of the people who are being considered for it.'

This was the basis for the formulation of the personnel specification as a way of matching people to jobs on the basis of comparative sets of data: defining job requirements and personal suitability along the same lines. This enables a classification approach to recruitment (see Paragraph 2.4 above) to be implemented, if suitable.

4.4 The *Seven Point Plan* put forward by Professor Rodger in 1951 draws the selector's attention to seven points about the candidate:

(a) physical attributes (such as neat appearance, ability to speak clearly and without impediment);

(b) attainment (including educational qualifications);

(c) general intelligence;

(d) special aptitudes (such as neat work, speed and accuracy);

(e) interests (practical and social);

(f) disposition (or manner: friendly, helpful and so on);

(g) background circumstances.

4.5 Munro Fraser's *Five Point Pattern of Personality* (1966) draws the selector's attention to:

(a) impact on others, including physical attributes, speech and manner;
(b) acquired knowledge or qualifications*, including education, training and work experience;
(c) innate ability, including mental, agility, aptitude for learning;
(d) motivation: individual goals, demonstrated effort and success at achieving them; and
(e) adjustment: emotional stability, tolerance of stress, human relations skills.

*Most personnel specifications include achievements in education, because there appears to be a strong correlation between management potential and higher education.

4.6 Note that the personnel specification includes job requirements in terms of a candidate's:

(a) capacities - what he is *capable* of; and
(b) inclinations - what he *will* do.

In other words, behavioural versatility must be accounted for, by considering not only the individual's mental and physical attributes, but his current attitudes, values, beliefs, goals and circumstances - all of which will influence his response to work demands.

4.7 Each feature in the specification should be classified as:

(a) essential - for instance, honesty in a cashier is essential whilst a special aptitude for conceptual thought is not;

(b) desirable - for instance, a reasonably pleasant manner should ensure satisfactory standards in a person dealing with the public

(c) contra - indicated-some features are actively disadvantageous, such as an inability to work in a team when acting as project leader.

PERSONNEL SPECIFICATION: Customer Accounts Manager

	ESSENTIAL	DESIRABLE	CONTRA-INDICATED
Physical attributes	Clear speech Well-groomed Good health	Age 25-40	Age under 25 Chronic ill-health and absence
Attainments	2 'A' levels GCSE Maths and English Thorough knowledge of retail environment	Degree (any discipline) Marketing training 2 years' experience in supervisory post	No experience of supervision or retail environment
Intelligence	High verbal intelligence		
Aptitudes	Facility with numbers Attention to detail and accuracy Social skills for customer relations	Analytical abilities (problem solving) Understanding of systems and IT	No mathematical ability Low tolerance of technology
Interests	Social: team activity		Time-consuming hobbies 'Solo' interests only
Disposition	Team player Persuasive Tolerance of pressure and change	Initiative	Anti-social Low tolerance of responsibility
Circumstances	Able to work late, take work home	Located in area of office	

Exercise 3

Do not continue until you have written out on a piece of paper a definition of each of the following terms as you now understand them.

(a) Job analysis
(b) Job description
(c) Job specification
(d) Personnel specification

Solution

Look back at paragraph 3.7 when you have made your own attempt. If you cheated, decide in which of (a) - (d) above the attribute of 'laziness' would have been identified!

5. JOB ADVERTISEMENT

5.1 The object of recruitment advertising is to home in on the target market of labour, and to attract interest in the organisation and the job.

In a way, it is already part of the *selection* process. The advertisement will be placed where suitable people are likely to see it (say, internally only - immediately pre-selecting members of the organisation - or in a specialist journal, pre-selecting those specialists). It will be worded in a way that further weeds out people who would not be suitable for the job (or for whom the job would not be suitable).

Advertising methods

5.2 The way in which a job is advertised will depend on:

(a) the type of organisation; and
(b) the type of job.

A factory is likely to advertise a vacancy for an unskilled worker in a different way to a company advertising for an ACIS for a company secretarial position. Managerial jobs may merit national advertisement, whereas semi- or un-skilled jobs may only warrant local coverage, depending on the supply of suitable candidates in the local area. Specific skills may be most appropriately reached through trade, technical or professional journals, for example for accountants or computer programmers.

In addition, there is the consideration of whether to advertise within the organisation (internal recruitment) or outside (external recruitment), or both. (The issues have already been discussed in relation to management succession.)

5.3 The choice of advertising medium will also depend on:

(a) the *cost* of advertising. It is more expensive to advertise in a national newspaper than on local radio, and more expensive to advertise on local radio than in a local newspaper etc;

(b) the *readership and circulation* (type and number of readers/listeners) of the medium, and its suitability for the number and type of people the organisation wants to reach;

(c) the *frequency* with which the organisation wants to advertise the job vacancy, and the duration of the recruitment process.

5.4 Some methods or media for advertising jobs are as follows.

(a) *In-house magazines and notice-boards*. An organisation might advertise vacancies for particular jobs through its own in-house magazine or journal, inviting applications from employees who would like a transfer or a promotion to the particular vacancy advertised. In-house notice boards are a traditional, and still much-used, method of advertising or 'posting' internal vacancies.

(b) *Professional and specialist newspapers or magazines*, such as *Personnel Management, Marketing* or *Computing*.

(c) Other national newspapers, especially for senior management jobs or vacancies for skilled workers, where potential applicants will not necessarily be found through local advertising. *Local newspapers* would be suitable for jobs where applicants are sought from the local area.

(d) *Local radio, television and cinema*. These are becoming increasingly popular, especially for large-scale campaigns, for large numbers of vacancies.

(e) *Job centres*. On the whole, vacancies for unskilled work (rather than skilled work or management jobs) are advertised through local job centres, although in theory any type of job can be advertised here.

(f) *School and university careers offices*. When an organisation recruits school leavers or graduates, it would be convenient to advertise vacancies through their careers officers. Suitable information should be made available. Ideally, the manager responsible for recruitment in an area should try to maintain a close liaison with careers officers. Some large organisations organise special meetings or *careers fairs* in universities and colleges (the so-called 'milk round'), as a kind of showcase for the organisation and the careers it offers. This type of work may become more important as the number of young people falls during the 1990s and competition for high-calibre recruits becomes more intense, especially in industries like banking which has traditionally drawn the cream of graduates into management trainee positions, and has drawn much of its more junior manpower from school-leavers.

Content of the advertisement

5.5 Preparation of the job information requires skill and attention in order to fulfil its objectives of attraction and preselection. It should be:

(a) concise, but comprehensive enough to be an accurate description of the job, its rewards and requirements;

(b) in a form that will attract the attention of the maximum number of the right sort of people;

(c) attractive, conveying a favourable impression of the organisation, but not falsely so: disappointed expectations will be a prime source of dissatisfaction for an applicant when he actually comes into contact with the organisation;

(d) relevant and appropriate to the job and the applicant. Skills, qualifications and special aptitudes required should be prominently set out, along with special features of the job that might attract - on indeed deter - applicants, such as shiftwork or extensive travel.

5.6 The advertisement, based on information set out in the job description, job and person specifications and recruitment procedures, should contain information about:

(a) the organisation: its main business and location, at least;

(b) the job: title, main duties and responsibilities and special features;

(c) conditions: special factors affecting the job;

(d) qualifications and experience (required, and preferred); other attributes, aptitudes and/or knowledge required;

(e) rewards: salary, benefits, opportunities for training, career development, and so on;

(f) application: how to apply, to whom, and by what date.

5.7 It should encourage a degree of *self-selection*, so that the target population begins to narrow itself down. The information contained in the advertisement should deter unsuitable applicants as well as encourage potentially suitable ones.

Exercise 4

'Anyone born much before 1950 is battling against stiff odds in the UK jobs market. In some sectors even those born in the early 1960s are running into trouble.

Evidence of widespread opposition by employers to recruiting older staff came this week, with the publication of a survey of job advertisements by the independent research group Industrial Relations Services. It found that almost a third of advertisers specified an age bar, an increase from a quarter four years ago. Of those that stated a preference, four out of five wanted someone under the age of 45.

A newspaper advertisement by financial service company Laurentian Milldon this week is not untypical. Advertising a sales job, it begins: "Are you aged 24-36, ambitious, energetic, a good communicator?" It finishes a touch unconvincingly: "We are an equal opportunity employer."

Ageism is against the law in the US - but unlike racism and sexism it is not outlawed in job selection in the UK. Inquiries by the FT this week confirm the cult of youth is on the rise in most sectors and at most levels within UK companies.' *Financial Times*, March 1993

Why do you think employers are opposed to recruiting older staff?

Solution

The article goes on to describe a number of 'excuses' offered by recruiters.

(a) Cost (although performance-based pay systems are taking over from age-based ones in many companies)

(b) Fear that the pay-back period on training will be too short

(c) A young customer base (on the supposition of an affinity between people of a common age)

(d) A 'young' organisational culture

(e) In IT recruitment, lack of relevant experience amongst older workers

(f) 'Image' - if this is the right word! Middle managers will often go for very young glamorous secretaries, particularly if they are recruiting themselves'

On a more positive note, one recruitment consultant said that his impression was that 'a lot of employers have found that they have discarded a lot of experience as they discarded older people'.

Other methods of reaching the labour market

5.8 Various *agencies* exist, through whom the employer can reach the public.

(a) *Government agencies* include the Manpower Services Commission (MSC), with its network of Job Centres, and regional offices of the Professional and Executive Register (PER) which act as agents.

(b) *Institutional agencies* exist to help their own members to find employment: for example, the career services of educational institutions such as schools and colleges, and the employment services of professional institutions and trade unions.

(c) *Private employment agencies* have proliferated in recent years. There is a wide range of agencies specialising in different grades and skill-areas of staff - clerical, technical, professional and managerial. Private agencies generally offer an immediate pool of labour already on their books, and many also undertake initial screening of potential applicants, so that the recruitment officer sees only the most suitable.

5.9 There are also more informal recruitment methods, not directly involving advertisement.

(a) Unsolicited applications are now frequently made to organisations, especially where there are few advertised vacancies. Some applicants may have heard about impending vacancies through the grapevine.

(b) Some vacancies may be filled purely by word-of-mouth, and on the recommendation of established workers. If the organisation has a strong tradition of sons following fathers (say, on the docks, or in mining areas), or if there is a strong cultural type of the acceptable worker, family members vouched for by employees may be taken on for appropriate vacancies. (Some national cultures support this practice more than others.)

(c) 'Head-hunting' has become increasingly popular. Informal approaches are made to successful executives currently employed elsewhere.

5.10 The role of the *recruitment consultant* is to perform the staffing function on behalf of the management of the client organisation: in other words, to fill vacancies with the type and calibre of staff required.

The tasks involved in this include:

(a) analysing, or being informed of, the requirements - the demands of the post, the organisation's preferences for qualifications, personality and so on;

(b) helping to draw up, or offering advice on, job descriptions, person specifications and other recruitment and selection aids;

(c) designing job advertisements;

(d) screening applications, so that those most obviously unsuitable are weeded out immediately;

(e) helping with short-listing for interview;

(f) advising on the constitution and procedures of the interview;

(g) offering a list of suitable candidates with notes and recommendations.

Much will depend on whether the consultant is employed to perform the necessary tasks, or merely to advise and recommend.

5.11 The decision of whether or not to use consultants will depend on a number of factors.

(a) *Cost.* This will particularly be a factor where the desired recruitment is at a low level, since a quality recruitment decision will not be so crucial, and the fees may not therefore be cost-effective.

(b) *The level of expertise and specialist techniques or knowledge which the consultant can bring to the process.* Consultants may be expert in using interview techniques, analysis of personnel specifications and so on. In-house staff, on the other hand, have experience of the particular field (some recruitment consultants specialise - say, in accountancy or computing - but not all) and of the culture of the organisation into which the recruits must fit.

(c) *The level of expertise, and specialist knowledge, available within the organisation.* Even if the consultant is not familiar with the field and organisation, there may be little choice: the cost of training in-house personnel in the necessary interview and assessment techniques may be prohibitive.

(d) Whether there is a *need for impartiality* which can only be filled by an outsider trained in objective assessment. If fresh blood is desired in the organisation, it may be a mistake to have staff selecting clones of the common organisational type.

(e) Whether the import of an outside agent will be *regarded as helpful* by in-house staff. Consultants may be regarded as the tools of the manager who employs them, or as meddlers and outsiders. On the other hand, an outsider will avoid internal office politics.

(f) Whether the *structure and politics of the organisation* are conducive to allowing in-house staff to make decisions of this kind. Consultants are not tied by status or rank and can discuss problems freely at all levels. They are also not likely to fear the consequences of their recommendations for their jobs or career prospects. On the other hand, the organisation may regard in-house selection as an important part of succession and a way of maintaining the organisational culture.

(g) *Time*. Consultants will need to learn about the job, the organisation and the organisation's requirements. The client will not only have to pay fees for this period of acclimatisation: it may require a post to be filled more quickly than the process allows.

(h) *Supply of labour*. If there is a large and reasonably accessible pool of labour from which to fill a post, consultants will be less valuable. If the vacancy is a standard one, and there are ready channels for reaching labour (such as professional journals), the use of specialists may not be cost effective.

6. THE EVALUATION OF RECRUITMENT PROCEDURES

6.1 Because the external advertisement of job vacancies is an expensive element of the recruitment process, it is important that its effectiveness should be monitored regularly.

6.2 Control information will include the nature and cost of each insertion, and the *response* to it: the number of replies and also their quality. It may be that a certain medium is too costly for the number of worthwhile responses it generates, perhaps for a certain type of job vacancy. Advertising in the national press, for example, may generate a large number of responses - but not of the required quality; meanwhile an insertion in a professional journal may, with a smaller circulation, generate a higher proportion of potentially suitable applicants.

The simple formula: $\dfrac{\text{recruitment costs}}{\text{number starting work}}$ provides an index: a falling index shows cost improvement

6.3 The effectiveness of the procedure must ultimately be judged by the success achieved by the candidates appointed under the existing recruitment and selection programme. This is a highly qualitative and subjective judgement: even if the appointee proves *not* to have been the Right Person for the job (according to various criteria), there is nothing to indicate that the recruitment/selection procedures were at fault. The Right Person may not have applied: she may not in fact have existed in the labour market at that time.

6.4 Proper records of failures, in particular, will nevertheless be useful. An analysis of failures, and the sequence of events preceding them, may indicate areas for review and corrective action where necessary. *Follow-up* appraisal and up-to-date records are therefore essential.

Chapter roundup

- The essential points in this chapter can be summarised in diagrammatic form.

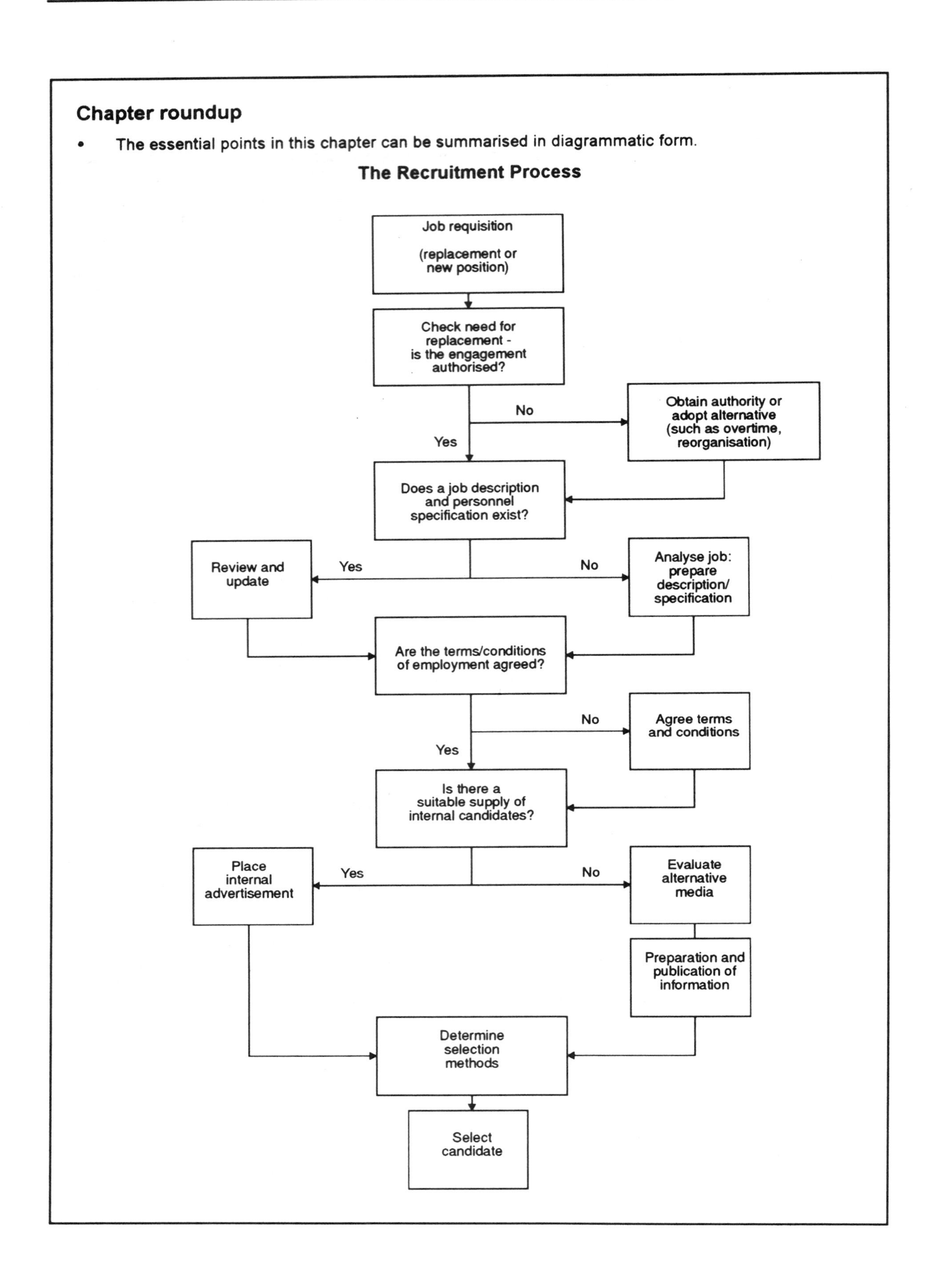

Test your knowledge

1 What factors influence the supply of labour in the external market? (see para 1.3)

2 What, broadly, is the distinction between the classification and selection approaches to recruitment? (2.4)

3 What information will a job description usually contain? (3.13)

4 Which of the following does Professor Rodger's 7-Point Plan personnel specification *not* specifically include?

A Physical attributes

B Interests

C Motivation

D Background circumstances (4.4)

5 What factors are relevant to the decision of where to advertise a job vacancy? (5.2, 5.3)

6 List *six* factors which should be considered when making a decision whether to employ recruitment consultants or to use internal staff for the recruitment process. (5.10, 5.11)

7 What are the uses of job descriptions (so far as the personnel function is concerned)? To what extent can it be argued that, in some organisations or situations, job descriptions are dysfunctional? (*ICSA December 1988)* (3.9-3.12)

Now try question 9 at the end of the text

Chapter 10

SELECTION

This chapter covers the following topics.

1. The selection process
2. Interviews
3. Selection testing
4. Once selection has been made

Signpost

- There is more on face-to-face communication in Chapter 16.
- Contracts of employment are described in more detail in Chapter 15.

1. THE SELECTION PROCESS

1.1 The selection of employees (at all levels, but particularly for key posts) must be approached systematically if it is to be efficient and effective. The recruiting officer must know what the organisation's requirements are, and must measure each potentially suitable candidate against those requirements. He should not waste time and resources investigating candidates who are clearly disqualified from the post by some area of unsuitability. ('If only ...' is not a useful assessment.) On the other hand, of those who are broadly suitable, summary rejection of someone who proves to be a considerable asset to another organisation is as much a danger as the incautious acceptance of a candidate who proves to be unsuitable.

Applications

1.2 Job advertisements usually ask candidates to fill in a *job application form*, or to send information about themselves and their previous job experience (their CV or *curriculum vitae*), usually with a covering letter briefly explaining why they think they are qualified to do the job.

An application form or CV will be used to find out relevant information about the applicant, in order to decide, at the initial sifting stage:

(a) whether the applicant is obviously unsuitable for the job; or
(b) whether the applicant might be of the right calibre, and worth inviting to interview.

1.3 The application form will be designed by the organisation, and if it is to be useful in the sifting process, it should fulfil the following criteria.

(a) It should ask questions which will elicit information about the applicant, which can be compared with the requirements of the job. For example, if the personnel specification requires a minimum of 2 'A' Level passes the application form should ask for details of the applicant's educational qualifications. Similarly, if practical and social interests are thought to be relevant, the applications form should ask for details of the applicant's hobbies and pastimes, membership of societies and sporting clubs or teams and so on.

(b) It should give applicants the opportunity to write about themselves, their career ambitions or why they want the job. By allowing applicants to write in their own words at length, it might be possible to obtain some information about their:

(i) neatness;
(ii) intelligence;
(iii) ability to express themselves in writing;
(iv) motivation; and even
(v) character.

Exercise 1

Think of four ways in which an application form could be *badly* designed. You may be able to do this from personal experience.

Solution

Here are four suggestions, derived from Torrington and Hall *Personnel Management* (1991).

(a) Boxes too small to contain the information asked for

(b) Forms so lengthy that prospective applicants either complete them perfunctorily or apply to another employer instead

(c) Forms with illegal or offensive questions

(d) Forms not properly tailored to the vacancy. For example a form asking about 'teaching experience' might be sent to an applicant for a non-teaching administrative post at a college.

Selection

1.4 A typical selection system will then include the following basic procedures.

(a) Take any initial steps required. If the decision to interview or reject cannot be made immediately, a standard letter of acknowledgement might be sent, as a courtesy, to each applicant. It may be that the job advertisement required applicants to write to the personnel manager with personal details and to request an application form: this would then be sent to applicants for completion and return.

(b) Set each application against key criteria in the job advertisement and specification. Critical factors may include age, qualifications, experience or whatever.

(c) Sort applications into 'possible', 'unsuitable' and 'marginal'.

(d) 'Possibles' will then be more closely scrutinised, and a shortlist for interview drawn up. Ideally, this should be done by both the personnel specialist and the prospective manager of the successful candidate, who will have more immediate knowledge of the type of person that will fit into the culture and activities of his department.

(e) Invite candidates for interviews (requiring them to complete an application form, if this has not been done at an earlier stage). Again, if large numbers of interviewees are involved, standard letters should be used (pre-printed, or prepared on a word processor using the direct mail facility).

(f) Reinforce interviews with selection testing, if suitable.

(g) Review un-interviewed 'possibles', and 'marginals', and put potential future candidates on hold, or in reserve.

APPLICATION FORM

Post applied for: .... Date:

Surname: Mr/Mrs/Miss/Ms First names:
Address:

Post Code: Telephone:

Age: Date of birth: / /
Nationality Marital status:

EDUCATION AND TRAINING

Place of education (including schools after 11 years)	Dates	Examinations passed/qualifications

EXPERIENCE

Name of employer and main business	Position held	Main duties	From	To

OTHER INFORMATION
Please note your hobbies and interests, and any other information you would like to give about yourself or your experience.

State of health (include any disability)

May we contact any of your previous employers? Yes ☐ No ☐

If yes, please give the names of any managers to whom we may apply.

If selected, I would be able to start from / / .

(h) Send standard letters to unsuccessful applicants, and inform them simply that they have not been successful. Reserves will be sent a holding letter: 'We will keep your details on file, and should any suitable vacancy arise in future ...'. Rejects should be briefly, but tactfully, dismissed: 'Thank you for your interest in the post of .... We have given your application careful consideration. I regret to inform you, however, that we have decided not to ask you to attend for an interview. The standard of application was very high ...'.

1.5 We will now go on to look at the two major selection techniques: interviews and tests.

2. INTERVIEWS

2.1 Despite frequent criticism for its inability to elicit reliable and meaningful information on which to base selection (discussed later), interviewing is a popular technique which gives the organisation a chance to assess the applicant (and particularly his interpersonal skills) directly, and gives the applicant a chance to learn more about the organisation, the people and the working environment.

The interview has a three-fold purpose.

(a) Finding the best person for the job.

(b) Making sure that the applicant understands what the job is and what the career prospects are. He must be allowed a fair opportunity to decide whether he wants the job.

(c) Making the applicant feel that he has been given fair treatment in the interview, whether he gets the job or not.

Preparation of the interview

2.2 The interview is a two way process, but the interviewer must have a clear idea of what the interview is setting out to achieve, and must be in sufficient control of the interview to make sure that every candidate is asked questions which cover the same ground and obtain all the information required.

The agenda and questions will be based on:

(a) the job description, and what abilities are required of the job holder;

(b) the personnel specification. The interviewer must be able to judge whether the applicant matches up to the personal qualities required from the job holder;

(c) the application form.

2.3 The interview process should be efficiently run to make a favourable impression on the candidates: they should be clearly informed when and where to come, whom to ask for, what to bring with them etc and should be expected by the receptionist or other receiving staff. A waiting room should be available, preferably with cloakroom facilities. Arrangements should have been made to welcome and escort candidates: they should not be placed under the extra stress of being left stranded, getting lost, or being ignored. Accommodation for the interview should be private and free from distractions/interruption.

2.4 The layout of the room should be planned. Most interviewers wish to put candidates at their ease, and so it would be inadvisable to put the candidate in a 'hot-seat' across a desk (a psychological barrier) from them. On the other hand, some interviewers might want to observe the candidate's reaction under severe pressure, and deliberately make the layout of the room uncomfortable and off-putting.

Conduct of the interview

2.5 The manner of the interviewers, the tone of their voice, and the way their early questions are phrased can all be significant in establishing the tone of the interview, and the ease with which the candidate can talk freely.

2.6 Questions should be paced and put carefully. The interviewer should not be trying to confuse the interviewee, plunging immediately into demanding questions or picking on isolated points; nor should he allow the interviewee to digress or gloss over important points. The interviewer must retain control over the information-gathering process.

2.7 A variety of questions may be used, to different effects.

(a) *Open questions* or open-ended questions ('Who...? What...? Where...? When...? Why...?') force interviewees to put together their own responses in complete sentences. This encourages the interviewee to talk, keeps the interview flowing, and is most revealing ('Why do you want to be in Personnel?')

(b) *Probing questions* are similar to open questions in their phrasing but aim to discover the deeper significance of the candidate's experience or achievements. ('What was it about HRM that particularly appealed to you?')

> In an article in *PM Plus* (August 1991) Alan Fowler describes the purpose of such questions as being 'to provide a clearer focus to too short or too generalised answers'. He adds that 'Poor interviewers to often let a candidate's general and fairly uninformative answer pass without a probe, simply because they are working through a list of prepared open questions.'

(c) *Closed questions* are the opposite, inviting only 'yes' or 'no' answers: ('Did you...?', 'Have you...?'). A closed question has the following effects.

(i) It elicits answers only to the question asked by the interviewer. This may be useful where there are small points to be established ('Did you pass your exam?') but there may be other questions and issues that he has not anticipated but will emerge if the interviewee is given the chance to express himself ('How did you think your studies went?').

(ii) It does not allow the interviewee to express his personality, so that interaction can take place on a deeper level.

(iii) It makes it easier for interviewees to conceal things ('You never *asked* me....').

(iv) It makes the interviewer work very hard.

(d) *Multiple questions* are just that: two or more questions are asked at once. ('Tell me about your last job? How did your knowledge of HRM help you there, and do you think you are up-to-date or will you need to spend time studying?') This type of question can be used to encourage the candidate to talk at some length, but not to stray too far from the point. It might also test the candidate's ability to listen and handle large amounts of information, but should be used judiciously in this case.

(e) *Problem solving questions* present the candidate with a situation and ask him to explain how he would deal with it. ('How would you motivate your staff to do a task that they did not want to do?'). Such questions are used to establish whether the candidate will be able to deal with the sort of problems that are likely to arise in the job, or whether he has sufficient technical knowledge (in which case a line manager rather than the personnel manager might be the best person to ask the questions and judge the responses).

(f) *Leading questions* lead the interviewee to give a certain reply. ('We are looking for somebody who likes detailed figure work. How much do you enjoy dealing with numbers?' or 'Don't you agree that...?'. 'Surely...?')

The danger with this type of question is that the interviewee will give the answer that he thinks the interviewer wants to hear, but it might legitimately be used to deal with a highly reticent or nervous candidate, simply to encourage him to talk.

2.8 The interviewer *must* listen to and evaluate the responses, to judge what the interviewee:

(a) wants to say;
(b) is trying *not* to say;
(c) is saying - but doesn't mean, or is lying about;
(d) is having difficulty saying.

In addition, the interviewer will have to be aware when he:

(a) is hearing something he needs to know;

(b) is hearing something he *doesn't* need to know;

(c) is hearing only what he expects to hear;

(d) is not hearing clearly - when his own attitudes, perhaps prejudices, are getting in the way of his response to the interviewee and his views.

2.9 The candidate should be given the opportunity to ask questions. Indeed, a well-prepared candidate should go into an interview knowing what questions he may want to ask. His choice of questions might well have some influence on how the interviewers finally assess him. Moreover, there is information that the candidate will need to know about the organisation and the job, and about:

(a) terms and conditions of employment (although negotiations about detailed terms may not take place until a provisional offer has been made); and

(b) the next step in the selection process - whether there are further interviews, when a decision might be made, or which references might be taken up.

Types of interview

2.10 *Individual* or *one-to-one interviews* are the most common selection method. They offer the advantages of direct face-to-face communication, and opportunity to establish *rapport* between the candidate and interviewer: each has to give his attention solely to the other, and there is potentially a relaxed atmosphere, if the interviewer is willing to establish an informal style.

The disadvantage of a one-to-one interview is the scope it allows for a biased or superficial decision.

(a) The candidate may be able to disguise lack of knowledge in a specialist area of which the interviewer himself knows little.

(b) The interviewer's perception may be selective or distorted (see Paragraph 2.13 below), and his lack of objectivity may go unnoticed and unchecked, since he is the sole arbiter.

(c) The greater opportunity for personal rapport with the candidate may cause a weakening of the interviewer's objective judgement: he may favour someone he got on with over someone who was unresponsive but better-qualified. Again, there will be no cross-check with another interviewer.

2.11 *Panel interviews* are designed to overcome the above disadvantages. A panel may consist of two or three people who together interview a single candidate: most commonly, a personnel manager and the departmental manager who will have responsibility for the successful candidate. This may be more daunting for the candidate (depending on the tone and conduct of the interview) but it has several advantages.

(a) The personnel and line specialists gather the information they need about the candidate at the same time, cutting down on the subsequent information-sharing stage, and making a separate subsequent interview with one or the other unnecessary.

(b) The questions each specialist wants put to the interviewee (related to their own field of activity and expertise) will be included, and the answers will be assessed by the specialist concerned: each will also be able to give the interviewee the information he wants, both about the employment aspects of the position, and about the departmental task-related aspects.

(c) The interviewers can discuss their joint assessment of the candidate's abilities, and their impressions of his behaviour and personality at the interview. They can thus gain a more complete picture, and can modify any hasty or superficial judgements. Personal bias is more likely to be guarded against, and checked if it does emerge.

2.12 Large formal panels, or *selection boards*, may also be convened where there are a number of individuals or groups with an interest in the selection. This has the advantage of allowing a number of people to see the candidate, and to share information about him at a single meeting: similarly, they can compare their assessments on the spot, without a subsequent effort at liaison and communication.

Offsetting these administrative advantages, however, there are severe drawbacks to the effectiveness of the selection board as a means of assessment.

(a) Questions tend to be more varied, and more random, since there is no single guiding force behind the interview strategy. The candidate may have trouble switching from one topic to another so quickly, especially if questions are not led up to, and not clearly put - as may happen if they are unplanned. Candidates are also seldom allowed to expand their answers, and so may not be able to do justice to themselves.

(b) If there is a dominating member of the board, the interview may have greater continuity - but he may also 'force' the judgements of other members, and his prejudices may be allowed to dominate the assessment.

(c) Some candidates may not perform well in a formal, artificial situation such as a board interview, and may find such a situation extremely stressful. The interview will thus not show the best qualities of someone who might nevertheless be highly effective in the work context and in the face of work-related pressures.

(d) The pressures of a board interview favour individuals who are confident, and who project an immediate and strong image: those who are articulate, dress well and so on. First impressions of such a candidate may cover underlying faults or shortcomings, while a quiet, overawed candidate may be dismissed, despite his strengths in other areas.

Exercise 2

'In some cases, the option of sequential interviewing may be open to managers. Instead of, say, four people on a panel spending an hour with each of four candidates, the members might each spend an hour alone with each candidate. The panel could then take its selection decision in the light of the information obtained at the separate interviews.'

W David Rees, *The Skills of Management* (1991)

Weigh up the pros and cons of 'sequential interviewing'.

Solution

Rees suggests the following points. You may not agree with all of them.

(a) The method need take no more time for the organisation: in fact single interviews may turn out to take less than an hour.

(b) Candidates would have to spend more time being interviewed but 'might not mind this if they felt it was a more effective method of selection'.

(c) It is normally easier to create rapport and coax information out of a person when you are seeing him alone.

(d) However, inexperienced interviewers will be far more 'exposed' than they would have been in a panel interview. They may prefer merely to observe.

(e) The method is liable to create more argument about the final decision: the interviewee may be flagging by the end of the session and give a totally different impression to different interviewers.

The limitations of interviews

2.13 Interviews are criticised because they fail to provide accurate predictions of how a person will perform in the job. The main reasons why this might be so are as follows.

(a) Limited scope and relevance.

> 'Among the qualities which neither the interview nor intelligence tests are able to assess accurately are the candidate's ability to get on with and influence his colleagues, to display qualities of spontaneous leadership and to produce ideas in a real-life situation.' *Plumbley*

(b) Errors of judgement by interviewers. These might be:

(i) the *halo effect* - a tendency for people to make an initial general judgement about a person based on a single obvious attribute, such as being neatly dressed, or well-spoken, or having a public school education. This single attribute will colour later perceptions, and might make an interviewer mark the person up or down on every other factor in their assessment;

(ii) *contagious bias* - a process whereby an interviewer changes the behaviour of the applicant by suggestion. The applicant might be led by the wording of questions or non-verbal cues from the interviewer, and change what he is doing or saying in response;

(iii) a possible inclination by interviewers to *stereotype* candidates on the basis of insufficient evidence. Stereotyping groups together people who are assumed to share certain characteristics, then attributes certain traits to the group as a whole, and then (illogically) assumes that each individual member of the supposed group will possess that trait: all women are weak, or whatever;

(iv) *incorrect assessment* of qualitative factors such as motivation, honesty or integrity. Abstract qualities are very difficult to assess in an interview;

(v) *logical error*. For example, an interviewer might decide that a young candidate who has held two or three jobs in the past for only a short time will be unlikely to last long in any job.

(c) Lack of skill and experience in interviewers. The problems with inexperienced interviewers are not only bias, but:

(i) inability to evaluate information about a candidate properly;

(ii) inability to compare a candidate against the requirements for a job or a personnel specification;

(iii) bad planning of the interview;

(iv) inability to take control of the direction and length of the interview;

(v) a tendency to talk too much in interviews, and to ask questions which call for a short answer;

(vi) a tendency to jump to conclusions on insufficient evidence, or to place too much emphasis on isolated strengths or weaknesses;

(vii) a tendency to act as an inquisitor and make candidates feel uneasy;

(viii) a reluctance to probe into facts and challenge statements where necessary.

2.14 While some interviewers will be experts from the personnel department of the organisation, it is usually thought desirable to include line managers in the interview team. They cannot be full-time interviewers, obviously: they have their other work to do. No matter how much training they are given in interview techniques, they will lack continuous experience, and probably not give interviewing as much thought or interest as they should. A simplified set of criteria, which is in non-personnel-specialist's terminology, and is clearly related to relevant items on the job specification, may help.

3. SELECTION TESTING

3.1 In some job selection procedures, an interview is supplemented by some form of selection test. The interviewers must be certain that the results of such tests are reliable, and that a candidate who scores well in a test will be more likely to succeed in the job. The test will have no value unless there is a direct relationship between ability in the test and ability in the job. The test should be designed to be *discriminating* (to bring out the differences in subjects), *standardised* (so that it measures the same thing in different people, providing a consistent basis for comparison) and *relevant* to its purpose.

3.2 There are four types of test commonly used in practice:

(a) intelligence tests;
(b) aptitude tests;
(c) proficiency tests; and
(d) personality tests.

Sometimes applicants are required to attempt several tests (a *test battery*) aimed at giving a more rounded picture than would be available from a single test.

3.3 (a) *Intelligence tests* aim to measure the applicant's general intellectual ability. They may test the applicant's memory, his ability to think quickly and logically and his skill at solving problems.

(b) *Aptitude tests* are designed to predict an individual's potential for performing a job or learning new skills. There are various accepted areas of aptitude, including clerical, numerical and mechanical.

(c) *Proficiency tests* are perhaps the most closely related to an assessor's objectives, because they measure ability to do the work involved. An applicant for an audio typist's job, for example, might be given a dictation tape and asked to type it. This is a type of attainment test, in that it is designed to measure abilities or skills already acquired by the candidate.

(d) *Personality tests* may measure a variety of characteristics, such as an applicant's skill in dealing with other people, his ambition and motivation or his emotional stability. They usually consist of questionnaires asking respondents to state their interest in or preference for jobs, leisure activities and so on. To a trained psychologist, such questionnaires may give clues about the dominant qualities or characteristics of the individuals tested, but wide experience is needed to make good use of the results.

3.4 This kind of testing must be used with care as it suffers from several limitations.

(a) There is not always a direct (let alone predictive) relationship between ability in the test and ability in the job: the job situation is very different from artificial test conditions.

(b) The interpretation of test results is a skilled task, for which training and experience is essential. It is also highly subjective (particularly in the case of personality tests), which belies the apparent scientific nature of the approach.

(c) Additional difficulties are experienced with particular kinds of test. For example:

(i) an aptitude test measuring arithmetical ability would need to be constantly revised or its content might become known to later applicants;

(ii) personality tests can often give misleading results because applicants seem able to guess which answers will be looked at most favourably;

(iii) it is difficult to design intelligence tests which give a fair chance to people from different cultures and social groups and which test the *kind* of intelligence that the organisation wants from its employees: the ability to score highly in IQ tests does not necessarily correlate with desirable traits such as mature judgement or creativity, merely mental agility. In addition, 'practice makes perfect': most tests are subject to coaching and practice effects.

(d) It is difficult to exclude bias from tests. Many tests (including personality tests) are tackled less successfully by women than by men, or by immigrants than by locally-born applicants because of the particular aspects chosen for testing. This is a particular problem in countries, such as the UK, where equal opportunities legislation makes it illegal to discriminate in employment on the basis of sex or race.

Research

Dr Ivan Robertson and Peter Makin (*Financial Times*, 9 June 1986).

The authors referred to a scale constructed by psychologists to describe how reliable various techniques are at predicting who will do well. The scale ranges from 1 (meaning a method that is right every time) to 0 (meaning a method that is no better than random chance). On this scale, interviews have been found to score only 0.2 - little better than tossing a coin. IQ tests score better at 0.4, and a similar level is achieved by the use of the so-called biodata technique, a questionnaire which applicants complete with details of their life-style and attitudes.

Despite these results, Robertson and Makin found that the great majority of the large companies they circularised used only interviews and references in their recruitment procedures. Only 4% of companies responding to their questionnaire said that they made use of the comparatively sophisticated technique of personality tests.

Group selection methods

3.5 Group selection methods might be used by an organisation as the final stage of a selection process for management jobs. They consist of a series of tests, interviews and group situations over a period of two days, involving a small number of candidates for a job. Typically, six or eight candidates will be invited to the organisation's premises for two days. After an introductory session to make the candidates feel at home, they will be given one or two tests, one or two individual interviews, and several group situations in which the candidates are invited to discuss problems together and arrive at solutions as a management team. Techniques in such programmes include:

(a) group role-play exercises, in which they can explore (and hopefully display) interpersonal skills and/or work through simulated managerial tasks;

(b) case studies, where candidates' analytical and problem-solving abilities are tested in working through described situations/problems, as well as their interpersonal skills, in taking part in (or leading) group discussions of the case study.

3.6 These group sessions might be thought useful because:

(a) they give the organisation's selectors a longer opportunity to study the candidates;

(b) they reveal more than application forms, interviews and tests alone about the ability of candidates to persuade others, negotiate with others, and explain ideas to others and also to investigate problems efficiently. These are typically management skills;

(c) they reveal more about how the candidates' personalities and attributes will affect the work team and his own performance. Stamina, social interaction with others (ability to co-operate and compete), intelligence, energy, self confidence or outside interests will not necessarily be meaningful in themselves (as analysed from written tests), but may be shown to affect performance in the work context.

Since they are most suitable for selection of potential managers who have little or no previous experience and two days to spare for the sessions, group selection methods are most commonly used for selecting university graduates for management trainee jobs.

Exercise 3

Think back to your own selection for your first job with your organisation. What procedures did you have to go through? What sort of interview did you have: was it well-conducted, looking back on it now? Did you have to take any tests?

Find out who does the selection interviewing in your office, what guidelines are laid down for them, and what training they get in interview technique.

Find out what selection tests (if any) are used in your organisation, and for what kinds of jobs.

---

4. ONCE SELECTION HAS BEEN MADE

4.1 Once an eligible candidate has been found, a provisional offer can be made, by telephone or in writing, subject to satisfactory references.

The organisation should be prepared for its offer to be rejected at this stage. An applicant may have received and accepted another offer; he may not have been attracted by his first-hand view of the organisation, and may have changed his mind; he may only have been testing the water in applying in the first place: gauging the market for his skills and experience for future reference, or seeking a position of strength from which to bargain with his present employer. A small number of eligible applicants should therefore be kept in reserve.

References

4.2 *References* provide further confidential information about the prospective employee. This may be of varying value, as the reliability of all but the most factual information must be in question. A reference should contain:

(a) straightforward factual information confirming the nature of the applicant's previous job(s), period of employment, pay, and circumstances of leaving;

(b) opinions about the applicant's personality and other attributes. These should obviously be treated with some caution. Allowances should be made for prejudice (favourable or unfavourable), charity (withholding detrimental remarks), and possibly fear of being actionable for libel (although references are privileged, as long as they are factually correct and devoid of malice).

At least two *employer* references are desirable, providing necessary factual information, and comparison of personal views. *Personal* references tell the prospective employer little more than that the applicant has a friend or two.

4.3 Written references save time, especially if a standardised letter or form has been pre-prepared. A simple letter inviting the previous employer to reply with the basic information and judgement required may suffice. If the recruiting officer wishes for a more detailed appraisal of the applicant's suitability, he may supply brief details of the post in question and ask for the previous employer's opinion - but it will be an ill-informed and subjective judgement. A standard form to be completed by the referee may be more acceptable, and might pose a set of simple questions about:

(a) job title;
(b) main duties and responsibilities;
(c) period of employment;
(d) pay/salary; and
(e) attendance record.

If a judgement of character and suitability is desired, it might be most tellingly formulated as the question: 'Would you re-employ this individual? (If not, why not?)'

4.4 Telephone references may be time-saving, if standard reference letters or forms are not available. They may also elicit a more honest opinion than a carefully prepared written statement. For this reason, a telephone call may also be made to check or confirm a poor or grudging reference which the recruiter suspects may be prejudiced.

The pen is more double-edged than the sword!

'I am delighted to write a reference for X, now that I no longer employ him.'

'The man who gets Y to work for him will be happy indeed.'

'I cannot speak too highly of Z's ability.'

'Things have not been the same since P left.'

Exercise 4

(a) At the end of a recent selection process one candidate was, in the view of everyone involved, outstanding. However, you have just received a very bad reference from her current employer. What do you do?

(b) For fun, rephrase the following comments in the way that you might expect to see them appear in a letter of reference

(i) Mr Smith is habitually late
(ii) Remains immature
(iii) Socially unskilled with clients
(iv) Is rather dull

Solution

(a) It is quite possible that her current employer is desperate to retain her. Disregard the reference, and seek one from a previous employer if possible.

(b) The phrases given are 'translations' by Adrian Furnham (*Financial Times*, December 1991) of the following.

(i) 'Mr Smith was occasionally a little lapse in time keeping'
(ii) 'Clearly growing out of earlier irresponsibility'
(iii) 'At her best with close friends'
(iv) 'Got a well deserved lower second'

Contracts of employment

4.5 Once the offer of employment has been confirmed and accepted, the contract of employment can be prepared, to include:

(a) job title;

(b) duties (usually with a clause to the effect that: 'The employee will perform such duties and will be responsible to such persons as the Company may from time to time direct.');

(c) date of commencement;

(d) salary or rate of pay, method and frequency of payment, shift- and overtime-rates;

(e) hours of work: including rest pauses, lunch break, overtime, shifts;

(f) holiday arrangements: days paid, qualifying period, accrual (if applicable), fixed dates (if any), public holidays;

(g) sick pay: duration, national insurance benefit deductions, requirement for medical certificates, maximum time lost before termination;

(h) pensions and pension schemes;

(i) notice of termination: from, and to, employee;

(j) grievance and disciplinary procedures, works rules, union or staff association membership;

(k) special terms: there may be an 'intellectual property' clause protecting patents or confidential information when the employee leaves

(l) provision for variation of the terms of the contract, subject to proper notice being given.

Chapter roundup

- The contents of this chapter may again be summarised in diagrammatic form: see the next page. Here, in addition, is a checklist for the recruitment/selection officer
  - Vacancy authorised?
  - Is there an up-to-date job description?
  - Conditions of employment (salary/holiday/pension) determined?
  - Notice of vacancy internally posted?
  - Job advertisement/agency brief agreed?
  - Interviewing arrangements agreed?
  - Short-listed candidates informed?
  - Unsuitable/reserve candidates informed?
  - References taken up?
  - Offer/rejection letters despatched?
  - Replies to offers accounted for?
  - Statement of terms prepared?
  - Placement/induction plans prepared?

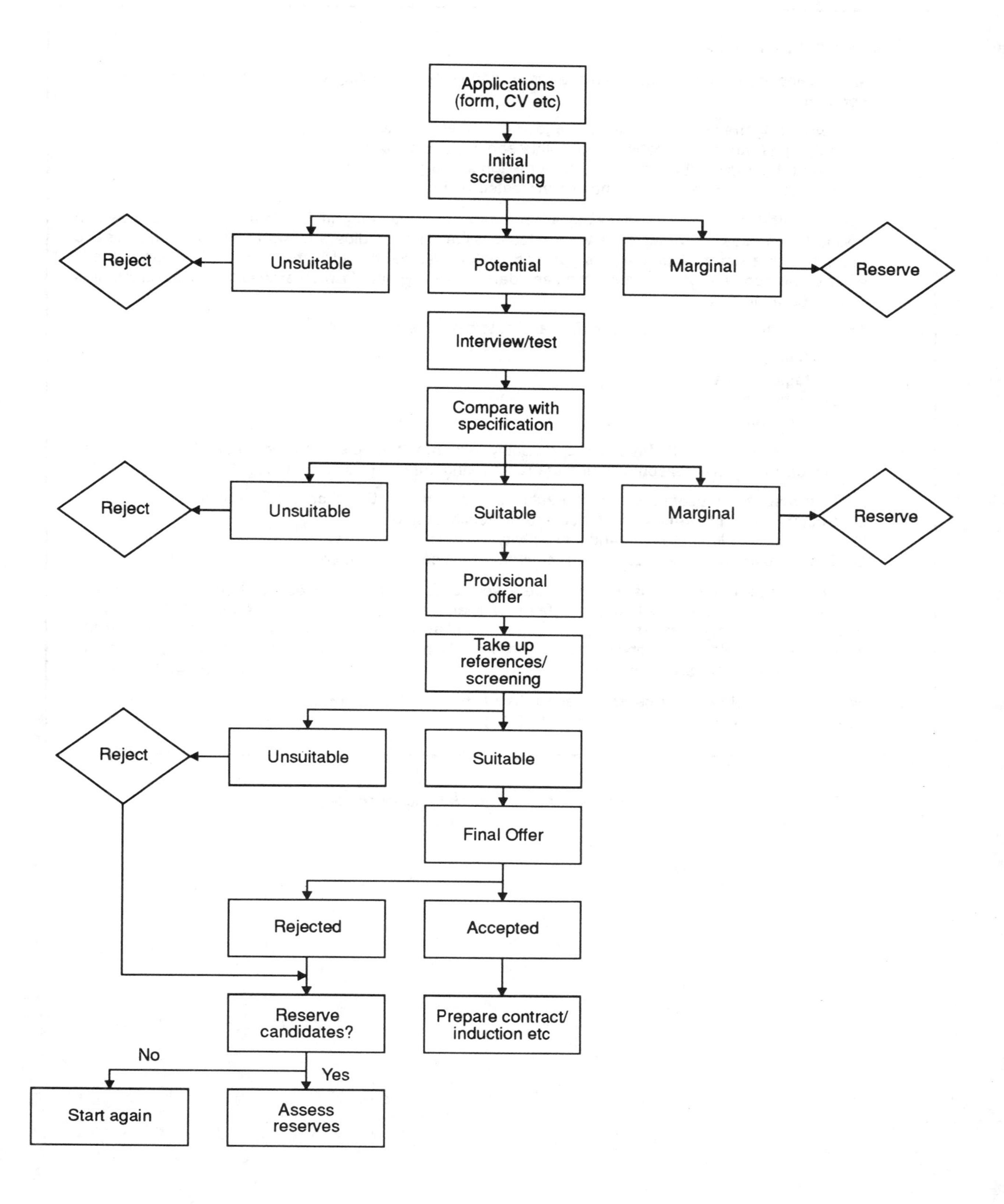
Applications (form, CV etc)
Initial screening
Reject
Unsuitable
Potential
Marginal
Reserve
Interview/test
Compare with specification
Reject
Unsuitable
Suitable
Marginal
Reserve
Provisional offer
Take up references/ screening
Reject
Unsuitable
Suitable
Final Offer
Rejected
Accepted
Reserve candidates?
Prepare contract/ induction etc
No
Yes
Start again
Assess reserves

Test your knowledge

1 A job selection interview has several aims. If you were conducting one, though, you should not be concerned with:

A comparing the applicant against the job/personnel specification
B getting as much information as possible about the applicant
C giving the applicant information about the job and organisation
D making the applicant feel he has been treated fairly (2.1)

2 Amon Leigh-Hewman is interviewing a candidate for a vacancy in his firm. He asks a question about the candidate's views on a work-related issue. The candidate starts to answer, and sees to his horror that Amon in pursing his lips and shaking his head slightly to himself. 'Of course, that's what some people say', continues the candidate, 'but I myself ...' Amon smiles. His next question is 'Don't you think that ...?'

Amon is getting a distorted view of the candidate because of:

A the halo effect
B contagious bias
C stereotyping
D logical error (2.13)

3 Selection tests such as IQ tests and personality tests may not be effective in getting the right person for the job for several reasons. Which of the following criticisms is false, though?

A Test results can be influenced by practice and coaching rather than genuine ability
B Subjects are able (and tend) to deliberately falsify results
C Tests do not eliminate bias and subjectivity
D Tests are generally less accurate predictors of success than interviews (3.4 + box)

4 'Even though the interview is known to be a very poor technique for establishing the suitability of candidates to fill a defined job, the faith of managers in the efficacy of the interview (when conducted by themselves) remains high and unshakable.' Is this an accurate observation? To the extent that the selection interview needs to be made more valid and reliable, what improvements can you suggest? *(ICSA June 1986)* (2.5 - 2.9, 2.13-2.14, 3.4 (box))

5 What are the relative merits and drawbacks of the panel and one-to-one interviews in personnel selection? *(ICSA June 1987)* (2.10 - 2.12)

Now try question 10 at the end of the text

Chapter 11

TRAINING AND DEVELOPMENT

This chapter covers the following topics.

Signpost

- Identifying the *need* for training and development is an aspect of performance assessment, which is discussed in the next chapter.
- Training and development also fulfil individual needs such as are described in Chapter 13 on motivation.
- Leadership was the subject of Chapter 6.

1. THE ORGANISATION'S APPROACH TO TRAINING

1.1 Providing the organisation with the most suitable human resources for the task and environment is an on-going process. It involves not only recruitment and selection, but the training and development of employees - prior to employment, or at any time during their employment, in order to help them meet the requirements of their current and potential future job.

Training, like selection, is concerned with:

(a) fitting people to the requirements of the job;

(b) securing better occupational adjustment; and

(c) in methodological terms, setting and achieving targets, and defining performance criteria against which the success of the process can be monitored.

1.2 The term 'industrial training', used in connection with training in a work context, includes a wide range of activities: commercial training, competence or skills training, management development, apprenticeships and so on.

The main purpose of industrial training is to raise competence and therefore performance standards. From the perspective of HR management, however, it is also concerned with personal development, helping individuals to expand and fulfil their potential (also, theoretically, motivating them to higher performance through the opportunity for personal growth).

1.3 This two-fold purpose of training may, however, encourage a certain conflict in the identification of training needs and priorities, where the perceived interests of organisation and individual differ. The organisation requires a fundamentally practical long-term training plan, explicitly geared to the requirements set out in the manpower plan.

> 'Training is to some extent a management reaction to change, eg changes in equipment and design, methods of work, new tools and machines, control systems, or in response to changes dictated by new products, services, or markets. On the other hand, training also induces change. A capable workforce will bring about new initiatives, developments and improvements - in an organic way, and of its own accord. Training is both a cause and an effect of change.'
>
> Bryan Livy: *Corporate Personnel Management*

1.4 Easy assumptions about training and, within each organisation, the training programme, should constantly be challenged, if the desired outcome from training is to be achieved.

(a) '*Training is a personnel department matter*.' Yes it is - but not exclusively. Line managers are conversant with the requirements of the job, and the individuals concerned; they are also responsible for the performance of those individuals. They should be involved in:

(i) training need and priority identification;

(ii) training itself. A specialist trainer may be used as a catalyst, but the experience of line personnel, supervisors and senior operatives will be invaluable in ensuring a practical and participative approach; and

(iii) follow-up of performance, for the validation of training methods.

In addition, experience and learning opportunities in the job itself are very important, and the individual should be encouraged in *self learning*, and in the 'ownership' of his own development at work.

(b) '*The important thing is to have a training programme.*' The view that training in itself is such a Good Thing that an organisation can't go wrong by providing some is a source of inefficiency. The individual needs and expectations of trainees must be taken into account: the purpose of training must be clear, to the organisation (so that it can direct training effort and resources accordingly) and to the individual, so that he feels it to be worthwhile and meaningful - without which the motivatory factor will be lost. If the individual feels that he is training in order to grow and develop, to find better ways of working, or to become part of the organisation culture, he will commit himself to learning more thoroughly than if he feels he is only doing it to show willing, to fulfil the manpower plan, or whatever. It is too easy, also, to run old or standard programmes, without considering that:

(i) the learning needs of current trainees may be different from past ones;

(ii) the requirements of the job may not all be susceptible to classroom or study methods: are the most relevant needs being met?

(iii) the training group may not be uniform in its needs: the training package may be off-target for some members.

(c) '*Training will improve performance.*' It *might* - and *should*, all other things being equal - but a training course is not a simple remedy for poor performance. Contingency theory must be applied to situations where employee performance is below the desired standard: an employee who is adequately *trained* to perform may still not be *able* or *willing* to do so, because of badly designed working methods or environment, faulty equipment, inappropriate supervision, poor motivation, lack of incentive (poor pay scales or promotion prospects), or non-work factors, such as health, domestic circumstances and so on. In particular, it must be remembered that performance is not just a product of the System, but a product, and manifestation, of human behaviour: training methods, and their expected results, must take into account human attitudes, values, emotions and relationships.

Exercise 1

What is your organisation's attitude to training, and has there been any noticeable change in recent years?

Have a look at your organisation's annual report or organisation manual, and see what its stated 'vision' and objectives for training are. Does the organisation see itself as committed to training? And in practice: does systematic training go on, are there equal opportunities for personal development, do managers in fact grumble that training is all cost and no benefit?

The systematic approach to training

1.5 According to the Department of Employment, training is 'the systematic development of the attitude/knowledge/skill/behaviour pattern required by an individual in order to perform adequately a given task or job.'

The application of systems theory to the design of training has gained currency in the West in recent years. A *training system* uses scientific methods to programme learning, from:

(a) the identification of training needs; this is a product of job analysis and specification (what is required, in order to do the job) and an assessment of the present capacities and inclinations of the individuals (their 'pre-entry' behaviour); to

(b) the design of courses, selection of methods and media; to

(c) the measurement of trained performance against pre-determined proficiency goals - the 'terminal behaviour' expected on the job.

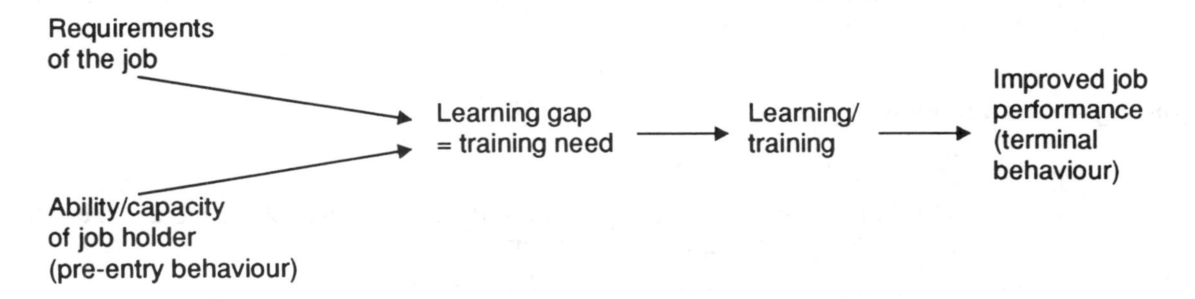

2. TRAINING NEEDS

2.1 The training needs of individuals and groups will obviously vary enormously, according to the nature of the job and particular tasks, the abilities and experience of the employees. As suggested earlier, training should not be a 'shot in the dark'; the homogeneity of the training group cannot be assumed; clear and obtainable objectives are essential to the efficiency and effectiveness of the training programme.

2.2 Some training requirements will be obvious and 'automatic'.

(a) If a piece of legislation is enacted which affects the organisation's operations, training in its provisions will automatically be indicated: so, for example, personnel staff will need to be trained if and when various EC Directives are enacted in UK law.

(b) The introduction of new technology similarly implies a training need.

(c) An organisation seeking accreditation for its training scheme, or seeking a British Standard (say, for quality systems - see below), will have certain training requirements imposed on them by the approving body.

BS 5750 Training requirements

2.3 The British Standard for Quality Systems (BS 5750) which many UK organisations are working towards (often at the request of customers, who perceive it to be a 'guarantee' that high standards of quality control are being achieved) includes training requirements. As the following extract shows, BS 5750 identifies training needs for those organisations registering for assessment, and also shows the importance of a systematic approach to ensure adequate control.

'The training, both by specific training to perform assigned tasks and general training to heighten quality awareness and to mould attitudes of all personnel in an organisation, is central to the achievement of quality.

The comprehensiveness of such training varies with the complexity of the organisation.

The following steps should be taken:

1 Identifying the way tasks and operations influence quality in total;

2 Identifying individuals' training needs against those required for satisfactory performance of the task;

3 Planning and carrying out appropriate specific training;

4 Planning and organising general quality awareness programmes;

5 Recording training and achievement in an easily retrievable form so that records can be updated and gaps in training can be readily identified.'

BSI, 1990

2.4 Some qualitative indicators might be taken as symptoms of a need for training: absenteeism, high labour turnover, grievance and disciplinary actions, crises, conflict, poor motivation and performance. Such factors will need to be investigated to see what the root causes are, and whether training *will* solve the problem.

Training needs analysis

2.5 Formal training needs analysis may be carried out.

Training needs should be identified as the gap between the requirements of the job and the actual current performance of the job-holders. In other words:

Required level of competence minus present level of competence = training need.

2.6 The present level of employees' competence (which includes not only skill and knowledge, but inclination as well) can be measured by an appropriate pre-test of skills, knowledge, performance, attitude and so on. Assessment interviews and the ongoing system of performance appraisal may also be used.

2.7 The required level of competence for the job can be determined by:

(a) job analysis;

(b) skills analysis, for more skilled jobs. Here, not only the task- and worker-oriented requirements of the job are identified, but also the skill elements of the task, such as:

(i) what sensory information (cues and stimuli) need to be recognised?
(ii) what senses (vision, touch, hearing etc) are involved?
(iii) what left-hand/right-hand/foot operations are required?
(iv) what counter-balancing operations are required?
(v) what interactions with other operatives are required?

(c) role analysis, for managerial and administrative jobs requiring a high degree of co-ordination and interaction with others;

(d) existing records, such as job specifications and descriptions, person specifications, the organisation chart (depicting roles and relationships) and so on.

Competence analysis

2.8 We have so far used the word 'competence' in its general sense of 'capability', but you should be aware that it now carries more technical connotations, as part of a revolution in the approach to vocational and professional qualifications and training.

2.9 Competence-based education and training focuses on the output of the learning process (what the trainee should be able to do at the end of it) rather than its input (a scheme of learning, or a syllabus of topics to be covered). The idea has been around for some time, having its roots in teacher education in the US, but has only gained currency in the UK in the 1980s. In 1986, the Manpower Services Commission launched a Standards Development Programme, while a review of vocational qualifications led to the establishment of the National Council for Vocational Qualifications (NCVQ), with responsibility for developing criteria for a new qualification framework based on *standards of competence,* against which candidates are assessed, rather than knowledge tested by examinations.

2.10 In practice, this has involved analysing real jobs in each occupational area in order to find out what 'acceptable performance' in the job entails. *Standards of competence* are formulated. They identify the *key roles* of the occupation and break them down into areas or *units of competence*. These in turn are formulated as statements of:

(a) the specific activities a job holder should be able to perform *(elements of competence)*;
(b) how well *(performance criteria)*;
(c) in what contexts and conditions (listed in a *range statement*); and
(d) with what 'underpinning' knowledge and understanding.

2.11 The NCVQ *accredits* suitably restructured qualifications, awarded by bodies such as training councils and professional bodies, as NVQs (National Vocational Qualifications) at a range of levels (1-5). Among the professional bodies, for example, the Association of Accounting Technicians (AAT) has already implemented a competence-based training and assessment scheme, leading to NVQ Level 2, 3 and 4 in progressive stages. NVQs are designed to be vocationally relevant and nationally recognised, since the standards of competence being assessed are devised by 'lead bodies' made up of highly qualified representatives of the occupation or profession nationwide. There are over 160 lead bodies: some are industry-specific (retail, construction and so on); some are more broadly occupational (training and development, accountancy and administration, marketing and so on).

The Personnel Standards Lead Body

2.12 'Under the chairmanship of David Sieff, director of corporate and government affairs at Marks and Spencer, the PSLB has defined the continuous improvement of people's contribution to organisational success as the key purpose of the personnel function.

Research on how that purpose can be met has now been completed. Before going on to detail standards of competence for personnel practitioners, the lead body presented the findings to an advisory forum, a consultation process that will be repeated at each stage in the production process.

Explaining the rationale behind the forum's composition, Christina Townsend [deputy chairman] says: "The work we are producing must be readily understood not only by personnel people but by those they work with and work for. In that context we see chief executives as extremely important. We don't expect them to take forward the details of our work but we do want them to say whether or not they agree with our definition of the key purpose of personnel and the changes that are likely to take place in the next few years". *'Personnel Management*, August 1992

2.13 In general terms, Connock *(HR Vision)* suggests that competence definition and analysis may be a useful approach within an organisation, as a way of assessing its future requirements, and providing data to underpin recruitment, training, appraisal, potential assessment, succession planning and reward strategies. Connock suggests that, being systematic and based on real-life, observable behaviour, competence analysis provides a thorough and objective picture of job requirements at different levels, and one that is relevant to the circumstances and values of the organisation. He does, however, recognise that achieving definitions of competence is a long and complex task, and tends therefore to be over-simplistic and quickly outdated.

Training objectives

2.14 The training department manager will have to make an initial investigation into the problem of the gap between job or competence requirements and current performance or competence. (As we noted earlier, it may be that shortcomings in the capacities and inclinations of employees would not be improved by training, but by a review of the work environment, systems and procedures, work methods, technology, industrial relations, leadership style, motivation and incentives.)

2.15 If it is concluded that the provision of training would improve work performance, training *objectives* can then be defined. They should be clear, specific and related to observable, measurable targets, ideally detailing:

(a) behaviour - what the trainee should be able to do;
(b) standard - to what level of performance; and
(c) environment - under what conditions (so that the performance level is realistic).

(Note that this corresponds directly to the approach used for standards of competence - Paragraph 2.9 above.)

2.16 Objectives are usually best expressed in terms of active verbs: at the end of the course the trainee should be able to describe ...., or identify .... or distinguish x from y .... or calculate .... or assemble .... and so on. It is insufficient to define the objectives of training as 'to give trainees a grounding in ....' or 'to encourage trainees in a better appreciation of ....': this offers no target achievement which can be quantifiably measured.

Exercise 2

Here is a quick test to see if you are paying attention.

(a) What three things is training concerned with?
(b) Is training a reaction to change or a cause of change?
(c) No organisation should be without a training programme. True or false?
(d) What is the systematic approach to training?
(e) Training needs should be identified as ... what?
(f) What is a standard of competence?
(g) How should training objectives be expressed?

The answers can all be found in the first two sections of this chapter.

---

3. THE LEARNING PROCESS

3.1 Having identified training needs and objectives, the manager will have to decide on the best way to approach training: there are a number of types and techniques of training, which we will discuss below.

There are different schools of learning theory which explain and describe the learning process in very different ways.

(a) Behaviourist psychology concentrates on the relationship between 'stimuli' (input through the senses) and 'responses' to those stimuli. 'Learning' is the formation of *new* connections between stimulus and response, on the basis of experience or 'conditioning': we modify our responses in future according to whether the results of our behaviour in the past have been good or bad. We get *feedback* on the results of our actions, which may be rewarding ('positive reinforcement') or punishing ('negative reinforcement') and therefore an incentive or a deterrent to similar behaviour in future. Trial-and-error learning, and carrot-and-stick approaches to motivation work on this basis.

(b) The cognitive (or 'information processing') approach argues that the human mind takes sensory information and imposes organisation and meaning on it: we interpret and rationalise. We use feedback information on the results of past behaviour to make rational decisions about whether to maintain successful behaviours or modify unsuccessful behaviours in future, according to our goals and our plans for reaching them.

3.2 Whichever approach it is based on, learning theory offers certain useful propositions for the design of effective training programmes, namely that:

(a) the individual should be *motivated* to learn. The advantages of training should be made clear, according to the individual's motives - money, opportunity, valued skills or whatever;

(b) there should be clear *objectives and standards* set, so that each task has some meaning. Each stage of learning should present a challenge, without overloading the trainee or making him lose confidence. Specific objectives and performance standards for each will help the trainee in the planning and control process that leads to learning, providing targets against which performance will constantly be measured;

(c) there should be timely, relevant *feedback* on performance and progress. This will usually be provided by the trainer, and should be concurrent - or certainly not long delayed. If progress reports or performance appraisals are given only at the year end, for example, there will be no opportunity for behaviour adjustment or learning in the meantime;

(d) positive and negative *reinforcement* should be judiciously used. Recognition and encouragement enhances the individual's confidence in his competence and progress: punishment for poor performance - especially without explanation and correction - discourages the learner and creates feelings of guilt, failure and hostility. Helpful or constructive criticism, however, is more likely to be beneficial;

(e) active *participation* is more telling than passive reception (because of its effect on the motivation to learn, concentration and recollection). If a high degree of participation is impossible, practice and repetition can be used to reinforce receptivity, but participation has the effect of encouraging 'ownership' of the process of learning and changing - committing the individual to it as his *own* goal, not just an imposed process.

Learning styles

3.3 It is also believed that the way in which people learn best will be different according to the type of person, that is, there are learning styles which suit different individuals.

Peter Honey and Alan Mumford have drawn up a popular classification of four learning styles.

(a) *Theorist*

This person seeks to understand underlying concepts and to take an intellectual, 'hands-off' approach based on logical argument. Such a person prefers training:

(i) to be programmed and structured;
(ii) to allow time for analysis; and
(iii) to be provided by teachers who share his preference for concepts and analysis.

Theorists find learning difficult if they have a teacher with a different style (particularly an activist style), material which skims over basic principles and a programme which is hurried and unstructured.

(b) *Reflector*

People who observe phenomena, think about them and then choose how to act are called reflectors. Such a person needs to work at his own pace and would find learning difficult if forced into a hurried programme with little notice or information.

Reflectors are able to produce carefully thought-out conclusions after research and reflection but tend to be fairly slow, non-participative (unless to ask questions) and cautious.

(c) *Activist*

These are people who like to deal with practical, active problems and who do not have much patience with theory. They require training based on hands-on experience.

Activists are excited by participation and pressure, such as making presentations and new projects. Although they are flexible and optimistic, however, they tend to rush at something without due preparation, take risks and then get bored.

(d) *Pragmatist*

These people only like to study if they can see its direct link to practical problems - they are not interested in theory for its own sake. They are particularly good at learning new techniques in on-the-job training which they see as useful improvements. Their aim is to implement action plans and/or do the task better. Such a person is business-like and realistic, but may discard as being impractical good ideas which only require some development.

3.4 The implications for management are that people react to problem situations in different ways and that, in particular, training methods should where possible be tailored to the preferred style of trainees.

The learning cycle

3.5 Another useful model is the *experiential learning cycle* devised by David Kolb. Kolb suggested that classroom-type learning is 'a special activity cut off from the real world and unrelated to one's life': a teacher or trainer directs the learning process on behalf of a passive learner. Experiential learning, however, involves doing, and puts the learner in an active problem-solving role: a form of *self-learning* which encourages the learner to formulate and commit himself to his own learning objectives.

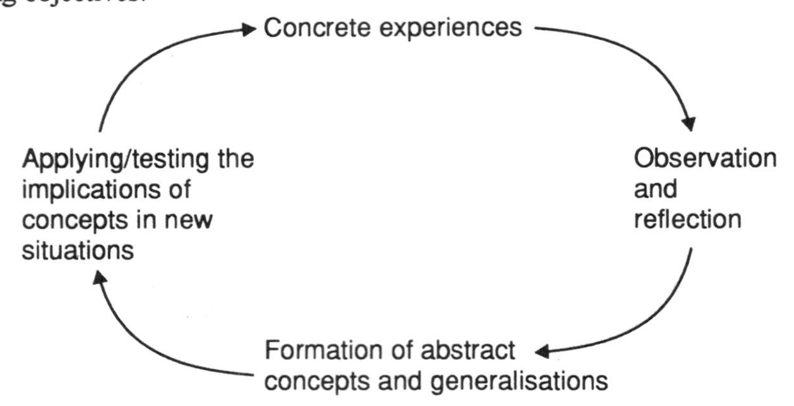

Say an employee interviews a customer for the first time (concrete experience). He observes his performance and the dynamics of the situation (observation) and afterwards, having failed to convince the customer to buy his product, he analyses what he did right and wrong (reflection). He comes to the conclusion that he had failed to listen to what the customer really wanted and feared, underneath his general reluctance: he realises that the key to communication is listening (abstraction/generalisation). In his next interview he applies his strategy to the new set of circumstances (application/testing). This provides him with a new experience with which to start the cycle over again.

Act

Analyse actions

Understand principles

Apply principles

3.6 This is the model for many of the modern approaches to training, and particularly management development, which recommend experiential learning - 'learning by doing', or self-learning.

In effect, it involves elements of *all* the learning styles identified by Honey and Mumford.

Exercise 3

Again this is a 'recap' exercise.

List five propositions from learning theory that may be helpful in the design of effective teaching principles.

---

4. TRAINING METHODS AND MEDIA

4.1 Training methods and media must next be evaluated, and a programme designed. There are a variety of options, discussed below, including:

(a) formal training and education; by internal or external residential courses, day courses or lectures, distance learning, programmed learning or computer-aided learning;

(b) on-the-job training - including induction, coaching and job-rotation;

(c) awareness oriented training - including T groups, assertiveness training and neuro-linguistic programming (NLP).

4.2 The training course should only go ahead if the likely benefits are expected to exceed the costs of designing and running it. The problem here is not so much in estimating costs, but in estimating the potential benefits.

(a) Costs will be the costs of the training establishment, training materials, the salaries of the staff attending training courses, their travelling expenses, the salaries of training staff, and so on.

(b) Benefits might be measured in terms of:

(i) quicker working and therefore reductions in overtime or staff numbers;
(ii) greater accuracy of work;
(iii) more extensive skills and versatility for labour flexibility;
(iv) enhanced job satisfaction.

As you will appreciate, the benefits are more easily stated in general terms than quantified in money terms.

Formal training and education

4.3 *Internal* courses are sometimes run by the training department of larger organisations. Skills may be taught at a technical level, related to the organisation's particular product and market, or in aspects such as marketing, teambuilding, interviewing or information technology management. Some organisations encourage the wider development of staff by offering opportunities to learn languages or other skills. Some of this is accomplished informally - for example, by encouraging foreign-language discussion groups in meal breaks.

4.4 One way of conveniently decentralising training (as well as quickly developing training courses to meet emerging needs) is the use of computer-based training (CBT) and Interactive Video (IV), through teaching equipment in offices or even trainees' homes. Training programmes may be developed by the organisation or by outside consultants - or bought 'off the shelf': the software (or 'courseware') can then be distributed, so that large numbers of dispersed staff can learn about new products or procedures quickly and simultaneously.

4.5 *External* courses vary, and may involve:

(a) day-release, which means that the employee attends a local college on one day of the week;

(b) evening classes, or 'distance learning' (a home study or 'correspondence' course, plus limited face-to-face teaching) which make demands on the individual's time outside work;

(c) full-time but brief introductory or revision courses for examinations of professional bodies (like the ICSA);

(d) a sponsored full-time course at a university or polytechnic for 1 or 2 years. This might be the case, for example, for a manager doing an MBA degree.

4.6 The disadvantages of formal training might be that:

(a) an individual will not benefit from formal training unless he/she wants to learn. The individual's superior may need to provide encouragement in this respect;

(b) if the subject matter of the training course does not relate to an individual's job, the learning will not be applied, and will quickly be forgotten. Many training managers provide internal courses without relating their content to the needs of individuals attending them. Equally, professional examinations often include subjects in which individuals have no job experience, and these are usually difficult to learn and are quickly forgotten afterwards;

(c) individuals may not be able to accept that what they learn on a course applies in the context of their own particular job. For example, a manager may attend an internal course on man-management which suggests a participatory style of leadership, but on returning to his job he may consider that what he has learned is not relevant in his case, because his subordinates are too young or too inexperienced;

(d) immediate and relevant *feedback* on performance and progress may not be available from the learning process, which will lower the learner's incentive;

(e) it does not suit activists or pragmatists.

On-the-job training

4.7 On-the-job training is very common, especially when the work involved is not complex. Trainee managers require more coaching, and may be given assignments or projects as part of a planned programme to develop their experience. Unfortunately, this type of training will be unsuccessful if:

(a) the assignments do not have a specific purpose from which the trainee can learn and gain experience;

(b) the trainee is a theorist or reflector and needs to get away from the pressures of the workplace to think through issues and understand the underlying principles before he can apply new techniques;

(c) the organisation is intolerant of any mistakes which the trainee makes. Mistakes are an inevitable part of on-the-job learning, and if they are punished the trainee will be reluctant to take further risks: he will be de-motivated to learn.

In addition, there may be real risks involved in throwing people in at the deep-end: the cost of mistakes or inefficiencies may be high and the pressure on learners great.

4.8 An important advantage of on-the-job training is that it takes place in the environment of the job itself, and in the context of the work group in which the trainee will have to operate. The style of supervision, personal relations with colleagues, working conditions and pressures, the culture of the office/shop floor and so on will be absorbed as part of the training process.

(a) There will be no re-adjustment to make, as there will be when a trainee transfers externally-acquired knowledge, skills and academic attitudes to the job, the work place and work group.

(b) The work group itself will be adapting to the dynamics of the trainee's situation: it will adjust to his new capacities and inclinations. All too often, a trainee returns from a course to find that he cannot apply his new-found ideas and skills: the supervisor is set in his ways, colleagues resent his perceived superiority, the tasks and methods stay the same. Learning within the work group also prevents the trainee from becoming isolated and introverted (concentrating on his own learning process).

(c) The perceived relevance of the training to the job and performance criteria is much greater, and therefore the training is felt to be of greater value.

(d) There is greater opportunity for relevant, performance-related *feedback*, which is an integral part of the learning process (both as the means by which the individual recognises the need for modified behaviour, and as a motivating factor, with the knowledge of progress and success). Feedback on performance in a non-work environment may still leave doubts in the trainee's mind as to the value of his progress when he returns to the job.

4.9 Different methods of on-the-job training include:

(a) *induction:* introducing a new recruit or transferred employee to his job;

(b) *coaching:* the trainee is put under the guidance of an experienced employee who shows the trainee how to do the job. This is sometimes called 'sitting with Nellie'. The length of the coaching period will depend on the complexity of the job and the previous experience of the trainee;

(c) *job rotation:* the trainee is given several jobs in succession, to gain experience of a wide range of activities. (Even experienced managers may rotate their jobs, to gain wider experience; this philosophy of job education is commonly applied in the Civil Service, where an employee may expect to move on to another job after a few years);

(d) *temporary promotion:* an individual is promoted into his superior's position whilst the superior is absent due to illness. This gives the individual a chance to experience the demands of a more senior position;

(e) *'assistant to' positions:* a junior manager with good potential may be appointed as assistant to the managing director or another executive director. In this way, the individual gains experience of how the organisation is managed 'at the top';

(f) *project or committee work:* trainees might be included in the membership of a project team or committee, in order to obtain an understanding of inter-departmental relationships, problem-solving and particular areas of the organisation's activity.

You will note that most of these are forms of experiential learning: learning by doing. We will look at the first two in more detail.

Induction training

4.10 From his first day in a job, a new recruit must be helped to find his bearings. There are limits to what any person can pick up in a short time, so that the process of getting one's feet under the table will be a gradual one. You will note from the following paragraphs that induction is an ongoing process.

4.11 On the first day, a manager or personnel officer should welcome the new recruit. The manager might discuss in broad terms what he requires from people at work, working conditions, pay and benefits, training opportunities and career opportunities. He should then introduce the new recruit to the person who will be his immediate supervisor.

4.12 The immediate supervisor should then take over the on-going process of induction. He should:

(a) pinpoint the areas that the recruit will have to learn about in order to start his job. Some things (such as detailed technical knowledge) may be identified as areas for later study or training, while others (say, some of the office procedures and systems with which the recruit will have to deal) will have to be explained immediately. A list of learning priorities should be drawn up, so that the recruit, and the supervisor, are clear about the rate and direction of progress required;

(b) explain first of all the nature of the job, and the goals of each task, of the recruit's job and of the department as a whole. This will help the recruit to work to specific targets and to understand how his tasks relate to the overall objectives of the department - or even the organisation as a whole;

(c) explain about hours of work and stress the importance of time-keeping. If flexitime is operated, he should explain how it works;

(d) explain the structure of the department: to whom the recruit will report, to whom he can go with complaints or queries and so on;

(e) introduce the recruit to the people in the office. He should meet the departmental manager and all the members of the immediate work team (and perhaps be given the opportunity to get to know them informally). One particular colleague may be assigned to the recruit as a *mentor* for their first few days, to keep an eye on him, answer routine queries, 'show him the ropes'. The layout of the office, procedures for lunch hours or holidays, rules about smoking or eating in the office and so on will then be taught informally;

(f) plan and implement an appropriate training programme for whatever technical or practical knowledge is required. Again, the programme should have a clear schedule and set of goals so that the recruit has a sense of purpose, and so that the programme can be efficiently organised to fit in with the activities of the department;

(g) coach and/or train the recruit; check regularly on his progress, as demonstrated by performance, and as reported by the recruits' mentor, and as perceived by the recruit himself. Feedback information on how he is doing will be essential to the learning process, correcting any faults at an early stage and building the confidence of the recruit;

(h) integrate the recruit into the culture of the office. Much of this may be done informally: he will pick up the prevailing norms of dress, degree of formality in the office, attitude to customers etc. However, the supervisor should try to sell the values and style of the office and should reinforce commitment to those values by rewarding evidence of loyalty, hard work and desired behaviour.

4.13 After three months, six months or one year the performance of a new recruit should be formally appraised and discussed with him. Indeed, when the process of induction has been finished, a recruit should continue to receive periodic appraisals, just like every other employee in the organisation.

Exercise 4

'Joining an organisation with around 8,500 staff, based on two sites over a mile apart and in the throes of major restructuring, can be confusing for any recruit. This is the situation facing the 20 to 30 new employees recruited each month by the Guy's and St Thomas' Hospital Trust, which was formed by the merger of the two hospitals in April.

In a climate of change, new employees joining the NHS can be influenced by the negative attitudes of other staff who may oppose the current changes. So it has become increasingly important for the trust's management executive to get across their view of the future and to understand the feelings of confusion new staff may be experiencing.' *Personnel Management Plus*, August 1993

See if you can design a 9 - 5 induction programme for these new recruits, in the light of the above. The programme is to be available to *all* new recruits, from doctors and radiographers to accountants, catering and cleaning staff and secretaries.

Solution

Here is the programme as published in *Personnel Management Plus*.

INDUCTION DAY

9.00	Welcome	
9.05	Introduction	*Ground rules and objectives for the day*
9.25	Presentation	*The history of Guy's and St Thomas' hospitals*
10.25	Presentation	*Talk on structure of the management team, trust board and executive*
10.45	Group exercise	*with chief executive Tim Matthews on patient care, funding, hospital processes and measuring the care provided*
12.20	Lunch	
1.15	Tour of Guy's	
2.30	Presentation	*Looking at trust with new eyes - suggestions for change*
2.20	Presentation	*Information on staff organisations*
3.10	Presentation	*Security issues, fire drills, health and safety (including handouts)*
3.30	Presentation	*Session on occupational health*
3.40	Presentation	*Local areas and staff benefits*
3.45	Tour of St Thomas'	
4.30	Presentation	*Facilities management and patient care*
4.55	Closing session	*Evaluation and finish*

Particularly important is the focus on patient care and the group exercises. 'Feedback from the participants shows that they enjoy the discussions and learn a lot more about their colleagues and the trust by participating rather than being talked at.'

Coaching

4.14 Essential steps in the *coaching* process are as follows.

(a) Establish learning targets. The areas to be learnt should be identified, and specific, realistic goals stated. These will refer not only to the timetable for acquiring necessary skills and knowledge, but to standards of performance to be attained, which should if possible be formulated by agreement with the trainee.

(b) Plan a systematic learning and development programme. This will ensure regular progress, appropriate stages for consolidation and practice. It will ensure that all stages of learning are relevant to the trainee and the task he will be asked to perform.

(c) Identify opportunities for broadening the trainee's knowledge and experience - by involving him in new projects, encouraging him to serve on interdepartmental committees, giving him new contacts, or simply extending his job, giving him more tasks, greater responsibility and so on.

(d) Take into account the strengths and limitations of the trainee, and take advantage of learning opportunities that suit his ability, preferred style and goals.

(e) Exchange feedback. The supervisor will want to know how the trainee sees his progress and his future. He will also need performance information, to monitor the trainee's progress, adjust the learning programme if necessary, identify further needs which may emerge, and plan future development for the trainee.

Awareness-oriented training

4.15 An 'alternative' approach to training is to encourage the trainee to gain insight into his own behaviour, and to try and change the negative or restricting attitudes that prevent the individual from attaining more effective performance.

Such approaches vary from the relatively straight-forward 'encounter group' principle, which allows people to analyse and practise interpersonal skills in a controlled group, to the more controversial concept of 'neuro-linguistic programming' (NLP).

T Groups

4.16 Group learning is not common in industry but is more common in organisations such as social services departments of local government authorities. The purpose of group learning is to:

(a) give each individual in a training group (or T group) a greater insight into his own behaviour;

(b) to teach an individual how he appears to other people, as a result of responses from other members of the group;

(c) to teach an understanding of intra-group processes, and how people inter-relate;

(d) to develop an individual's skills in taking action to control such intra-group processes.

Assertiveness training

4.17 Assertiveness may be described as clear, honest and direct communication. It is not to be confused with 'bossiness' or aggression. Aggressive behaviour is competitive and directed at 'beating' someone else: assertion is based on equality and co-operation. Assertion is a simple affirmation that every individual has certain rights and can stand by them in the face of pressures from other people; that there is middle ground between being powerful and powerless, between the role of 'top dog' and 'door mat'. It means:

(a) not being dependent on the approval of others;
(b) not feeling guilty if you do not put other people's needs first all the time;
(c) having the confidence to receive criticism openly and give it constructively;
(d) avoiding conflict without having to give up your own values and wants;
(e) being able to express your own values and feelings without guilt or fear;
(f) making clear requests for what you want.

You can see that much of this is to do with awareness of one's own feelings, and attitudes about one's role and rights as a person. Training is therefore partly a matter of identifying, challenging and changing attitudes.

4.18 Assertiveness training commonly uses group role-play exercises to:

(a) test individuals' natural reactions in situations;

(b) analyse what the assertive alternative would be: and

(c) allow individuals to *practise* being assertive without the pressures of real-life scenarios. There are practical techniques and skills which can be taught - as well as habits of mind which can be developed - in order to achieve an assertive approach to situations such as asking for what you want, saying 'no', and giving and receiving criticism.

4.19 Assertiveness training is popularly seen as a prime means of remedying under-achievement in women, or of helping women to avoid exploitation at work. It is likely to be part of a 'Women Into Management' or similar training and education programme. The techniques and insights involved are likely to be of benefit to men as well, but it has been recognised that it is primarily women who are disadvantaged in western society by the failure to distinguish between assertion and aggression, submission and conflict-avoidance.

(a) For example, it has been suggested that women are more prone to the 'compassion trap' - a sense of obligation to put everyone else's needs before your own all the time.

(b) Women also suffer from attitudes towards money, particularly the sense that it is not 'feminine' to know about or talk about (let alone argue about) money. This discomfort combined with a lack of self-esteem and self-assertion can make negotiating salary levels a nightmare for many women, who may be aware that they are not paid what they are worth.

(c) The same kind of attitudes to power and authority make it difficult for women to criticise, confront and direct male subordinates and colleagues.

(d) The failure of victims of sexual harassment to come forward may indicate another area in which assertiveness could be of value.

Neuro-linguistic Programming (NLP)

4.20 NLP is a technique which emerged in the USA some 20 years ago. It is based on:

(a) identifying and breaking down the behaviour patterns found in excellent performers; and

(b) communicating these to people who wish to emulate their performance; in a way that

(c) overcomes the restricted thinking processes and limiting self-beliefs that typically hold those people back.

4.21 Typical NLP techniques include:

(a) the development of detached self-awareness or self-consciousness, so that trainees become able constantly and objectively to monitor their own behaviour. This is achieved through 'disassociation' - the attainment of a detached state in which they can observe and evaluate themselves dispassionately and as it were from 'outside' the situation;

(b) conditioning, such as that used by behaviourist psychologists. Positive and negative responses can be evoked by stimuli with which they have become associated in the mind: if you have spent many happy times with a friend, for example, the stimulus (the sight of him) will become associated with the response (happiness), and you will tend thereafter to feel happy at the sight of your friend, or even thinking about him, or visualising his face. A trainee can thus be conditioned to recall a 'trigger' stimulus (a person, image, place or event) in order to summon up associated feelings (of confidence, calmness or energy, say) which may be useful. This technique is called 'anchoring';

(c) developing the ability to establish rapport with others through 'matching' behaviour. The principle of rapport is that people feel comfortable with people who are like themselves in some way. 'Matching' is a technique of observing the behaviour of others and adapting your behaviour to theirs in some way - say, in the amount of gesturing they use, the volume of their voice, or the heartiness of their manner;

(d) developing the ability to shift perspectives - to view things from another person's point of view - using mental 'mapping' and visualisation techniques. This can be used to help people with work relationships;

(e) the 'mental' rehearsal' of events and plans. This operates as a form of practice, and also gives the mind positive suggestions (as if the event has already successfully taken place), giving confidence and therefore - usually - enhanced actual performance;

(f) using words as powerful tools of suggestion and understanding.

4.22 In a feature in *Personnel Management*, July 1992, a NLP devised for IT consultants at KPMG Management Consulting was described. A two-day course aimed 'to develop mental processing techniques, to refine information gathering techniques and to aid the development of new behavioural skills, including the ability to develop rapport with others'.

NLP: training breakthrough?

'Staff trained by NLP techniques acquire flexibility of thinking, choice of behaviour and control over their feelings; they are better equipped to handle themselves and others and to produce operational results. NLP also enables people to measure their results through sensory processes, giving them a very real sense of what they are achieving.'

Or deception?

'This raises ... questions about the extent to which people should appear to be something they are not in order to achieve a particular result. Much the same question could be levelled at job applicants; how far should they go at a selection interview to persuade the interviewer they are an appropriate choice and would fit into the organisation?

NLP does not aim to resolve people's ethical problems; it aims to give them better quality information and more flexibility so they may make their own choices about behaviour.'

Or manipulation?

'Possibly because of the speed with which NLP produces results, some people have regarded it warily and are concerned that it is potentially manipulative. In reality these fears are unfounded; of course, NLP could be misused, but so could any other technique you might consider. The concern tends to arise when there is a confusion between influencing and manipulation. We all influence others all the time; it is not possible to exist and not exert an influence on those around us.'

Carol Harris - director of the Association for NLP, *Personnel Management*, July 1992.

5. VALIDATION AND EVALUATION OF TRAINING SCHEMES

5.1. Implementation of the training scheme is not the end of the story. The scheme should be validated and evaluated.

(a) *Validation* means observing the results of the course, and measuring whether the training objectives have been achieved.

(b) *Evaluation* means comparing the actual costs of the scheme against the assessed benefits which are being obtained. If the costs exceed the benefits, the scheme will need to be re-designed or withdrawn.

5.2 There are various ways of validating a training scheme.

(a) *Trainee reactions to the experience:* asking the trainees whether they thought the training programme was relevant to their work, and whether they found it useful. This form of monitoring is rather inexact, and it does not allow the training department to measure the results for comparison against the training objective.

(b) *Trainee learning:* measuring what the trainees have learned on the course, perhaps by means of a test at the end of a course.

(c) *Changes in job behaviour following training:* studying the subsequent behaviour of the trainees in their jobs to measure how the training scheme has altered the way they do their work. This is possible where the purpose of the course was to learn a particular skill.

(d) *Organisational change as a result of training:* finding out whether the training has affected the work or behaviour of other employees not on the course - seeing whether there has been a general change in attitudes arising from a new course in, say, computer terminal work. This form of monitoring would probably be reserved for senior managers in the training department.

(e) *Impact of training on organisational goals:* seeing whether the training scheme (and overall programme) has contributed to the overall objectives of the organisation. This too is a form of monitoring reserved for senior management, and would perhaps be discussed at board level in the organisation. It is likely to be the main component of a cost-benefit analysis.

5.3 Validation is thus the measurement of terminal behaviour (trained work performance) in relation to training objectives.

6. MANAGEMENT TRAINING AND DEVELOPMENT

6.1 You might subscribe to the trait theory of leadership, that some individuals are born with the personal qualities to be a good manager, and others aren't. There might be some bits of truth in this view, but very few individuals, if any, can walk into a management job and do it well without some guidance, experience or training.

6.2 In every organisation, there should be some arrangement or system whereby:

(a) managers gain *experience*, which will enable them to do another more senior job in due course of time;

(b) subordinate managers are given *guidance* and *counselling* by their bosses;

(c) managers are given suitable *training* and *education* to develop their skills and knowledge; and

(d) managers are enabled to plan their future and the opportunities open to them in the organisation.

If there is a planned programme for developing managers, it is called a *management development programme*. Note, however, that even such a programme requires the support of the manager himself: self-learning - creating and using learning opportunities in the job - is an important part of management development.

6.3 Drucker has suggested that management development should be provided for all managers, not just the ones who are considered promotable material. 'The promotable man concept focuses on one man out of ten - at best one man out of five. It assigns the other nine to limbo. But the men who need management development the most are not the balls of fire who are the ... promotable people. They are those managers who are not good enough to be promoted but not poor enough to be fired. Unless they have grown up to the demands of tomorrow's jobs, the whole management group will be inadequate, no matter how good ... the promotable people. The first principle of manager development must therefore be the development of the entire management group'.

On the other hand, Handy noted that '... it remains true that career planning in many organisations is not a development process so much as a weeding-out process'.

6.4 The *Financial Times*, 13 November 1991, put forward four possible views of management training. (Note: these are extreme 'types' - not real people!)

(a) *Cynics* believe that it is a waste of time, and despise the resulting Smart-Alecs and 'course junkies'. They believe that management practices are either learnt through experience, or cannot be taught. Many believe that people are basically 'untrainable' anyway: 'They certainly do not take seriously the "proof" that training works, arguing that what can be taught is not important and, equally, what is teachable is not relevant.' Moreover, they subscribe to the view that: 'Those who can, manage; those who can't, become management trainers.'

(b) *Sceptics* are less hostile but not entirely convinced. They believe that training can help - but that not all training courses are clear or helpful in practice, and even if they are, the benefits tend to wear off. 'Back in the work place, the idealistic practices are ignored or even punished and hence discontinued. Most believe the solution lies in selecting people who are already well-trained or at least trainable.'

(c) *Enthusiasts* 'simply cannot see how people are expected to manage without being explicitly taught and trained'. They both reward training attendance, and use it as a reward. They take training needs audits, course appraisal and follow-up very seriously.

(d) *Naive proponents* 'are proselytisers of the near miraculous benefits of such-and-such a course, test, guru or concept. If only, they argue, people were to go on a course, understand and live its message, all would be well with the organisation.' They innocently embrace personal testimony and glossy brochure claims for courses.

6.5 It is therefore worth getting clear in your own mind why management training and development are needed.

(a) The prime objective of management development is improved performance capacity - from the managers *and* those they manage.

(b) Management development secures management succession: a pool of promotable individuals in the organisation.

(c) An organisation should show an interest in the career development of its staff, so as to motivate them and encourage them to stay with the firm, especially if it intends to rely on internal promotion from lower ranks to fill senior management positions.

(d) Individual managers should be encouraged to 'own' their own careers and personal development plans.

Exercise 5

'Good managers make good managers.' Discuss the validity of this proposition with particular reference to the benefits to be derived from formal management training/development programmes.

(ICSA December 1986)

Solution

The statement 'good managers make good managers' may be taken to mean either:

(a) that individuals with potential for promotion should be promoted. Those who are successful in one management position are likely to be successful in higher positions: they have experience, their abilities and aptitudes have been brought out under pressure, they have tasted the satisfactions available through responsibility etc; or

(b) that the main influence on management development is the way in which they themselves are managed - ie the management of managers: formal training may be considered of limited relevance.

Either interpretation could be discussed. Reference should be made to the benefits of training programmes, and particularly to their effectiveness in addressing problems such as application of theoretical knowledge to the work situation, adjustment to work after the 'euphoria' of training exercises and so on.

Management development, education and training

6.6 Approaches to management development fall into three main categories.

(a) *Management education* - study for an MBA (Masters in Business Administration) degree or DMS (Diploma in Management Studies), for example.

(b) *Management training* - largely off-the-job formal learning activities.

(c) *Experiential learning* - learning by doing.

The last of these categories has gained in popularity in recent years, with methods such as Management by Objectives being claimed to offer a higher degree of relevance to performance, and 'ownership' by the trainee.

Constable and McCormick, however, in their influential report 'The Making of British Managers', suggest that there is still a place for a systematic education and training programme featuring study for professional qualifications and design of in-house training courses. Such a programme is the basis of efforts to ensure that managers are (and can demonstrate that they are) properly trained, through the *Management Charter Initiative* (MCI), a government backed body which seeks to improve standards of management training through the competence-based approach.

6.7 The report by Constable and McCormick found from a survey of UK employers that:

(a) most employers regard innate ability and experience as the two key ingredients of an effective manager. But education and training help, especially in broadening the outlook of managers with only functional experience previously, and without experience of general management;

(b) there was agreement that it would be both inappropriate and impossible to make management a controlled profession similar to accountancy and law. However, making a managerial career more similar to the professions and having managers acquire specific *competences* appropriate to each stage of their career, were seen as beneficial.

6.8 A useful distinction between management education, training and development was given by Constable and McCormick.

(a) '*Education* is that process which results in formal qualifications up to and including post-graduate degrees.'

(b) *Training* is 'the formal learning activities which may not lead to qualifications, and which may be received at any time in a working career'; for example, a course in manpower forecasting or counselling skills;

(c) '*Development* is broader again: job experience and learning from other managers, particularly one's immediate superior, are integral parts of the development process.'

6.9 Recommendations of the report 'The Making of British Managers' included the following.

'Senior management

1 Create an atmosphere within the organisation where continuing management training and development is the norm.

2 Utilise appraisal procedures which encourage management training and development.

3 Encourage individual managers, especially by *making time available* for training.

4 Provide support to local educational institutes to provide management education and training (E & T).

5 Integrate in-house training courses into a wider system of management E & T. *Make the subject matter of in-house courses relevant to managers' needs*. Work closely with academic institutions and professional institutions (eg the ICSA or IPM) to ensure that the 'right' programmes are provided.

Individual managers

1 Actively want and seek training and development. 'Own' their own career

2 Recognise what new skills they require, and seek them out positively

3 Where appropriate, join a professional institute and seek to qualify as a professional member'

6.10 Although these recommendations focus on 'E and T', it is important to realise that this no longer implies bookwork, academic and theory-based studies. W A G Braddick *(Management for Bankers)* suggests the kind of shift in focus that has occurred in management development methods in recent years. (The notes are ours.)

			Notes
Principles	→	specifics	(Every organisation is unique)
Precepts	→	analysis/diagnosis	(Address the issues)
Theory-based	→	action-centred	(Understand it - but do it)
Academic	→	real time problems	(Tackle 'live' problems)
Functional focus	→	issue and problem focus	(Deal with 'whole' activities)
Excellent	→	team members and individual leaders	(Develop people - together)
Patient	→	agent	(Learn actively, take control)
One-off	→	continuous	(Keep learning)

6.11 Thus management education and training now tends to focus on the real needs of specific organisations, and to be grounded in practical skills. In-house programmes and on-the-job techniques have flourished, as have techniques of off-the-job learning which simulate real issues and problems: case study, role play, desk-top exercises, leadership exercises and so on.

6.12 Constable and McCormick reported in 1987 that: 'The total scale of management training is currently at a very low level. The general situation will only improve when many more companies conscientiously embrace a positive plan for management development. This needs to be accompanied by strong demand on the part of individual managers for continuing training and development throughout their careers.'

6.13 On the other hand, a report in *Personnel Management Plus* (September 1992) suggests that the UK now leads spending on management training, with a rise of 15% (compared to 14% in Germany and only 2% in Belgium, for example). Of total spending, 37% went on internally-run programmes, 35% on in-house programmes run by consultants, and 28% on external courses through training organisations. 'Managing people' and 'Communication skills' were the most popular courses.

The HR view?

'Management training is now a huge industry. Consultancy companies, magazines and human resource departments are exclusively dedicated to teaching people how to manage more effectively and efficiently.

The newly inducted, the freshly promoted, the diagnosed incompetent as well as high flyers are sent on courses on such subjects as time management, communication skills and finance for non-specialists.

But do these courses work? Can they be measured in some way and be shown to have a desirable effect?

A lot of sweat, tears and ink has been spilled over this apparently simple question. Hard-headed types want evidence that the expense is justified by increased productivity and revenue.

On the other hand, human resources managers seem happy enough if they get the feeling, through ratings on a feedback form, that participants have 'enjoyed' the course.

Financial Times 13 November 1991

Career development

6.14 Note that management development includes *career development* and succession planning by the organisation. This will require attention to a number of matters.

(a) The types of experience a potential senior manager will have to acquire. It may be desirable for a senior manager, for example, to have experience of:

(i) both line and staff/specialist management - in order to understand how authority is effectively exercised in both situations, and the potentially conflicting cultures/objectives of the two fields;

(ii) running a whole business unit (of whatever size) in order to develop a business, rather than a functional or sub-unit perspective. This is likely to be a vital transition in a manager's career, from functional to general management;

(iii) dealing with head office, from a subsidiary management position - in order to understand the dynamics of centralised/decentralised control;

(iv) international operations - if the organisation is in (or moving into) the international arena, say, in post-1992 Europe. Understanding of cultural differences is crucial to effective strategic and man management;

(v) other disciplines and organisations. Some consultancies and banks, for example, offer secondment with business organisations, development agencies or the civil service. Organisations often encourage potential high-fliers (and sometimes personnel specialists) to gain experience in different areas of the business.

(b) The individual's guides and role models in the organisation. It is important that an individual with potential should measure himself against peers - assessing weaknesses and strengths - and emulate role models, usually superiors who have already 'got what it takes' and proved it. Potential high fliers can be fast tracked by putting them under the guidance of effective motivators, teachers and power sources in the organisation.

At any stage in a career, a *mentor* will be important. The mentor may occupy a role as the employee's teacher/coach/trainer, counsellor, role model, protector or sponsor/ champion, spur to action or improvement, critic and encourager. In May 1993 the FT reported that 40% of British companies now have a mentoring scheme, and a further 20% are thinking about creating one.

(c) The level of opportunities and challenges offered to the developing employee. Too much responsibility too early can be damagingly stressful, but if there is not *some* degree of difficulty, the employee may never explore his full potential and capacity.

Exercise 6

'Does getting wet, cold and generally miserable in the countryside help you to become a better manager? Supporters of outward bound courses believe it does, but critics question the value of having highly-paid and specialised executives tramping around the woods honing boy scout-level skills.

Is outdoor training not just a way of keeping ageing physical exercise teachers, sadistic ex-corporals and overpaid consultants employed? And is it just an expensive fad in training, no better or worse than classroom teaching?' *Financial Times*, January 1993

Give your own views on these questions, in the light of everything you have read so far in this Study Text.

Solution

The FT article goes on to give three arguments in favour of this type of training.

(a) Experimental versus theoretical learning. Since the 1960s, when encounter groups thrived, trainers have claimed that real learning occurs when people are put in difficult, novel and problematic situations and not when they study elaborate theories or abstract ideas.

(b) Emotions not ideas. Most training courses are about ideas, concepts, skills and models. They involve brain work and traditional classroom activities. But outdoor trainers claim that modern management is as much about self-confidence and courage.

(c) Team membership not leadership. There are plenty of leadership courses but not too many for those who have to follow. Despite mouthing platitudes about teams and team work, Anglo-Americans come from an individualistic, not a collective, culture. Team work does not come easily or naturally.

Chapter roundup

- A systematic approach to training can be illustrated in a flowchart as follows.

Training

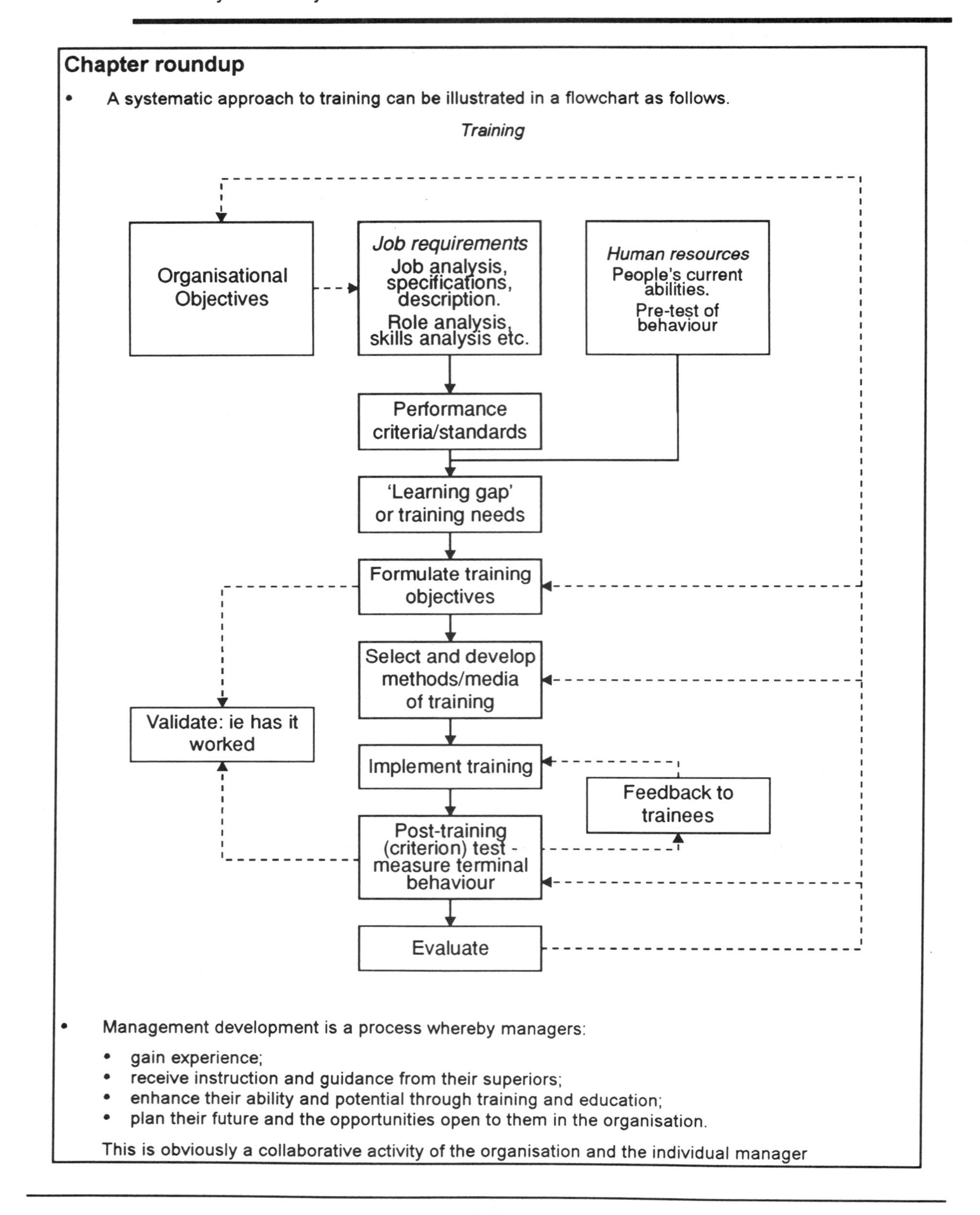

- Management development is a process whereby managers:
  - gain experience;
  - receive instruction and guidance from their superiors;
  - enhance their ability and potential through training and education;
  - plan their future and the opportunities open to them in the organisation.

  This is obviously a collaborative activity of the organisation and the individual manager

Test your knowledge

1 How do 'training needs' arise in an organisation? How would you carry out a 'training needs analysis' if required to do so? *(ICSA June 1987)* (see paras 2.2 - 2.7)

2 Outline the role of:

(a) motivation;
(b) feedback; and
(c) 'reinforcement'

in the learning process. (3.1, 3.2)

3 Draw the 'experiential learning curve' designed by Kolb. Choose any task you have done today, and show how the learning cycle would help you to do it better next time. (3.5)

4 What are the advantages and disadvantages of 'on-the job' training methods? (4.7, 4.8)

5 Outline the process of:

(a) induction; (4.12)
(b) coaching. (4.14)

6 What are the objectives of management development? (6.5)

7 Distinguish between management education, training and development, with examples of methods for each. (6.6)

8 What is a mentor? (6.14)

Now try question 11 at the end of the text

Chapter 12

PERFORMANCE APPRAISAL AND COUNSELLING

This chapter covers the following topics.

1. The purposes of appraisal
2. Appraisal procedures
3. Problems with appraisal
4. The identification of potential
5. Counselling

Signpost

- Performance appraisal can identify the need for training, discussed in the previous chapter.
- Motivation is dealt with in Chapter 13; interviewing and questioning were explained in Chapter 10; leadership qualities were discussed in Chapter 6.

1. THE PURPOSES OF APPRAISAL

1.1 The general purpose of any staff appraisal system is to improve the efficiency of the organisation by ensuring that the individuals within it are performing to the best of their ability and developing their potential for improvement. Within this overall aim, staff appraisals are used in practice for:

(a) *reward review* - measuring the extent to which an employee is deserving of a bonus or pay increase as compared with his peers;

(b) *performance review*, for planning and following-up training and development programmes, ie identifying training needs, validating training methods and so on; and

(c) *potential review*, as an aid to planning career development and succession, by attempting to predict the level and type of work the individual will be capable of in the future.

1.2 In his book *Human resource management* (1988) George Thomason identifies the variety of objectives of appraisals.

(a) Establishing what actions are required of the individual in a job in order that the objectives for the section or department are realised.

(b) Establishing the key or main results which the individual will be expected to achieve in the course of his or her work over a period of time.

(c) Assessing the individual's level of performance against some standard, to provide a basis for remuneration above the basic pay rate.

(d) Identifying the individual's levels of performance to provide a basis for informing, training and developing him.

(e) Identifying those persons whose performance suggests that they are promotable at some date in the future and those whose performance requires improvement to meet acceptable standards.

(f) Establishing an inventory of actual and potential performance within the undertaking to provide a basis for manpower planning.

(g) Monitoring the undertaking's initial selection procedures against the subsequent performance of recruits, relative to the organisation's expectations.

(h) Improving communication about work tasks between different levels in the hierarchy.

1.3 Whatever the purpose of appraising staff in a particular situation, the review should be a systematic exercise, taken seriously by assessor and subject alike. It may be argued that such deliberate stock-taking is unnecessary, since managers are constantly making judgements about their subordinates and (should be) giving their subordinates feedback information from day to day.

However, it must be recognised that:

(a) managers may obtain random impressions of subordinates' performance (perhaps from their more noticeable successes and failures), but rarely form a coherent, complete and objective picture;

(b) they may have a fair idea of their subordinates' shortcomings - but may not have devoted time and attention to the matter of improvement and development;

(c) judgements are easy to make, but less easy to justify in detail, in writing, or to the subject's face;

(d) different assessors may be applying a different set of criteria, and varying standards of objectivity and judgement, which undermines the value of appraisal for comparison, as well as its credibility in the eyes of the appraisees;

(e) unless stimulated to do so, managers rarely give their subordinates adequate feedback on their performance, especially if the appraisal is a critical one.

1.4 There is clearly a need for a system which tackles the basic problems of:

(a) the formulation and appreciation of desired traits and standards against which individuals can be consistently and objectively assessed. Assessors must be aware of factors which affect their judgements;

(b) recording assessments. Managers should be encouraged to utilise a standard and understood framework, but still allowed to express what they consider important, and without too much form-filling; and

(c) getting the appraiser and appraisee together, so that both contribute to the assessment and plans for improvement and/or development.

Exercise 1

Why is appraisal necessary if employees are being given day to day feedback on their performance?

Solution

This is just to check that you were paying attention in the last few paragraphs.

---

2. APPRAISAL PROCEDURES

2.1 A typical system would therefore involve:

(a) identification of *criteria* for assessment, perhaps based on job analysis, performance standards, person specifications and so on;

(b) the preparation by the subordinate's manager of an *appraisal report*;

(c) an *appraisal interview*, for an exchange of views about the results of the assessment, targets for improvement, solutions to problems and so on;

(d) review of the assessment by the assessor's own superior, so that the appraisee does not feel subject to one person's prejudices. Formal appeals may be allowed, if necessary to establish the fairness of the procedure;

(e) the preparation and implementation of *action plans* to achieve improvements and changes agreed; and

(f) *follow-up:* monitoring the progress of the action plan.

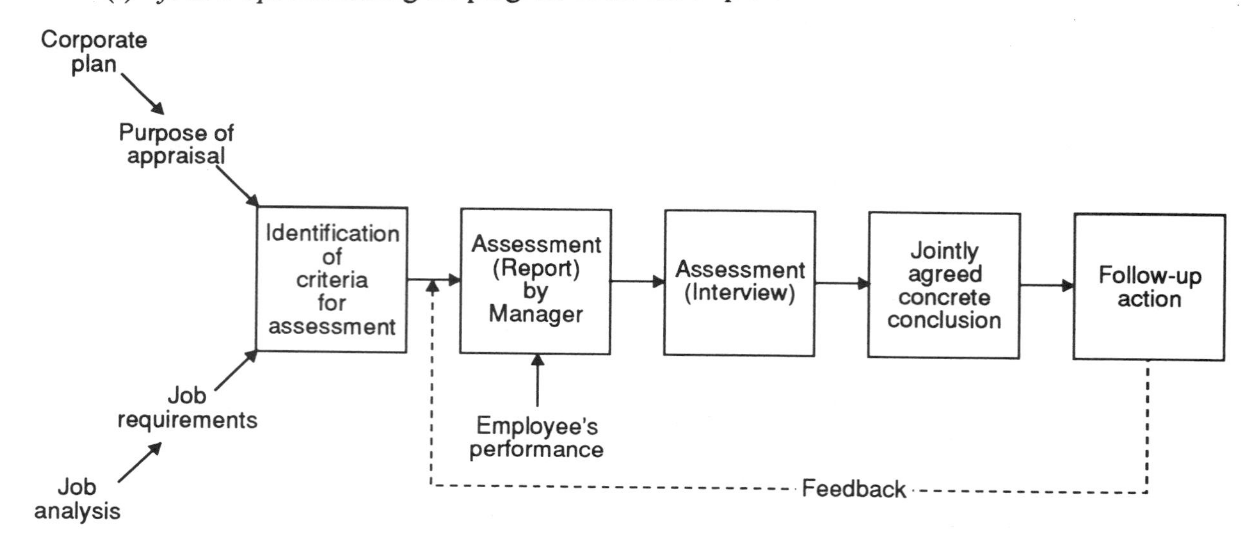

2.2 There may not need to be standard forms for appraisal - and elaborate form-filling procedures should be avoided - as long as managers understand the nature and extent of what is required, and are motivated to take it seriously. Most systems, however, provide for appraisals to be recorded, and report forms of various lengths and complexity may be designed for standard use.

The appraisal report

2.3 The basis of appraisal must first be determined. Assessments must be related to a common standard, in order for comparisons to be made between individuals: on the other hand, they should be related to meaningful performance criteria, which take account of the critical variables in each different job. A blanket approach may provide a common standard, but may not offer a significant index for job performance.

In particular, there is the question of whether *personality* or *performance* is being assessed: in other words, what the individual is, or what he does. According to Livy, 'Personal qualities have questionable validity as a measure of performance, and may introduce unreliability since they are prone to ambiguity and have moral connotations ... In practical terms, this ... has encouraged the use of results-based appraisals (such as Management By Objectives) and the development of job-related performance criteria.'

It is now generally considered that personality is only relevant where there is reason to suspect that it is a factor in poor performance: as long as the employee is achieving the desired results, personality characteristics are not a useful measurement.

2.4 Various appraisal techniques have been formulated.

(a) *Overall assessment*. This is the simplest method, simply requiring the manager to write in narrative form his judgements about the appraisee, possibly with a checklist of personality characteristics and performance targets to work from. There will be no guaranteed consistency of the criteria and areas of assessment, however, and managers may not be able to convey clear, effective judgements in writing. Kay Rowe studied several such schemes and concluded: 'A few suggested careful thought and a conscientious effort to say something meaningful, but the vast majority were remarkable for their neutrality. Glib, generalised, enigmatic statements abounded'.

(b) *Guided assessment.* Assessors are required to comment on a number of specified characteristics and performance elements, with guidelines as to how terms such as 'application', 'integrity' and 'adaptability' are to be interpreted in the work context. This is a more precise, but still rather vague method.

(c) *Grading*. Grading adds a comparative frame of reference to the general guidelines, whereby managers are asked to select one of a number of levels or degrees to which the individual in question displays the given characteristic. These are also known as *rating scales*, and used to be much used in standard appraisal forms. Their effectiveness depends to a large extent on:

(i) the relevance of the factors chosen for assessment. These may be nebulous personality traits, for example, or clearly defined job factors such as job knowledge, performance against targets, or decision-making;

(ii) the definition of the agreed standards of assessment. Grades A-D might simply be labelled 'Outstanding - Satisfactory - Fair - Poor', in which case assessments are subject to much variation and subjectivity. They may, on the other hand, be more closely related to work priorities and standards, using definitions such as 'Performance is good overall, and superior to that expected in some important areas', or 'Performance is broadly acceptable, but employee needs training in several major areas and/or motivation is lacking.'

Numerical values may be added to ratings to give rating scores. Alternatively a less precise *graphic scale* may be used to indicate general position on a plus/minus scale.

Factor: job knowledge

(d) *Behavioural incident methods*. These concentrate on employee behaviour, which is measured against typical behaviour in each job, as defined by common *'critical incidents'* of successful and unsuccessful job behaviour reported by managers. Time and effort are required to collect and analyse reports and to develop the scheme, and it only really applies to large groups of people in broadly similar jobs. However, it is firmly rooted in observation of real-life job behaviour, and the important aspects of the job, since the analysis is carried out for *key tasks*, (those which are identified as critical to success in the job and for which specific standards of performance *must* be reached).

The behavioural equivalent of the graphic scale (illustrated above) for a manager's key task of 'marketing initiative' might appear as, for example:

Produces no new ideas for marketing. Appears apathetic to competitive challenge	Produces ideas when urged by head office. Ideas not clearly thought out nor enthusiastically applied	Produces ideas when urged by head office and gives full commitment to new programmes	Spontaneously generates new ideas for marketing and champions them through head office approval. Ideas related to identified needs and effective in practice

(e) *Results-orientated schemes*. All the above techniques may be used with more or less results-orientated criteria for assessment - but are commonly based on trait or behavioural appraisal. A wholly results-orientated approach (such as Management by Objectives) sets out to review performance against specific targets and standards of performance agreed in advance by manager and subordinate together. The advantages of this are that:

(i) the subordinate is more involved in appraisal of his own performance, because he is able to evaluate his success or progress in achieving specific, jointly-agreed targets;

(ii) the manager is therefore relieved, to an extent, of his role as critic, and becomes a counsellor. A primarily *problem-solving* approach (what does the employee require in order to do his job better?) may be adopted;

(iii) learning and motivation theories suggest that clear and known targets are important in modifying and determining behaviour.

The effectiveness of the scheme will still, however, depend on the targets set (are they clearly defined? realistic?) and the commitment of both parties to make it work. The measurement of success or failure is only part of the picture: reasons for failure and opportunities arising from success must be evaluated.

2.5 You will learn more about Management by Objectives in your later studies. Briefly, on an *individual* level, it involves the following.

(a) Identification of job aims

(b) Identification of key results which are critical to achieving job aims, and the key tasks involved in obtaining them

(c) Agreement on performance standards for the key tasks

(d) Setting short-term goals and action plans (as part of an improvement plan, to achieve agreed standards)

(e) Progress monitoring and control

2.6 We will look at some of the pitfalls facing managers in making their assessments later. However, at this written stage of the assessment process, managers will need guidance, to help them make a relevant, objective and helpful report. Training sessions may be arranged. Most large organisations with standard review forms also issue detailed *guidance notes* to aid assessors with the written and discussion elements.

Some notes which might refer to the written report element might include the following.

(a) The immediate superior of each individual should write his report.

(b) The report should be based on a job description prepared or agreed in consultation with the individual being reported on, as a fair reflection of his duties and responsibilities.

(c) Spaces for 'comments' on this form should be used to explain grades awarded, and to describe special circumstances which have affected performance (such as ill-health).

(d) Grades should reflect actual performance during the past twelve months. There should be no reference to previous reports.

(e) Reports are strictly confidential: their contents may only be disclosed to persons concerned with them in the course of official work.

(f) Each report should be countersigned by a departmental manager or other appropriate executive, with additional comment where appropriate.

(g) Report writers have a continuous responsibility to give performance information to their subordinates, positive and negative. No grading 'below average' should be recorded without the subordinate having received prior warning and opportunity to correct shortcomings.

Personnel Appraisal: Employees in Salary Grades 5-8

Date of Review	Time on Position Yrs. Mths.	S.G.	Age Yrs.	Name
Period of Review	Position Title			Area

Important: Read guide notes carefully before proceeding with the following sections

Section One	N/A	U	M	SP	E	O	Section Two	1	2	3	4	5
Performance Factors							Personal Characteristics					
Administrative Skills							Initiative					
Communications – Written							Persistence					
Communications – Oral							Ability to work with others					
Problem Analysis							Adaptability					
Decision Making							Persuasiveness					
Delegation							Self-Confidence					
Quantity of Work							Judgement					
Development of Personnel							Leadership					
Development of Quality Improvements							Creativity					

Section Three Highlight Performance Factors and particular strengths/weaknesses of employee which significantly affect Job Performance

Overall Performance Rating (Taking into account ratings given)

Prepared by: Signature .... Date .... Position Title ....

Section Four Comments by Reviewing Authority

I R Review Initial

Signature .... Date .... Position Title .... Date

Section Five Supervisor's Notes on Counselling Interview

Signature .... Date .... Position Title ....

Section Six Employees Reactions and Comment

Signature .... Date ....

Performance Classification

Outstanding performance is characterised by high ability which leaves little or nothing to be desired.

Personnel rated as such are those who regularly make significant contributions to the organisation which are above the requirements of their position. Unusual and challenging assignments are consistently well handled.

Excellent performance is marked by above-average ability, with little supervision required.

These employees may display some of the attributes present in **'outstanding'** performance, but not on a sufficiently consistent basis to warrant that rating. Unusual and challenging assignments are normally well handled.

Satisfactory Plus performance indicates fully adequate ability, without the need for excessive supervision.

Personnel with this rating are able to give proper consideration to normal assignments, which are generally well handled. They will meet the requirements of the position. **'Satisfactory plus'** performers may include those who lack the experience at their current level to demonstrate above-average ability.

Marginal performance is in instances where the ability demonstrated does not fully meet the requirements of the position, with excessive supervision and direction normally required.

Employees rated as such will show specific deficiencies in their performance which prevent them from performing at an acceptable level.

Unsatisfactory performance indicates an ability which falls clearly below the minimum requirements of the position.

'Unsatisfactory' performers will demonstrate marked deficiencies in most of the major aspects of their responsibilities, and considerable improvement is required to permit retention of the employee in his current position.

Personal Characteristics Ratings

1. – Needs considerable improvement – substantial improvement required to meet acceptable standards.
2. – Needs improvement – some improvement required to meet acceptable standards.
3. – Normal – meets acceptable standards.
4. – Above normal – exceeds normally acceptable standards in most instances.
5. – Exceptional – displays rare and unusual personal characteristics.

Exercise 2

Study the appraisal form on the previous page. This is based on a genuine example reproduced in a book published in 1982. Can you suggest any ways in which the form could be improved for the 1990s?

Solution

Amongst the points you might have made are the following.

(a) Some of the wording has an old-fashioned ring to it ('salary grades 5-8', 'reviewing authority')

(b) Many organisations have now dispensed with rating scales on the grounds that they are contentious, and they don't adequately distinguish attributes that are very significant from those that are less so.

(c) The term 'weaknesses' is used whereas many organisations would use a term like ' areas for improvement'.

There are other points to be made. Alternatively you may think the form is fine as it is, and these objections are merely faddish 'political correctness'.

Interview and counselling

2.7 The extent to which any discussion or counselling interview is based on the written report varies in practice. For certain purposes - say, mutually agreed programmes for further training of the individual under review - the report may be distributed to the appraisee in advance of the interview, so that he has a chance to make an independent assessment for discussion with his manager.

2.8 Maier *(The Appraisal Interview)* identifies three types of approach to appraisal interviews.

(a) The *tell and sell* method. The manager tells the subordinate how he has been assessed, and then tries to 'sell' (gain acceptance of) the evaluation and the improvement plan. This requires unusual human relations skills in order to convey constructive criticism in an acceptable manner, and to motivate the appraisee to alter his behaviour.

(b) The *tell and listen* method. The manager tells the subordinate how he has been assessed, and then invites him to respond. The manager therefore no longer dominates the interview throughout, and there is greater opportunity for counselling as opposed to pure direction. The employee is encouraged to participate in the assessment and the working out of improvement targets and methods: it is an accepted tenet of behavioural theory that participation in problem definition and goal setting increases the individual's commitment to behaviour and attitude modification. Moreover, this method does not assume that a change in the employee will be the sole key to improvement: the manager may receive helpful feedback about how job design, methods, environment or supervision might be improved. Again, however, the interviewer needs to be a talented and trained listener and counsellor.

(c) The *problem-solving* approach. The manager abandons the role of critic altogether, and becomes a counsellor and helper. The discussion is centred not on the assessment, but on the employee's work problems. The employee is encouraged to think solutions through, and to commit himself to the recognised need for personal improvement. This approach encourages intrinsic motivation through the element of self-direction, and the perception of the job itself as a problem-solving activity. It may also stimulate creative thinking on the part of employee and manager alike, to the benefit of the organisation's adaptability and methods. Again, the interviewer will require highly-developed skills, and the attitudes of both parties to the process will need to be got right.

Example: appraisal interviews

2.9 The Behavioural Sciences Research Division of the Civil Service Department carried out a survey of appraisal interviews given to 252 officers in a government department, in 1973. Findings included the following.

(a) Interviewers have difficulty with negative performance feedback, and tend to avoid it if possible.

(b) Negative performance feedback, however, is significant in the effective conduct and follow-up of interviews. It is more likely to bring forth positive post-appraisal action, and is favourably received by appraisees, who feel it is the most useful function of the whole process, if handled frankly and constructively.

(c) The most common fault of interviewers is talking too much. The survey recorded the preference of appraisees for a 'problem-solving' style of participative interview, over a one-sided 'tell and sell' style.

2.10 Cuming suggests that, unlike the conventionally-used attitude surveys which concentrate on the compensatory aspects of employment (on the basic assumption that work is intrinsically unpleasant, and that management's task is to render it more satisfactory by pay, fringe benefits, friendly treatment, pride in the company and so on), the appraisal interview should ask positive and thought-provoking questions such as:

(a) Do you fully understand your job? Are there any aspects you wish to be made clearer?

(b) What parts of your job do you do best?

(c) Could any changes be made in your job which might result in improved performance?

(d) Have you any skills, knowledge, or aptitudes which could be made better use of in the organisation?

(e) What are your career plans? How do you propose achieving your ambitions in terms of further training and broader experience?

Follow-up

2.11 After the appraisal interview, the manager may complete his report, with an overall assessment, assessment of potential and/or the jointly-reached conclusion of the interview, with recommendations for follow-up action.

The manager should then discuss the report with the counter-signing manager (usually his own superior), resolving any problems he has had in making the appraisal or report, and agreeing on action to be taken. The report form may then go to the management development adviser, training officer or other relevant people as appropriate for follow-up.

2.12 *Follow-up* procedures will include:

(a) informing appraisees of the results of the appraisal, if this has not been central to the review interview. Some people argue that there is no point making appraisals if they are not openly discussed, but unless managers are competent and committed to reveal results in a constructive, frank and objective manner, the negative reactions on all sides may outweigh the advantages;

(b) carrying out agreed actions on training, promotion and so on;

(c) monitoring the appraisee's progress and checking that *he* has carried out agreed actions or improvements;

(d) taking necessary steps to help the appraisee to attain improvement objectives, by guidance, providing feedback, upgrading equipment, altering work methods or whatever.

Upward appraisal

2.13 A notable modern trend, adopted in the UK by companies such as BP, British Airways Central TV and others, is upward appraisal, whereby employees are not rated by their superiors but by their subordinates. The followers appraise the leader.

2.14 The advantages of this method were set out as follows in an article by Adrian Furnham in the *Financial Times* (March 1993).

(a) Subordinates tend to know their superior better than superiors know their subordinates. They see their bosses and know their moods, foibles and preferences, their adequacies, skills, strengths and limitations and things they do and do not like doing.

(b) As all subordinates rate their managers statistically, these ratings tend to be more reliable - the more subordinates the better. Instead of the biases of individual managers' ratings, the various ratings of the employees can be converted into a representative view. If the employees have very differing views of their bosses this can present problems, but represents very significant data meriting further investigation.

(c) Subordinates' ratings have more impact because it is more unusual to receive ratings from subordinates. It is also surprising to bosses because, despite protestations to the contrary, information often flows down organisations more smoothly and comfortably than it flows up. When it flows up it is qualitatively and quantitatively different. It is this difference that makes it valuable.

2.15 Problems with the method include fear of reprisals, vindictiveness, and extra form processing. Some bosses in strong positions might refuse to act, even if a consensus of staff suggested that they should change their ways.

Exercise 2

Look up the procedures manual of your organisation, and read through your appraisal procedures. Also get hold of any documentation related to them; the appraisal report form and notes, in particular.

How effective do you think your appraisal procedures are? Measure them against the criteria given above. How do you *feel* about appraisal interviews?

If you can get hold of an appraisal report form, have a go at filling one out for yourself - a good exercise in self-awareness!

3. PROBLEMS WITH APPRAISAL

3.1 In theory, such appraisal schemes may seem very fair to the individual and very worthwhile for the organisation, but in practice the system often goes wrong.

(a) Appraisal interviews are often defensive on the part of the subordinate, who believes that criticism may herald financial disadvantage or lost promotion opportunity, and on the part of the superior, who cannot reconcile the role of judge and critic with the human relations aspect of the interview and may in any case feel uncomfortable about 'playing God' with the employee's future.

(b) The superior might show conscious or unconscious bias in his report or may be influenced by his rapport with the interviewee, in the face-to-face phase. Systems without clearly-defined standard criteria will be particularly prone to the subjectivity of the assessor's judgements.

(c) The manager and subordinate may both be reluctant to devote time and attention to appraisal. Their experience in the organisation may indicate that the exercise is a waste of time (especially if there is a lot of form-filling) with no relevance to the job, and no reliable follow-up action.

(d) The organisational culture may simply not take appraisal seriously: interviewers are not trained or given time to prepare, appraisees are not encouraged to contribute, or the exercise is perceived as a 'nod' to Human Relations with no practical results.

> Kay Rowe conducted a study of appraisal systems, based on 1,440 completed appraisal forms from six organisations. In a famous article ('An appraisal of appraisals', *Journal of Management Studies*, 1964) which seems to have lost none of its relevance with age she concluded that:
>
> (a) appraisers are reluctant to appraise;
> (b) interviewers are reluctant to interview; and
> (c) there is no follow up.

Appraisal: good or bad for motivation?

3.2 The affect of appraisal on motivation is a particularly tricky issue.

(a) Feedback on performance is regarded as vital in motivation, because it enables an employee to evaluate his achievement and make future calculations about the amount of effort required to achieve objectives and rewards. Even negative feedback can have this effect - and is more likely to spur the employee on to post-appraisal action.

(b) Agreement of challenging but attainable targets for performance or improvement also motivates employees by clarifying goals and the value (and 'cost' in terms of effort) of incentives offered.

(c) A positive approach to appraisal allows employees to solve their work problems and apply creative thinking to their jobs.

3.3 However, people rarely react well to criticism - especially at work, where they may feel that their reward or even job security is on the line. In addition, much depends on the self-esteem of the appraisee.

(a) If the appraisee has a high self-image, he may be impervious to criticism: he will be able to deflect it - and the greater the criticism, the harder he will work to explain it away. If such a person is *not* criticised, he will be confirmed in his behaviour and sense of self-worth, which will motivate him to continue as he is: this is fine if he is doing a good job, but *not* if he is doing a bad job, and simply being given a 'soft' appraisal.

(b) If the appraisee has a low self-image, he may be encouraged by low levels of criticism, and this may help to improve his performance. Heavy criticism of a person of low self-esteem can, however, be psychologically damaging.

Improving the system

3.4 The appraisal scheme should itself be assessed (and regularly re-assessed) according to the following general criteria.

(a) *Relevance*

(i) Does the system have a useful purpose, relevant to the needs of the organisation and the individual?

(ii) Is the purpose clearly expressed and widely understood by all concerned, both appraisers and appraisees?

(iii) Are the appraisal criteria relevant to the purposes of the system?

(b) *Fairness*

(i) Is there reasonable standardisation of criteria and objectivity throughout the organisation?

(ii) Is there reasonable objectivity?

(c) *Serious intent*

(i) Are the managers concerned committed to the system - or is it just something the personnel department thrusts upon them?

(ii) Who does the interviewing, and are they properly trained in interviewing and assessment techniques?

(iii) Is reasonable time and attention given to the interviews - or is it a question of 'getting them over with'?

(iv) Is there a genuine demonstrable link between performance and reward or opportunity for development?

(d) *Co-operation*

(i) Is the appraisal a participative, problem-solving activity - or a tool of management control?

(ii) Is the appraisee given time and encouragement to prepare for the appraisal, so that he can make a constructive contribution?

(iii) Does a jointly-agreed, concrete conclusion emerge from the process?

(iv) Are appraisals held regularly?

(e) *Efficiency*

(i) Does the system seem overly time-consuming compared to the value of its outcome?

(ii) Is it difficult and costly to administer?

3.5 The personnel function has a role to play in encouraging line management to carry out systematic appraisal, and in overcoming some of the causes of managerial reluctance to appraise.

(a) Education in the potential benefits of and constructive approaches to appraisal may help, starting to build a culture where appraisal is perceived to be a primary problem-solving tool and keystone of managerial effectiveness.

(b) Personnel should design, and instruct managers in the use of, workable procedures and documentation for appraisal.

(i) Standard review forms, based on relevant and specific standards of assessment, might be provided.

(ii) Grades and ratings should be related to work priorities and standards and clearly defined.

(iii) Results-oriented schemes might be designed, relieving the manager to an extent of his role as critic and encouraging co-operative problem-solving and post-appraisal action.

(iv) Training sessions may be organised to help assessors. Most large organisations with standard review forms also issue detailed guidance notes to aid assessors.

3.6 However, a change in attitudes and practice such as would be required in many organisations will not be easy to achieve, particularly since part of the attitude problem is likely to be the feeling that appraisal is being imposed on line managers by the personnel department, which does not understand the operational difficulties. Ultimately, the cultural change may have to come from the personnel department's own practices: the personnel manager will have to lead by example, to show that it can be done.

Exercise 3

In an article in *Personnel Management* in September 1993 Clive Fletcher writes about 'the break-up of the traditional, monolithic approach' to performance appraisal in response to modern trends in organisations. Given that this Study Text has frequently referred to current developments, how do you think that these may be reflected in the field of performance appraisal?

Solution

This is by no means an easy question. Fletcher identifies trends such as performance management, which 'places performance appraisal in a central role in a more integrated and dynamic set of HR policies'; delayering of management levels and increased uses of matrix or project management raising the question of *who* appraises; employee empowerment, requiring increased use of self-appraisal; upward appraisal, whereby subordinates appraise their managers; the adoption of the 'competence' framework for describing performance; TQM and continuous improvement, which implies that performance should attempt to go *beyond* the current limitations of the system. You may have had other ideas.

4. THE IDENTIFICATION OF POTENTIAL

4.1 The review of potential is the use of appraisal to forecast the direction in which an individual is progressing, in terms of his career plans and skill development, and at what rate. It can be used as feedback to the individual, to indicate the opportunities open to him in the organisation in the future. It will also be vital to the organisation in determining its management succession plans.

4.2 Information for potential assessment will include:

(a) strengths and weaknesses in existing skills and qualities;

(b) possibilities and strategies for improvement, correction and development;

(c) the goals, aspirations and attitudes of the appraisee, with regard to career advancement, staying with the organisation and handling responsibility;

(d) the opportunities available in the organisation, including likely management vacancies, job rotation/enrichment plans and promotion policies for the future.

4.3 No single review exercise will mark an employee down for life as 'promotable' or otherwise. The process tends to be an on-going one, with performance at each stage or level in the employee's career indicating whether he might be able to progress to the next step. However, this approach based on performance in the current job is highly fallible: hence the 'Peter principle' of L J Peter, who pointed out that managers tend to be promoted from positions in which they have proved themselves competent until one day they reach a level at which they are no longer competent - promoted 'to the level of their own incompetence'!

4.4 Moreover, the management succession plan of an organisation needs to be formulated in the long term: there is a long lead time involved in equipping a manager with the skills and experience needed at senior levels and the organisation must develop people if it is to fill the shoes of departing managers without crisis.

4.5 Some idea of *potential* must therefore be built into appraisal. It is impossible to predict with any certainty how successful an individual will be in what will, after all, be different circumstances from anything he has experienced so far. However, some attempt can be made to:

(a) determine key *indicators of potential*: in other words, elements believed to be essential to management success; and/or

(b) *simulate* the conditions of the position to which the individual would be promoted, to assess his performance.

Key indicators of potential

4.6 Various research studies (by employing organisations and by theorists) have been carried out into exactly what makes a successful senior manager (and could be identified in junior people to indicate that they might *become* successful senior managers).

The following are some of the factors identified.

(a) General effectiveness (track record in task performance and co-worker satisfaction).

(b) Administrative skills (planning and organising, making good decisions).

(c) Interpersonal skills or intelligence (being aware of others, making a good impression, persuading and motivating).

(d) Intellectual ability or analytical skills (problem-solving, mental agility).

(e) Control of feelings (tolerance of stress, ambiguity and so on).

(f) Leadership (variously defined, but demonstrated in follower loyalty and commitment).

(g) Imagination and intuition (for creative decision-making and innovation).

(h) 'Helicopter ability' (the ability to rise above the particulars of a situation, to see the whole picture, sift out key elements and conceive strategies - the ability to 'see the wood for the trees').

(i) Orientation to work (being motivated by work rather than non-work satisfactions; self-starting, rather than needing to be motivated by others).

(j) Team work (ability and willingness to co-operate with others).

(k) Taste for making money (empathy with the profit motive: ambition for self and business).

(l) 'Fit' (having whatever mix of all the above skills, abilities and experience the business organisation needs - being in the right place at the right time).

4.7 Various techniques can be used to measure these attributes, including:

(a) written tests (for intellectual ability);

(b) simulated desk-top tasks or case studies (for administrative skills, analytical and problem-solving ability);

(c) role play (for negotiating or influencing skills, conflict resolution or team working);

(d) leadership exercises (testing the ability to control the dynamics of a team towards work-related objectives);

(e) personality tests (for work orientation, motivation and so on);

(f) interviews (for interpersonal skills, attitudes and so on);

(g) presentations or speeches (for communication skills).

4.8 Note the use of *simulated* activity, such as case study or role play, to give potential managers experience of managerial tasks. An alternative approach might be to offer them *real* experience (under controlled conditions) by appointing them to assistant or deputy positions or to committees or project teams, and assessing their performance. This is still no real predictor of their ability to handle the *whole* job, on a continuous basis and over time, however, and it may be risky, if the appraisee fails to cope with the situation.

Assessment centres

4.9 *Assessment centres* are an increasingly used approach, growing out of the War Office Selection Board methods during the Second World War.

The purpose of the method is to assess potential and identify development needs, through various *group* techniques. It is particularly useful in the identification of executive or supervisory potential, since it uses simulated but realistic management problems, to give participants opportunities to show potential in the kind of situations to which they would be promoted, but of which they currently have no experience.

4.10 Trained assessors - usually line managers two levels above the participants, and perhaps consultant psychologists - use a variety of games, simulations, tests and group discussions and exercises. Observed by the assessors, participants may be required to answer questionnaires about their attitudes, complete written tests, prepare speeches and presentations, participate in group role-play exercises, work through simulated supervisory tasks, and undertake self-appraisal and peer-rating. They are assessed on a range of factors, such as assertiveness, energy, initiative and creativity, stress-tolerance, sensitivity, abilities in persuasion, communication, and decision-making.

An assessment report is then compiled from the assessors' observations, test scores and the participant's self-assessment. This is discussed in a feedback counselling interview.

4.11 Advantages of assessment centres include:

(a) a high degree of acceptability and user confidence; avoidance of single-assessor bias;

(b) reliability in predicting potential success (if the system is well-conducted);

(c) the development of skills in the assessors, which may be useful in their own managerial responsibilities;

(d) benefits to the assessed individual, including experience of managerial/supervisory situations, opportunity for self-assessment and job-relevant feedback, and opportunities to discuss career prospects openly with senior management.

4.12 The cost of the scheme must, however, be considered. A well-run centre with trained and practised assessors requires considerable expense of managerial time, and the time of participants, as well as the fixed costs of setting up the programme, whether it is designed internally or bought 'off the shelf'.

In addition, an article in *Personnel Management* (June 1991) suggests that there are a number of 'design shortcomings' which frequently occur, including:

(a) inadequate specification of the target competencies against which participants are assessed;

(b) exercises which bear little relation *either* to the competencies being assessed, or to the organisation's cultural practices;

(c) little or no assessor training and selection;

(d) inadequate selection and briefing of candidates, so that some are 'lost' from the start;

(e) inefficient programming and scheduling of the events;

(f) inadequate or non-existent follow-up action, feedback, counselling and implementation of recommendations.

Example: Rover

4.13 Rover's approach to ACs was reported in *Personnel Management* (June 1991). 'An internal study, comparing appointments made using ACs with those resulting from more traditional interview approaches, suggested that decisions arising from ACs were more robust and reliable.'

However, Rover experienced problems with:

(a) the absence of learning and development experience for trainees (as opposed to observation opportunities for the employer)

(b) the acceptability and relevance of simulations to the participants

(c) the lack of clarity and visibility in assessment criteria

(d) the subjectivity of assessments

(e) the bulk of recorded data resulting from ACs, obscuring true assessment of outcomes.

4.14 Rover revised their AC system in 1989, and applied it to reorganisation in the personnel function itself. Major features of the new programme included the following.

(a) To establish clear and rigorous assessment criteria, a competence-based model was adopted, using a 'map' of eight themes or dimensions of competence: strategy, business, resource management, communication, quality, professional consultancy and organisation development.

The map was developed with senior personnel managers using brainstorming and workshop sessions, and later refined in discussion. Its concise nature and use of language culturally and organisationally specific to Rover generated interest and established credibility with senior managers and, subsequently, with participants.

(b) It was decided to give learning and development equal priority to assessment. Rover made the assessment criteria freely available to participants and offered them enrolment by self-assessment. Thus simulations and exercises were developed which would observe and measure agreed competence benchmarks and provide a challenging and satisfying development experience: for example, an exercise involving personnel strategy development for Rover.

(c) Scores in all tests were built up into individual profiles, which could be used to identify relative strengths and development needs: personal development plans were later agreed with each individual.

'In conclusion, we found that the shift from assessment to development centre approaches has been very successful. Using competence-based criteria gives greater visibility and credibility to the process and offers spin-off benefits in performance appraisal, training objectives and for succession and development planning purposes. Challenging exercises that reflect real business issues are more acceptable to participants and offer more realistic assessment.'

Exercise 4

'Performance appraisal is a waste of time if it consists principally of an assessment of performance over the past twelve months; it should concentrate more or less exclusively on planning for the future.' Discuss. *(ICSA December 1986)*

Solution

According to the examiner, 'the quotation was intended to convey the view that appraisal does not serve a very useful purpose if it consists of little more than an assessment of what the individual has done over the past review period, without using that experience on which to construct performance improvement objectives by capitalising on strengths and seeking to eradicate or minimise weaknesses'.

(a) There are justifications for performance review: (1.1)

(b) Past performance must be used in the determination of future targets and present potential: (4.2)

(c) However, the primary purpose must not be purely to praise or blame, but for learning, development and motivation (4.2-4.4)

(d) This purpose will *still* not be achieved, however, without careful planning and conduct of the appraisal, objective-setting and so on. (4.12, 4.13)

5. COUNSELLING

5.1 The subject of counselling was touched upon earlier in this chapter. 'Counselling can be defined as a purposeful relationship in which one person helps another to help himself. It is a way of relating and responding to another person so that that person is helped to explore his thoughts, feelings and behaviour with the aim of reaching a clearer understanding. The clearer understanding may be of himself or of a problem, or of the one in relation to the other.' *(Rees)*

5.2 The need for workplace counselling can arise in many different situations, including the following.

(a) During appraisal, as mentioned above
(b) In grievance or disciplinary situations
(c) Following change, such as promotion or relocation
(d) On redundancy or dismissal
(e) As a result of domestic or personal difficulties
(f) In cases of sexual harassment or violence at work

5.3 Be aware that this is by no means an exhaustive list, and that quite apart from the fact that you will have to deal with such situations in your everyday working life, such scenarios are commonly the stuff of case-study type questions in examinations.

5.4 Most of what follows is derived from the IPM's October 1992 *Statement on Counselling in the Workplace*.

The counselling process

5.5 The counselling process has three stages.

(a) *Recognition and understanding*. Often it is the employee who takes the initiative, but managers should be aware that the problem raised initially may be just the tip of the iceberg. (*Personnel Management Plus*, February 1993, cites a case where an employee came forward with a problem about pension contributions, and mentioned, as he was about to leave 'By the way - my wife wants a divorce'.)

(b) *Empowering*. This means enabling the employee to recognise their own problem or situation and encouraging them to express it.

(c) *Resourcing*. The problem must then be managed, and this includes the decision as to who is best able to act as counsellor. A specialist or outside resource may be better than the employee's manager.

The role of counselling in organisations

5.6 The IPM statement makes it clear that effective counselling is not merely a matter of pastoral care for individuals, but is very much in the organisation's interests.

(a) Appropriate use of counselling tools can prevent under performance, reduce labour turnover and absenteeism and increase commitment from employees. Unhappy employees are far more likely to seek employment elsewhere.

(b) Effective counselling demonstrates an organisation's commitment to and concern for its employees and so is liable to improve loyalty and enthusiasm among the workforce.

(c) The development of employees is of value to the organisation, and counselling can give employees the confidence and encouragement necessary to take responsibility for self and career development.

(d) Workplace counselling recognises that the organisation may be contributing to the employees' problems and therefore it provides an opportunity to reassess organisational policy and practice.

Counselling skills

5.7 Remember that the aim of counselling is to help the employee to help himself. Counsellors need to be observant enough to note behaviour which may be symptomatic of a problem, be sensitive to beliefs and values which may be different from their own (for example religious beliefs), be empathetic to the extent that they appreciate that the problem may seem overwhelming to the individual, and yet remain impartial and refrain from giving advice. Counsellors must have the belief that the individual has the resources to solve their own problems, albeit with passive or active help.

5.8 Interviewing skills are particularly relevant. Open questioning, listening actively (probing, evaluating, interpreting and supporting), seeing the problem from the individual's point of view, and above all being genuinely and sincerely interested are skills identified in the IPM Statement.

Confidentiality

5.9 There will be situations when an employee cannot be completely open unless he is sure that his comments will be treated confidentially. However, certain information, once obtained by the organisation (for example about fraud or sexual harassment) calls for action. In spite of the drawbacks, therefore, the IPM statement is clear that employees must be made aware when their comments will be passed on to the relevant authority, and when they will be treated completely confidentially.

The interview

5.10 The checklist below contains much useful advice for meeting and interviewing people generally, not merely in counselling situations.

Counselling checklist

Preparation

- choose a place to talk which is quiet, free from interruption and not open to view
- research as much as you can before the meeting and have any necessary papers readily available
- make sure you know whether the need for counselling has been properly identified or whether you will have to carefully probe to establish if a problem exists
- allow sufficient time for the session. (If you know you must end at a particular time, inform the individual of this)
- decide if it is necessary for the individual's department head to be aware of the counselling and its purpose
- give the individual the option of being accompanied by a supportive colleague
- if you are approaching the individual following information received from a colleague, decide in advance the extent to which you can reveal your source

- consider how you are going to introduce and discuss your perceptions of the situation
- be prepared for the individual to have different expectations of the discussion, eg the individual may expect you to solve the problem - rather than come to terms with it himself/herself
- understand that the individual's view of the facts of the situation will be more important than the facts themselves and that their behaviour may not reflect their true feelings

Format of discussion

- welcome the individual and clarify the general purpose of the meeting
- assure the individual that matters of confidentiality will be treated as such
- the individual may be reticent through fear of being considered somewhat of a risk in future and you will need to give appropriate reassurances in this regard
- be ready to prompt or encourage the individual to move into areas he/she might be hesitant about
- encourage the individual to look more deeply into statements
- ask the individual to clarify statements you do not quite understand
- try to take the initiative in probing important areas which may be embarrassing/emotional to the individual and which you both might prefer to avoid
- recognise that some issues may be so important to the individual that they will have to be discussed over and over again, even though this may seem repetitious to you
- if you sense that the individual is becoming defensive, try to identify the reason and relax the pressure by changing your approach
- occasionally summarise the conversation as it goes along, reflecting back in your own words (not parrot phrasing) what you understand the individual to say
- sometimes emotions may be more important than the words being spoken, so it may be necessary to reflect back what you see the individual feeling
- at the close of the meeting, clarify any decisions reached and agree what follow-up support would be helpful

Overcoming dangers

- if you take notes at an inappropriate moment, you may set up a barrier between yourself and the individual
- realise you may not like the individual and be on guard against this
- recognise that repeating problems does not solve them
- be careful to avoid taking sides
- overcome internal and external distractions. Concentrate on the individual and try to understand the situation with him/her
- the greater the perceived level of listening, the more likely the individual will be to accept comments and contributions from you
- resist the temptation to talk about your own problems, even though these may seem similar to those of the individual

Source: IPM Statement on Counselling in the Workplace

Exercise 5

Javed is a member of the section which you lead. His work is normally well above average and he knows it. You find him mildly arrogant and have difficulty in liking him, although he seems to have a great deal of respect for you. Frankly you think he is a bit of a crawler.

Of late you have noticed that Javed's work is slipping and he seems to keep himself to himself more than usual. One day he comes to you with a problem that he would normally deal with himself, and he is obviously distressed when you send him away to solve the problem on his own. On your guard, now, you observe that none of the other team members are co-operating with Javed and one or two rather catty remarks are being made behind his back.

You decide to have a counselling session with Javed. How do you conduct the session?

Discipline and grievance handling

5.11 These matters are not mentioned in your syllabus and so a few words will suffice. A *grievance* occurs when an individual thinks that he is being wrongly treated, usually by a colleague or supervisor. Many organisations have formal grievance procedures, stipulating who should be approached in the first instance, and who next if that approach fails, and so on. Generally, all the principles of effective counselling apply in this situation.

5.12 Disciplinary situations arise when an employee breaks organisational 'rules' excessively, for example repeated lateness, absenteeism, inadequate work performance, breaking safety rules and so on. ACAS have produced guidelines for disciplinary action consisting of progressively more serious steps.

(a) Informal talk
(b) Oral warning or reprimand
(c) Written or official warning
(d) Disciplinary layoffs, or suspension
(e) Demotion
(f) Discharge

These steps recognise that misbehaviour at work may be a cry for help (in which case counselling is highly appropriate), but ultimately should not be tolerated.

Employee assistance programmes (EAPs)

5.13 Recognising the difficulty of providing an effective counselling service in-house and also the special skills involved, a notable modern trend is to use outsiders for employee support.

5.14 Companies such as EAR, Focus, and ICAS provide Employee Assistance Programmes (EAPs), offering a 24-hour telephone line with instant access to a trained counsellor. Meetings can be face-to-face if the employee wants, and their immediate families are also covered by the scheme. The providers offer thorough briefing on the scheme for all employees and management information and consultancy for the employers.

5.15 About 80% of the top 500 US companies use such schemes, and around 150 UK companies are estimated to have taken them up since the late 1980s. The cost on average is between £15 and £30 per employee, and companies such as Mobil Oil, Whitbread, and Glaxo report considerable benefits. (*Personnel Management Plus*, February 1993; *Financial Times*, June 1993.)

Chapter roundup

- *Performance appraisal*

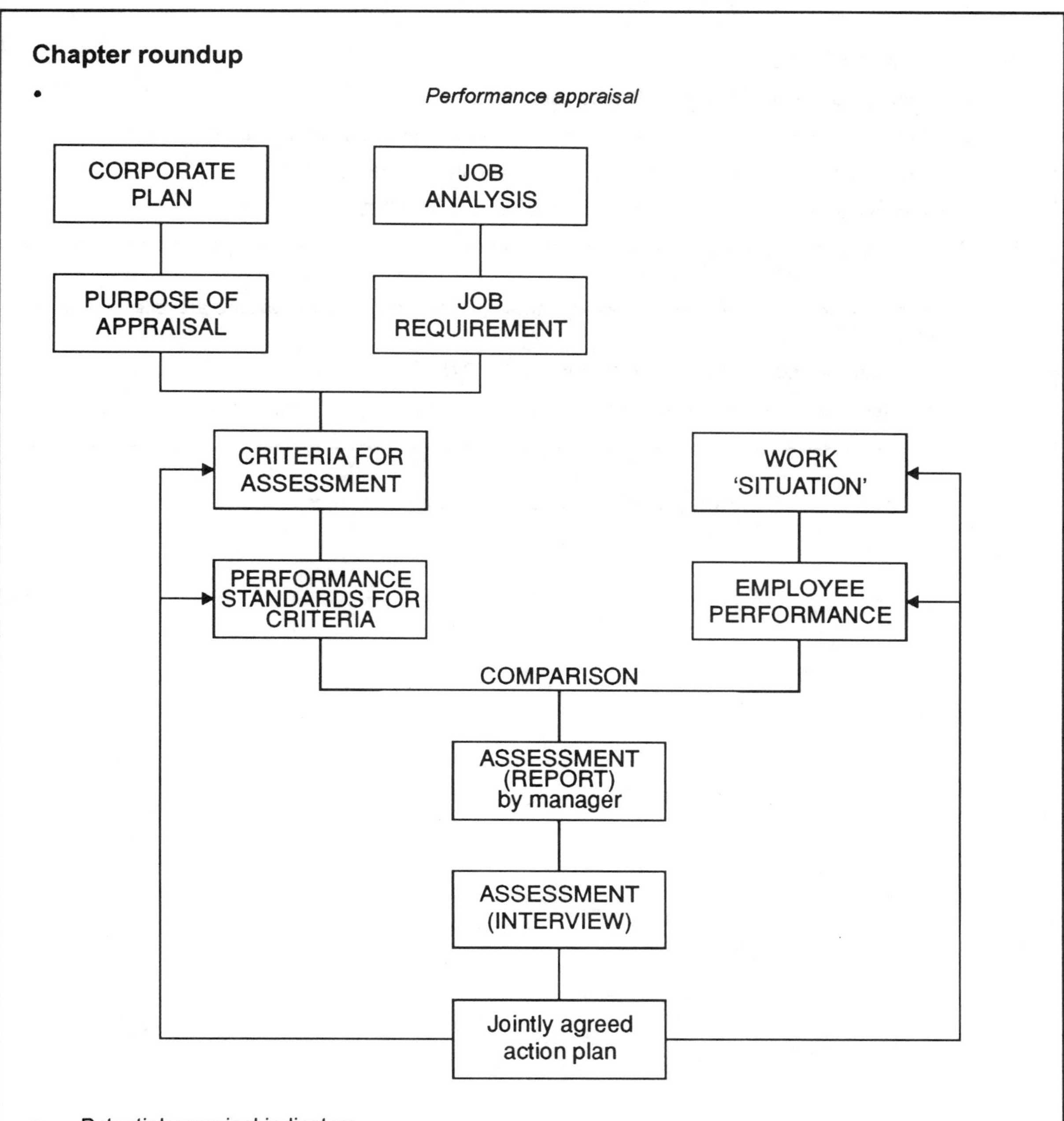

- Potential appraisal indicates:
  - the individual's promotability (present and likely future);
  - the individual's training and development needs;
  - the direction and rate of progress of the individual's development;
  - the future (forecast) management resource of the organisation;
  - the management recruitment, training and development needs of the organisation.
- In counselling one person helps another to help himself. Simple in principle, this can be one of the most difficult tasks of management.

Test your knowledge

1 What are the purposes of appraisal? (see paras 1.1, 1.2)

2 What bases or criteria of assessment might an appraisal system use? (2.3, 2.4)

3 Outline a results-oriented approach to appraisal, and its advantages. (2.4)

4 What follow-up procedures should be used after an appraisal? (2.12)

5 What kinds of criticism might be levelled at appraisal schemes by a manager who thought they were a waste of time? (3.1, 3.3)

6 What techniques might be used to measure an employee's potential to become a successful senior manager? (4.7, 4.10)

7 How can counselling be of benefit to an organisation? (5.6)

8 List six steps that it may be necessary to follow when handling a disciplinary situation. (5.12)

Now try question 12 at the end of the text

Chapter 13

MOTIVATION AND REWARDS

This chapter covers the following topics.

1. Definitions
2. Need theories
3. Two factory theory
4. Expectancy theory
5. Motivation and rewards
6. Motivation and performance
7. Pay as a motivator

Signpost

- The word ' motivation' has been used in its commonly understood sense throughout this text. We now go into a little more depth in view of its relevance to performance, discussed in the previous chapter, and to rewards, discussed in this chapter.
- The work environment was discussed in Chapter 4; organisation theories were covered in Chapter 2 (human relations, contingency theory and so on). The chapters on teamwork and leadership (Chapters 5 and 6) are of some relevance to the question of participation.
- The rate of pay that is fair for a job is discussed in the next chapter on job evaluation.

1. DEFINITIONS

1.1 The words 'motives' and 'motivation' are commonly used in different contexts to mean:

(a) goals, or outcomes that have become desirable for a particular individual. These are more properly 'motivating factors' - since they give people a reason for behaving in a certain way (in pursuit of the chosen goal): thus we say that money, power or friendship are 'motives' for doing something;

(b) the mental process of choosing desired outcomes, deciding how to go about them, assessing whether the likelihood of success warrants the amount of effort that will be necessary, and setting in motion the required behaviours. Our motivation to do something will depend on this calculation of the relationship between needs/goals, behaviour and outcome;

(c) social process by which the behaviour of an individual is influenced by others. 'Motivation' in this sense usually applies to the attempts of organisations to get workers to put in more effort by offering them certain rewards (financial and non-financial) if they do so.

1.2 One way of grouping the major theories of motivation is by distinguishing between:

(a) *content theories*; and
(b) *process theories*.

Content theories assume that human beings have an innate package of motives which they pursue; in other words, that they have a set of needs or desired outcomes. Maslow's need hierarchy theory and Herzberg's two-factor theory are two of the most important approaches of this type.

Process theories explore the process through which outcomes become desirable and are pursued by individuals. This approach assumes that man is able to select his goals and choose the paths towards them, by a conscious or unconscious process of calculation. Expectancy theory, and Handy's motivation calculus, are theories of this type. They take a contingency approach, by stressing the number of variables that influence the individual's decision in each case: there is no best way to motivate people.

2. NEED THEORIES

2.1 Need theories are content theories that suggest that the desired outcome of behaviour in individuals is the *satisfaction of innate needs*.

2.2 The American psychologist Abraham Maslow argued that man has seven innate needs. Maslow's categories are:

- physiological needs - avoiding cold and hunger, etc
- safety needs - freedom from threat, but also security, order, predictability
- love needs - for relationships, affection, sense of belonging
- esteem needs - for competence, achievement, independence, confidence and their reflection in the perception of others: recognition, appreciation, status, respect
- self-actualisation needs - for the fulfilment of personal potential: 'the desire to become more and more what one is, to become everything that one is capable of becoming'
- freedom of inquiry and expression needs - for social conditions permitting free speech encouraging justice, fairness and honesty
- knowledge and understanding need - to gain and order knowledge of the environment, to explore, learn, experiment

According to Maslow, the last two needs are the channels through which we find ways of satisfying all the other needs: they are the basis of satisfaction. The first two needs are essential to human survival. Satisfaction of the next two is essential for a sense of adequacy and psychological health. Maslow regarded self-actualisation as the ultimate human goal, although few people ever reach it.

2.3 David McClelland, writing in the 1950s, identified three types of motivating needs (in which you will recognise some of Maslow's categories).

(a) *The need for power*. People with a high need for power usually seek positions of leadership in order to influence and control.

(b) *The need for affiliation*. People who need a sense of belonging and membership of a social group tend to be concerned with maintaining good personal relationships.

(c) *The need for achievement*. People who need to achieve have a strong desire for success and a strong fear of failure.

Maslow's hierarchy of needs

2.4 In his motivation theory, Maslow put forward certain propositions about the motivating power of needs. (Note that the model was not constructed specifically around people's needs at work, although it can be applied to them as well as general life needs.) He suggested that Man's needs can be arranged in a 'hierarchy of relative pre-potency'. This means that there are 'levels' of need, each of which is dominant until satisfied; only then does the next level of need become a motivating factor.

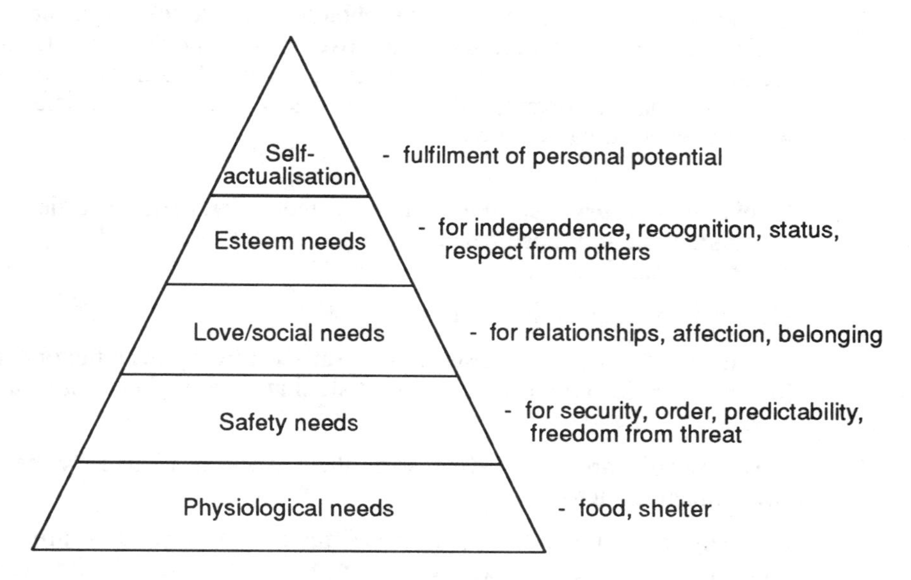

2.5 There is a certain intuitive appeal to Maslow's theory. After all, you are unlikely to be concerned with status or recognition while you are hungry or thirsty - primary survival needs will take precedence. Likewise, once your hunger is assuaged, the need for food is unlikely to be a motivating factor. Unfortunately, research does not bear out the proposition that needs become less powerful as they are satisfied, except at this very primitive level of primary needs (such as hunger and thirst).

Theory - not fact

2.6 The hierarchy of needs is only a theory - *not* an established or empirical fact - and there are various major problems associated with it.

(a) Empirical verification for the hierarchy is hard to come by.

(b) 'Maslow may simply have reflected American middle class values and the pursuit of the good life, and may not have hit on fundamental universal truths about human psychology.' (Buchanan and Huczynski, *Organisational Behaviour*). Several research studies have indicated that cultural patterns affect work behaviour and the success of management techniques: what works in one context may simply not work in another.

(c) It is difficult to predict behaviour using the hierarchy: the theory is too vague. It is impossible to define how much satisfaction has to be achieved before the individual progresses to the next level in the hierarchy. Different people emphasise different needs (and some are clearly able to suppress their basic physiological and safety needs for the sake of a perceived 'higher cause', or for the sake of other people). Also, the same need may cause different behaviour in different individuals.

(d) Application of the theory in work contexts presents various difficulties. (Since Maslow did not base his model in the work context, this is not a serious criticism of the model's validity - but does pose problems for managers attempting to derive practical techniques from it.) The role of pay is problematic, since it arguably acts as an instrument of, or 'stands in' for, other rewards - status, recognition, independence and so on. Moreover, as Drucker notes, a want changes in the act of being satisfied: 'incentives' such as remuneration, once regularly provided, come to be perceived as 'entitlements', and their capacity to create dissatisfaction, to become a deterrent to performance, outstrips their motivatory power. Self actualisation, too, is difficult to offer employees in practice, since its nature is so highly subjective.

3. TWO-FACTOR THEORY

3.1 In the 1950s, the American psychologist Frederick Herzberg interviewed 203 Pittsburgh engineers and accountants and asked two 'critical incident' questions. The subjects were asked to recall events which had made them feel good about their work, others which made them feel bad about it. Analysis revealed that the factors which created satisfaction were different from those which created dissatisfaction.

3.2 In his book *Work and the nature of man* Herzberg identified the factors which cause job dissatisfaction and those which cause job satisfaction. He distinguished between 'hygiene factors' and 'motivator factors'.

He saw two needs of individuals:

(a) the need to avoid unpleasantness, satisfied by 'hygiene factors'; and
(b) the need for personal growth, satisfied at work by 'motivator factors' only.

3.3 'When people are dissatisfied with their work it is usually because of discontent with the environmental factors'.

Herzberg calls these things 'hygiene' factors because they are essentially preventative. They prevent or minimise dissatisfaction but do not give satisfaction, in the same way that sanitation minimises threats to health, but does not give good health. They are also called 'maintenance' factors.

Satisfaction with environmental factors is not lasting. In time dissatisfaction will occur. For example an individual might want a pay rise which protects his income against inflation. If he is successful in obtaining the rise he wants, he will be satisfied for the time being, but only until next year's salary review.

3.4 *Hygiene factors* include:

(a) company policy and administration;
(b) salary;
(c) the quality of supervision;
(d) interpersonal relations;
(e) working conditions;
(f) job security.

3.5 *Motivator factors* create job satisfaction and are effective in motivating an individual to superior performance and effort. These factors give the individual a sense of self-fulfilment or personal growth, and consist of:

(a) status (although this may be a hygiene factor as well as a motivator factor);
(b) advancement;
(c) gaining recognition;
(d) being given responsibility;
(e) challenging work;
(f) achievement;
(g) growth in the job.

3.6 In suggesting means by which motivator satisfactions could be supplied, Herzberg encouraged managers to study the job itself (the type of work done, the nature of tasks, levels of responsibility) rather than conditions of work. 'Dissatisfaction arises from environment factors - satisfaction can only arise from the job.'

If there is sufficient challenge, scope and interest in the job, there will be a lasting increase in satisfaction and the employee will work well; productivity will be above 'normal' levels. The extent to which a job must be challenging or creative to a motivator-seeker will depend on each individual, his ability, his expectations and his tolerance for delayed success.

3.7 Herzberg specified three typical means whereby jobs can be redesigned to improve motivation:

(a) *job enrichment*, or 'the planned process of up-grading the responsibility, challenge and content of the work';

(b) *job enlargement*, the process of increasing the number of operations in which a worker is engaged and so moving away from narrow specialisation of work; and

(c) *job rotation*, or the planned operation of a system whereby staff members exchange positions with the intention of breaking monotony in the work and providing fresh job challenge.

Exercise 1

What 'intrinsic' or internal satisfactions and rewards do you value in your own job? (as opposed to the 'extrinsic' or external things the organisation can give you like pay comfortable working conditions). Why do you work: what do you get out of your work that you most value: what would you miss most if you didn't work?

Try putting each of these satisfactions into • Herzberg's categories of hygiene and motivator factors, and • Maslow's categories of needs.

4. EXPECTANCY THEORY

4.1 The expectancy theory of motivation is a process theory, based on the assumptions that human beings are purposive and rational (in other words, aware of their goals and capable of directing their behaviour towards those goals).

Essentially, the theory states that the strength of an individual's motivation to do something will depend on the extent to which he expects the results of his efforts, if successfully achieved, to contribute towards his personal needs or goals.

4.2 In 1964 Victor Vroom, another American psychologist, worked out a formula by which human motivation could actually be assessed and measured, based on an expectancy theory of work motivation.

Vroom suggested that the strength of an individual's motivation is the product of two factors:

(a) the strength of his preference for a certain outcome. Vroom called this *valence*. It may be represented as a positive or negative number, or zero - since outcomes may be desired, avoided or considered with indifference; and

(b) his expectation that that outcome will in fact result from a certain behaviour. Vroom called this *subjective probability*: it is only the individual's expectation, and depends on his perception of the link between behaviour and outcome. As a probability, it may be represented by any number between 0 (no chance) and 1 (certainty).

4.3 In its simplest form, the expectancy equation therefore looks like this.

Force or strength of motivation to do x.	=	Valence (strength of preference for outcome y).	×	Expectation (that doing x will result in y)

4.4 This is what you would expect: if either valence or expectation have a value of zero, there will be no motivation.

(a) An employee may have a high expectation that behaviour x (say, increased productivity) will result in outcome y (say, promotion) - because of past experience, or a negotiated productivity deal, for example. However, if he is indifferent to that outcome (perhaps because he doesn't want the responsibility), **V = 0**, and he will not be motivated to more productive behaviour.

(b) If the employee has a great desire for promotion - but doesn't believe that more productive behaviour will secure it for him (say, because he has been passed over previously) **E = 0**, and he will still not be highly motivated.

(c) If **V** = -1, (because the employee actively fears responsibility and doesn't want to leave his work group), the value for motivation will be negative, and the employee may deliberately *under*-produce.

4.5 Expectancy theory attempts to measure the strength of an individual's motivation to act in a particular way. It is then possible to compare 'F' values for a range of different behaviours, to discover which behaviour the individual is most likely to adopt. It is also possible to compare 'F' values for different individuals, to see who is most highly motivated to behave in the desired (or undesirable) way.

4.6 Handy's motivation calculus is similar: 'we decide how much "E" (which stands for effort, energy, excitement, expenditure, etc) to invest by doing a calculation,' weighing up the strength or salience of the need, the expectancy that 'E' will lead to a particular result, and the instrumentality of that result in reducing the need.

Exercise 2

Analyse the force of your motivation to study for the *Organisation and the Human Resource* examination in terms of expectancy theory.

5. MOTIVATION AND REWARDS

5.1 You will notice that not all the rewards or incentives that an organisation can offer its employees are directly related to monetary rewards. The satisfaction of any of the employee's wants or needs may be seen as a reward for past or future performance.

Different individuals have different goals, and get different things out of their working life: in other words they have different *orientations* to work. Any one or combination of the needs identified by Maslow and others may be the reason why a person works, or is motivated to work well.

(a) The human relations school regarded work relationships as the main reward offered to the worker.

(b) Later writers suggested a wide range of motivations, including:

   (i) job satisfaction, interest and challenge in the job itself - rewarding work;

   (ii) participation in decision-making - responsibility and involvement;

   (iii) the culture of the organisation, which itself can offer a range of psychological and physical rewards.

(c) Pay has always occupied a rather ambiguous position, but since people need money to live, it will certainly be *part* of the reward 'package' an individual gets from his work.

Job satisfaction

5.2 According to Herzberg, job satisfaction is offered by various factors at work which make employees feel good about their work and jobs, and is a source of motivation and higher productivity. As mentioned briefly, earlier, Herzberg suggested three forms of job design which can enhance job satisfaction: job enrichment, enlargement and rotation.

5.3 *Job enrichment* is planned, deliberate action to build greater responsibility, breadth and challenge of work into a job. It is thus a 'vertical' extension of the job design, which might include:

(a) removing controls;
(b) increasing accountability;
(c) creating natural work units;
(d) providing direct feedback;
(e) introducing new tasks; or
(f) allocating special assignments.

5.4 It would be wrong, however, to suppose that job enrichment alone will automatically make employees more productive.

> 'Even those who want their jobs enriched will expect to be rewarded with more than job satisfaction. Job enrichment is not a cheaper way to greater productivity. Its pay-off will come in the less visible costs of morale, climate and working relationships'. *(Handy)*

5.5 *Job enlargement* is frequently confused with job enrichment, though it should be clearly defined as a separate technique. Job enlargement, as the name suggests, is the attempt to widen jobs by increasing the number of operations in which a job holder is involved. This has the effect of lengthening the time cycle of repeated operations: by reducing the number of repetitions of the same work, the dullness of the job should also be reduced. Job enlargement is therefore a 'horizontal' extension of an individual's work.

Arguably, job enlargement is limited in its ability to improve motivation since, as Herzberg points out, to ask a worker to complete three separate tedious, unchallenging tasks is unlikely to motivate him more than asking him to fulfil one single tedious, unchallenging task.

5.6 *Job rotation* might take two forms.

(a) An employee might be transferred to another job after a period in an existing job, in order to give him new interest and challenge.

(b) Job rotation might be regarded as a form of training. Trainees might be expected to learn a bit about a number of different jobs by spending 6 months or 1 year in each job before being moved on. The employee is regarded as a trainee, rather than as an experienced person holding down a demanding job.

No doubt you will have your own views about the value of job rotation as a method of training or career development. It is interesting to note Drucker's view: 'The whole idea of training jobs is contrary to all rules and experience. A man should never be given a job that is not a real job, that does not require performance from him.'

It is generally accepted that the value of job rotation as a motivator is limited.

5.7 There are some points to note about the assessment of job satisfaction in general.

(a) There is little evidence that a satisfied worker actually works harder - so increased productivity *per se* will not imply satisfaction on the part of the work force. They may be motivated by fear, or work methods may have been improved.

(b) There is, however, support for the idea that satisfied workers tend to be loyal, and stay in the organisation.

(i) *Labour turnover* (the rate at which people leave an organisation) may therefore be an indication of dissatisfaction in the workforce - although there is a certain amount of natural wastage.

(ii) *Absenteeism* may also be an indication of dissatisfaction, or possibly of genuine physical or emotional distress.

(c) There is also evidence that satisfaction correlates with mental health - so that symptoms of stress or psychological failure may be a signal to management that all is not well.

5.8 Job enrichment and enlargement became popular techniques throughout the 1970s. The former, in particular, contributed to the 'quality of working life' movement. However, the world economic recession of the 1980s diverted management theorists' attention away from such issues. As Buchanan and Huczynski note: 'The quality of working life is less important when there is little work to be had.'

Nevertheless, the theories have offered managers ideas about what their subordinates look for and get out of their work, and what variables can be manipulated to give them greater challenge and satisfaction in their work. Relatively simple managerial changes, such as giving more direct feedback on performance, or reducing the number of formal controls on employee behaviour, can enhance the employee's experience of work.

Participation as a means of motivation

5.9 There is a theory that if a superior invites his subordinates to participate in planning decisions which affect their work, if the subordinates voluntarily accept the invitation, and results about actual performance are fed back regularly to the subordinates so that they can make their own control decisions, then the subordinate will be motivated:

(a) to be more efficient;
(b) to be more conscious of the organisation's goals;
(c) to raise his planning targets to reasonably challenging levels;
(d) to be ready to take appropriate control actions when necessary.

5.10 What exactly does participation involve and why might it be a good thing? Handy commented that: 'Participation is sometimes regarded as a form of job enlargement. At other times it is a way of gaining commitment by workers to some proposal on the grounds that if you have been involved in discussing it, you will be more interested in its success. In part, it is the outcome of almost cultural belief in the norms of democratic leadership. It is one of those 'good' words with which it is hard to disagree.'

5.11 The advantages of participation should perhaps be considered from the opposite end: what would be the disadvantages of *not* having participation? The answer to this is that employees would be told what to do, and would presumably comply with orders. However, their compliance might not be enthusiastic, and they might not be psychologically committed to their work.

5.12 Participation can involve employees and make them committed to their task, but only if:

(a) participation is genuine. It is very easy for a boss to pretend to invite participation from his subordinates but end up issuing orders. If subordinates feel the decision has already been taken, they might resent the falsehood of management efforts to discuss the decision with them;

(b) the efforts to establish participation are continual and pushed over a long period of time and with a lot of energy. However, 'if the issue or the task is trivial, or foreclosed, and everyone realises it, participative methods will boomerang. Issues that do not affect the individuals concerned will not, on the whole, engage their interest' *(Handy)*;

(c) the purpose of the participation of employees in a decision is made quite clear from the outset. If employees are consulted to make a decision, their views should carry the decision. If, however, they are consulted for advice, their views need not necessarily be accepted;

(d) the individuals really have the abilities and the information to join in decision-making effectively;

(e) the manager wishes for participation from his subordinates, and does not suggest it because he thinks it is the Done Thing.

> 'It is simply naive to think that participative approaches are always more effective than authoritarian styles of management or vice versa. The critics as well as the advocates of participative management would therefore be wise to direct their energies towards identifying the situations in which a variety of decision-making styles are effective, rather than towards universalistic claims for the applicability or otherwise of any single approach.' *(Hopwood)*

Responsibility for quality

5.13 *Quality circles* seem to have emerged first in the United States, but it was in Japan that they were adopted most enthusiastically. The modern success story of Japanese industry has prompted Western countries to imitate many of the Japanese working methods, with the result that quality circles are now re-appearing in American and West European companies.

A quality circle consists of a group of employees, perhaps about eight in number, which meets regularly to discuss problems of quality and quality control in their area of work, and perhaps to suggest ways of improving quality. The quality circle has a leader or supervisor who directs discussions and possibly also helps to train other members of the circle.

5.14 Quality circles are not random groups of employees. To make the system work a number of factors must be considered.

(a) A quality circle is a *voluntary* grouping. There is no point in coercing employees to join, because the whole point is to develop a spontaneous concern for quality amongst workers.

(b) Quality circles do not function automatically. Training may be needed in methods of quality control, problem solving techniques and methods of communication.

(c) The right leader must be chosen. The person required is one who is capable of directing discussions and drawing out contributions from each member of the circle.

5.15 Ideally, quality circles should be given more responsibility than merely suggesting or even championing improvements; commitment may be increased if the members of quality circles have responsibility for implementing their recommendations. In practice quality circles may become 'talk shops' for problem-solving, inter-disciplinary communication and suggestion/idea generation. Even so, their value should not be underestimated, particularly since they represent a genuine attempt to encourage an impetus towards quality at the lower levels of the organisation hierarchy: most initiatives aimed at improving quality - and quality awareness among staff - are activated and driven from above, by management. A bottom up approach such as quality circles may encourage greater ownership by staff of quality values.

5.16 Benefits claimed to arise from the use of quality circles include:

(a) greater motivation and involvement of employees;
(b) improved productivity and quality of output;
(c) greater awareness of problems by operational staff;
(d) greater awareness of quality and service issues, market and individual customer needs etc.

Culture as a motivator

5.17 Culture, as we have already discussed, is the shared value system of an organisation. Drucker speaks of the 'spirit of performance', which is the 'creation of energy' in the organisation.

Peters and Waterman likewise argue that employees can be 'switched on' to extraordinary loyalty and effort if:

(a) *the cause is perceived to be in some sense great*. They call this 'reaffirming the heroic dimension' of work. Commitment comes from believing that a task is inherently worthwhile: devotion to quality values, and to the *customer*, and his needs and wants, is an important motivator in this way. 'Owing to good luck, or maybe even good sense, those companies that emphasise quality, reliability, and service have chosen the *only* area where it is readily possible to generate excitement in the average down-the-line employee. They give people pride in what they do. They make it possible to love the product.' Shared values and good news swapping - a kind of folklore of past success and heroic endeavour - create a climate where intrinsic motivation is a driving force;

(b) *they are treated as winners*. 'Label a man a loser and he'll start acting like one.' Repressive control systems and negative reinforcement break down the employee's self-image. Positive reinforcement, good news swapping, attention from management and so on. enhance the employee's self-image and create positive attitudes to work and to the organisation;

(c) *they can satisfy their dual needs* to:

(i) be a conforming, secure part of a successful team; and
(ii) be a 'star' in their own right.

6. MOTIVATION AND PERFORMANCE

6.1 You may be wondering whether motivation is really so important. It could be argued that if a person is employed to do a job, he will do that job and no question of motivation arises. If the person doesn't want to do the work, he can resign. The point at issue, however, is the *efficiency* with which the job is done. It is suggested that if individuals can be motivated, by one means or another, they will work more efficiently (and productivity will rise) or they will produce a better quality of work. There is some debate as to what the actual effects of improved motivation are, whether efficiency or quality, but it has become widely accepted that motivation is beneficial to the organisation.

6.2 Barnard suggested that management needs to understand what motivates employees and act to encourage such motivation; otherwise, many employees will tend to act in a negative way, contrary to the aims of the organisation. 'If all those who may be considered potential contributors to an organisation are arranged in order of willingness to serve it, the scale descends from possibly intense willingness through neutral or zero willingness to intense opposition or hatred. The preponderance of persons in a modern society always lies on the negative side with reference to any existing or potential organisation.'

6.3 The case for job satisfaction as a factor in efficiency is not proven. You should be clear in your own minds that although it seems obviously a Good Thing to have employees who enjoy their work and are interested in it, there is no reason why the organisation should want a satisfied work force unless it makes the organisation function better: it is good for human reasons, but it is not necessarily relevant to organisational efficiency or effectiveness.

6.4 It is another point of debate whether intrinsic satisfaction motivates employees to improved performance, or whether it works more the other way around, and the perception of success and achievement from good performance is itself an important source of satisfaction.

Exercise 3

Ivan Robertson, ***Motivation: strategies, theory and practice*** (1992) concludes that there is no universal motivator that can galvanise the whole of a workforce into action: 'far from being alike in capacity, some people are intrinsically and perhaps even genetically more motivated than others, and so put greater effort into anything they are called upon to do.' (*Financial Times*, January 1993)

Michael Dixon, the author of the *Financial Times* piece, went on to point out to Professor Robertson that the fact that companies have no way to turn everyone on, needn't mean they have no ways of turning everyone *off*! Professor Robertson agreed that there probably were several examples of universal *de*-motivators.

How many can you think of?

Solution

Professor Robertson suggests providing no feedback at all, or better, scrambling feedback so that the links between efforts and rewards appear entirely random. Arbitrary controls, as exemplified by management reacting to an obviously isolated incident by clapping a permanent straitjacket on everybody were another example. You can probably think of more.

7. PAY AS A MOTIVATOR

7.1 Pay has a central - but ambiguous - role in motivation theory. It is not mentioned explicitly in any need list, but it may be the means to an infinite number of specific ends, offering the satisfaction of many of the various needs. Individuals may also, however, have needs unrelated to money, which money cannot satisfy, or which the pay system of the organisation actively denies. So to what extent is pay an inducement to better performance, ie a motivator or incentive?

7.2 An employee needs income to live. The size of his income will affect his standard of living. However, people tend not to be concerned to *maximise* their earnings. They may like to earn more, but are probably more concerned:

(a) to earn *enough* pay; and

(b) to know that their pay is *fair* in comparison with the pay of others both inside and outside the organisation.

7.3 Equity (perceived fairness of pay in relation to the job and to the pay of others) is often more important than maximising income, once the individual has enough pay to maintain a satisfactory lifestyle. Yet the Economic Man model - which assumes that people will adjust their effort if offered money, persists in payment-by-results schemes, bonuses, profit-sharing and other 'cash' incentives.

Payment systems then have to tread the awkward path between *equity* (an objective rate for the job, preserved differentials and so on) and *incentive* (an offered reward to stimulate extra effort and attainment by particular individuals and groups).

7.4 The Affluent Worker research of Goldthorpe, Lockwood et al (1968) illustrated an *instrumental* orientation to work (the attitude that work is not an end in itself, but a means to other ends). The highly paid Luton car assembly workers experienced their work as routine and dead-end. The researchers concluded that they had made a rational decision to enter employment offering high monetary reward rather than intrinsic interest: they were getting out of their jobs what they most wanted from them.

The Luton researchers, however, did not claim that all workers have an instrumental orientation to work, but suggested that a person will seek a suitable balance of:

(a) the rewards which are important to him; and
(b) the deprivations he feels able to put up with.

Even those with an instrumental orientation to work have limits to their purely financial aspirations, and will cease to be motivated by money if the deprivations - in terms of long working hours, poor conditions, social isolation or whatever - become too great: in other words, if the 'price' of pay is too high.

High taxation rates may also weight the deprivation side of the calculation; workers may perceive that a great deal of extra effort will in fact earn them little extra reward.

7.5 Pay is one of Herzberg's hygiene rather than motivator factors. It gets taken for granted, and so is more usually a source of dissatisfaction than satisfaction. (Lawler suggested that in the absence of information about how much colleagues are earning, individuals guess their earnings and usually over-estimate. This then leaves them dissatisfied because they resent earning less than they *think* their colleagues are getting!)

7.6 However, pay is the most important of the hygiene factors, according to Herzberg. It is valuable not only in its power to be converted into a wide range of other satisfactions (perhaps the only way in which organisations can - at least indirectly - cater for individual employee's needs and wants through a common reward system) but also as a consistent measure of worth or value, allowing employees to compare themselves and be compared with other individuals or occupational groups inside and outside the organisation.

7.7 We should also consider pay from the organisation's point of view. Wages and salaries are:

(a) a cost, which appears in the cost of the product or service to the market;

(b) an investment: money spent on one factor of production (labour) in the hope of a return; and

(c) a potentially crucial environmental variable, as an incentive and motivator, a source of job satisfaction or dissatisfaction, political power or conflict.

7.8 The objectives of pay from the organisation's point of view are:

(a) to attract and retain labour of a suitable type and quality;
(b) to fulfil perceived social responsibilities; and
(c) to motivate employees to achieve and maintain desired levels of performance.

Exercise 4

How do you feel personally about pay as a motivator? Where would you draw the line between extra money and the hardships required to earn it?

If you are in a supervisory or managerial position, examine your attitudes to motivating the staff in your section: do you assume that 'they'll do it because they are paid to do it'? What non-monetary satisfactions and encouragement can you (and do you) offer your staff to keep them working well? What about your own superior: what are his/her attitudes and how does he/she try to motivate you? Does it work? (If not, why not?)

Salary review

7.9 Salary reviews may be carried out as a general exercise, when all or most salaries have to be increased to keep pace with market rates, cost of living increases or negotiated settlements. *General reviews* are often carried out annually (government regulations permitting) during inflationary periods: this may or may not create problems in financing individual merit awards as well.

7.10 *Individual* salary reviews are carried out to decide on merit awards. Again, these are usually held annually - with interim reviews, possibly, for trainees and younger staff who are making fast progress. Some companies phase reviews throughout the year rather than hold them all at once; this is more difficult to administer but does diffuse the tension of a general review period.

7.11 Guidelines for salary review will be necessary to minimise the subjectivity of discretionary payments. The total cost of all merit increases, or minimum/maximum amounts for increases, might be specified.

A *salary review budget* will determine the increase that can be allocated for awards, as a percentage of payroll costs for the department. The size of the budget will depend on:

(a) how average salaries in each grade differ from the target salary (the mid-point): ideally, they should correspond, but may be too high or low. A high ratio indicates that *earnings drift* has taken place, and salaries have moved towards the upper end of the scale - which may or may not correspond to the merits of the staff concerned; and

(b) the amount the company estimates it will be able to pay, based on forecast revenue, profit, and labour cost savings elsewhere (perhaps from highly-paid employees leaving, and recruits entering at lower-paid levels: this is called *salary attrition*).

> 'Salary structures, job evaluation schemes, progression policies and salary review procedures all aim to make salary administration a scientific process. But they cannot entirely succeed. Salary administration is as much art as science and, inevitably, there are problems which can only be solved by exercising judgement in the light of circumstances.'
>
> Armstrong, *Handbook of Personnel Management Practice*

Problems of perceived fairness

7.12 Perceived equity and fair differentials are important elements in any salary structure, if it is to achieve its aim of attracting, holding (and to some extent, motivating) employees. Apart from the subjective aspects of assessment, grading and discretionary increment systems, there are typically problems associated with:

(a) *distortion of the salary structure*, as discussed above, by factors such as salary attrition and drift. Jobs should be meticulously graded and re-graded: non-merit-related awards should be controlled, and where averages appear to be dropping, the situation should be explained to staff;

(b) *the squeezing of differentials*. Some employees may benefit from negotiated increases, overtime and bonuses, which their superiors may not get. This upward pressure may be a cause of dissatisfaction to higher-grade employees who feel that their particular contribution is not being acknowledged. The only solution is to maintain differentials between the target salary for the supervisorial grade, and the average earnings (with overtime) of the subordinates. Panic measures, creating a knock-on effect on all other grades, should be avoided: the overall span of salary levels should be wide enough to allow for 15-20% differentials to be maintained between each grade;

(c) *salary limits*. There may be a problem motivating individuals who have reached the top of their grade scale but cannot be promoted out of the grade. The situation must be made quite clear to employees: secrecy will only deprive the employee of a perceived goal (promotion) which will be a further source of demotivation. Special bonuses may also be allowed, in exceptional circumstances;

(d) *market rates*. Where market rates exist and can be determined, they will influence salary levels. However, they may (say, where a category of staff is in short supply) indicate salaries higher than internal job evaluation would: differentials are upset, and employees see apparent injustice. It may not be possible to recruit and retain suitable staff without sacrificing equity to some extent. Jobs subject to such market pressures should, however, be noted as exceptions in the salary structure or 'red-circled': adjustments may therefore be made later, as a result of regular audits of market rates, and anomalies may disappear as market rates deflate.

Performance related pay

7.13 Pay (or part of it) is related to output (in terms of the number of items produced, or time taken to produce a unit of work), or results achieved (performance to defined standards in key tasks, according to plan).

The most common individual payment by results (PBR) scheme for wage earners is straight piecework: payment of a fixed amount per unit produced, or operation completed.

For managerial and other salaried jobs, however, a form of management by objectives will probably be applied so that:

(a) key results can be identified and specified, for which merit awards (on top of basic salary) will be paid;

(b) there will be a clear model for evaluating performance and knowing when or if targets have been reached and payments earned;

(c) the exact conditions and amounts of awards can be made clear to the employee, to avoid uncertainty and later resentment.

7.14 For service and other departments, a PBR scheme may involve *bonuses* for achievement of key results, or *points schemes*, where points are awarded for performance on various criteria (efficiency, cost savings, quality of service and so on) and certain points totals (or the highest points total in the unit, if a competitive system is used) win cash or other awards.

7.15 *Personnel Management*, November 1990, reported research into the benefits and problems of performance-related pay.

	Black & Decker	*Komatsu UK*	*Birds Eye Walls*	*Co. A*	*Co. B*	*Co. C*	*Co. D*	*Co. E*
	1 Benefits of PRP cited							
Improves commitment and capability	Yes	Yes	Yes	Yes	Yes	Yes	Yes	Yes
Complements other HR initiatives	Yes	Yes	Yes	Yes	Yes	Yes		Yes
Improves business awareness	Yes	Yes	Yes		Yes	Yes		
Better two-way communications	Yes	Yes	Yes		Yes	Yes		Yes
Greater supervisory responsibility		Yes	Yes	Yes	Yes			Yes

	Black & Decker	*Komatsu UK*	*Birds Eye Walls*	*Co. A*	*Co. B*	*Co. C*	*Co. D*	*Co. E*
	2 Potential problems cited							
Subjectivity			Yes			Yes	Yes	Yes
Supervisors' commitment and ability	Yes	Yes	Yes		Yes	Yes	Yes	Yes
Translating appraisals into pay	Yes	Yes	Yes	Yes		Yes		Yes
Divisive/against team working			Yes	Yes		Yes	Yes	Yes
Union acceptance/employee attitudes			Yes	Yes	Yes		Yes	Yes

'In the wrong hands, PRP can do more harm than good, so organisations considering PRP should consider carefully whether it is appropriate for them ... Other payment systems which do not seek to directly link individual performance and reward may be more suited to the aims of the business.'

Suggestion schemes

7.16 Another variant on performance-based pay is the *suggestion scheme*, where payments or prizes are offered to staff to come up with workable ideas on improving efficiency or quality, new marketing initiatives or solutions to production problems. The theory is that there is in any case motivational value in getting staff involved in problem-solving and planning, and that staff are often in the best position to provide practical and creative solutions to their work problems or the customer's needs - but that an added incentive will help to overcome any reluctance on the part of staff to put forward ideas (because it is seen as risky, or doing management's job for them, or whatever).

Wherever possible, the size of the payment should be related to the savings or value added as a result of the suggestion - either as a lump sum or percentage. Payments are often also made for a 'good try' - an idea which is rejected but considered to show initiative, effort and judgement on the part of the employee.

Suggestion schemes usually apply only to lower grades of staff, on the grounds that thinking up improvements is part of the supervisor's or manager's normal job, but with the increase of worker empowerment and 'bottom up' quality initiatives, such as quality circles, they are becoming more widespread in various forms.

7.17 Whichever system is used, results-oriented payments should:

(a) offer real incentives, sufficiently high after tax to make extraordinary effort worthwhile, perhaps 10-30% of basic salary;

(b) relate payments to criteria over which the individual has control (otherwise he will feel helpless to ensure his reward, and the expectancy element in motivation will be lacking);

(c) make clear the basis on which payments are calculated, and all the conditions that apply, so that individuals can make the calculation of whether the reward is worth the extra level of effort;

(d) be flexible and sensitive enough to reward different levels of achievement in proportion, and with provision for regular review, and adaptation to the changing needs of the particular organisation.

Bonus schemes

7.18 *Bonus schemes* are supplementary to basic salary, and have been found to be popular with entrepreneurial types, usually in marketing and sales. Bonuses are both incentives and rewards.

7.19 Group incentive schemes typically offer a bonus for a group (equally, or proportionately to the earnings or status of individuals) which achieves or exceeds specified targets. Offering bonuses to a whole team may be appropriate for tasks where individual contributions cannot be isolated, workers have little control over their individual output because tasks depend on each other, or where team-building is particularly required.

It may enhance team-spirit and co-operation as well as provide performance incentives, but it may also create pressures within the group if some individuals are seen to be 'not pulling their weight'.

7.20 Long-term, large-group schemes may be applied *factory-wide*, as an attempt to involve all employees in the organisation of production. Typically, bonuses would be calculated monthly on the basis of improvements in output per man per hour against standard, or value added (to the cost of raw materials and parts by the production process).

Value added schemes work on the basis that improvements in productivity (indicated by a fall in the ratio of employment costs to sales revenue) increases value added, and the benefit can be shared between employers and employees on an agreed formula. So if sales revenue increases and labour costs (after charges for materials, utilities and depreciation have been deducted) stay the same, or sales revenue remains constant but labour costs decrease, the balance becomes available. There has been an increase in such schemes in recent years (for example, at ICI).

(a) *Advantages* of factory-wide schemes

   (i) Increasing employee identification with the organisation's objectives

   (ii) Stimulating interest in productivity

   (iii) Encouraging joint consultation, which may be a source of job satisfaction, and may help overcome resistance to change

   (iv) Facilitating equality of treatment for indirect as well as direct product workers (cleaners, stores assistants and so on).

(b) *Disadvantages*

   (i) Need for detailed and comprehensive costing systems which must be communicated to all employees

   (ii) Failure to provide direct incentives, since reward has become divorced from individual effort.

7.21 Other collective incentive schemes include:

(a) *the Scanlon Plan* (1947). This is based on collective bargaining, and operated by a joint management-union committee. It includes a suggestion plan, to stimulate, implement and monitor improvements in production and so reductions in labour costs, the benefits of which are shared by union members and management; and

(b) *the Rucker Plan* (1955), which uses added value, and involves a joint productivity committee to achieve cost reductions, the savings again being shared.

Profit-sharing schemes and employee shareholders

7.22 Profit-sharing schemes offer employees (or selected groups of them) bonuses, perhaps in the form of shares in the company, related directly to profits. The formula for determining the amounts may vary, but in recent years, a straightforward distribution of a percentage of profits above a given target has given way to a value-added related concept. The profit formula itself is not easily calculated - profit levels being subject to accounting conventions - so care will have to be taken to publish and explain the calculations to employees if the scheme is not to be regarded with suspicion or as simply another fringe benefit.

7.23 Profit sharing is in general based on the belief that all employees can contribute to profitability, and that that contribution should be recognised. If it is, the argument runs, the effects may include profit-consciousness and motivation in employees, commitment to the future prosperity of the organisation and so on.

7.24 The actual incentive value and effect on productivity may be wasted, however, if the scheme is badly designed.

(a) A perceivedly significant sum should be made available to employees - once shareholders have received appropriate return on their investment - say, 10% of basic pay.

(b) There should be a clear, and not overly delayed, link between effort/performance and reward. Profit shares should be distributed as frequently as possible - consistent with the need for reliable information on profit forecasts and targets and the need to amass a significant pool for distribution.

(c) The scheme should only be introduced if profit forecasts indicate a reasonable chance of achieving the above: profit sharing is welcome when profits are high, but the potential for disappointment is great.

(d) The greatest effect on productivity arising from the scheme may in fact arise from its use as a focal point for discussion with employees, about the relationship between their performance and results, and areas and targets for improvement. Management must be seen to be committed to the principle.

Examples: profit sharing and employee shareholders

7.25 Several hundred companies already operate some sort of profit-sharing scheme and according to a survey by the British Institute of Management, published in November 1978, almost all of them considered their scheme to be fairly, or very 'successful'. (The criteria by which 'success' is judged, however, will presumably vary from firm to firm.)

The main reasons given in the survey for the success of a profit-sharing scheme were that:

(a) it encouraged given employees to identify with the company;
(b) it provided an incentive to work harder;
(c) it made staff profit-conscious;
(d) employees were able to earn more money;
(e) it helped to retain staff; and
(f) it was generally liked by employees.

It appears that profit related pay schemes have become increasingly popular. The number has tripled since 1990, stimulated by government support and extra tax incentives.

7.26 However, *Personnel Management*, May 1991, reported the results of a study of an employee share ownership scheme (operated as a voluntary Save as You Earn-related scheme) in a Midlands factory.

'In the event, we concluded there had been no change in attitudes which we felt should be attributed to the scheme,' they report. 'If this is correct, it is hard to see how it could have had any effect on the behaviour of the employees and any significant advantage to the firm.'

There were also very few joiners. Researchers say many workers on low incomes are unwilling to make a five-year or seven-year savings commitment. They warn that this could lead to such schemes becoming the preserve of higher-paid staff, intensifying the 'them' and 'us' attitudes of British industry.

7.27 Share ownership has little effect on class divisions, according to a survey of employees in two privatised utilities carried out by Leicester University. This revealed that only 10 per cent believed that 'them and us' attitudes were replaced with a sense of common purpose because of share ownership.

Although 80 per cent of the sample of nearly 450 employees were employee shareholders, 65 per cent said that it made no difference to how careful they were with the company's equipment, and 70 per cent felt it did not make people work harder.

Problems with cash incentives

7.28 There are a number of difficulties associated with incentive schemes based on monetary reward.

(a) Increased earnings simply may not be an incentive to some individuals. An individual who already enjoys a good income may be more concerned with increasing his leisure time, for example.

(b) Workers are unlikely to be in complete control of results. External factors, such as the general economic climate, interest rates and exchange rates may play a part in *profitability* in particular. In these cases, the relationship between an individual's efforts and his reward may be indistinct.

(c) Greater specialisation in production processes means that particular employees cannot be specifically credited with the success of their particular products. This may lead to frustration amongst employees who think their own profitable work is being adversely affected by inefficiencies elsewhere in the organisation.

(d) Even if employees *are* motivated by money, the effects may not be altogether desirable. An instrumental orientation may encourage self-interested performance at the expense of teamwork: it may encourage attention to output at the expense of quality, and the lowering of standards and targets (in order to make bonuses more accessible).

Workers remain suspicious that if they achieve high levels of output and earnings, management will alter the basis of the incentive rates to reduce future earnings. Work groups therefore tend to restrict output to a level that they feel is 'fair' and 'safe'.

Non-cash incentives

7.29 Incentive and recognition schemes are increasingly focused not on cash, but on non-cash awards. According to a feature in *Personnel Management*, September 1992: 'Traditionally aimed at sales people, gifts and travel incentives have been spreading slowly to other areas and are now used to add interest to quality schemes and encourage money saving ideas... to enable managers to show gratitude to staff for such things as continuous improvement and teamwork...to lift morale'.

(a) British Telecom - in the wake of large scale voluntary redundancies - launched an up-beat 'Living our values' initiative, including the awarding of gifts to employees exemplifying the organisation's values and being role models to others.

(b) British Aerospace has preserved its quality awards, despite job losses. Teams receive gold pens, watches, ties or scarves for a PSB (Problem Solution Benefit).

(c) ICL used to offer symbolic awards of bronze, silver and gold medals, but has now replaced these with a gift catalogue (called the 'Excellence Collection') from which nominees choose rewards they value: 'Change is essential if recognition schemes are going to succeed... This is one of the problems with this kind of programme. If you don't update it from time to time, it just gets tired.' (*Personnel Management*, September 1992).

(d) Abbey Life's top performers are given the opportunity to attend conventions in exotic foreign locations, with partners (and without an onerous work content): length of stay and luxury of location depend on performance.

(e) Trusthouse Forte 'has launched a drive to cut employee turnover through an incentive scheme which awards air mileage in return for staff loyalty ... In addition to the basic retention programme, THF is also offering further incentives to staff, including 500 miles for the employee of the month and 1,000 for employee of the year, with another 200 miles for staff receiving a complimentary letter from a guest.' *Personnel Management* (February 1991)

7.30 Such schemes can be regarded by some staff as manipulative, irrelevant ('awards are being made for things that are part of normal duties: no special effort required') or just plain gimmicky. (The general secretary of the staff union at Sun Alliance has been quoted as saying: 'I have worked for a firm which rewarded its top salespeople with a cruise. I can't imagine anything worse than being trapped on a yacht with a lot of other life assurance salesmen'!) However, it is generally considered that such schemes can be effective as incentives, team-building exercises, and, perhaps more fundamentally, ways of expressing recognition of achievement - without which staff may feel isolated, undervalued or neglected.

Chapter roundup

- Motivation is a term used in different contexts to refer to:
  - goals or outcomes that have become desirable for a particular individual, as in: 'he is motivated by money';
  - the mental process of choosing a goal and deciding whether and how to achieve it, as in: 'he is motivated to work harder';
  - the social process by which the behaviour of an individual is influenced by others, as in: 'the manager motivates his team'.
- *Content* theories suggest that man has a package of needs: the best way to motivate an employee is to find out what his needs are and offer him rewards that will satisfy those needs.
  - Abraham Maslow identified seven innate needs of all individuals and arranged them in a hierarchy, suggesting that an individual will be motivated to satisfy each category, starting at the bottom before going on to seek higher order satisfactions.
  - Frederick Herzberg identified two basic need systems: the need to avoid unpleasantness and the need for personal growth. He suggested factors which could be offered by organisations to satisfy both types of need: 'hygiene' and 'motivator' factors respectively.
- Process theories do not tell managers what to offer employees in order to motivate them, but help managers to understand the dynamics of employees' decisions about what rewards are worth going for. They are generally variations on the *expectancy* model: F = V × E.
- Various means have been suggested of improving job satisfaction but there is little evidence that a satisfied worker actually works harder. Likewise, participation in decisions, involvement in product quality or enthusing employees about the corporate culture may or may not have an impact.
- Pay is the most important of the 'hygiene' factors. Salary structures most commonly consist of a graded structure, with a range (or scale) of salaries within each grade. Many organisations operate incentive schemes and offer other forms of benefits

Test your knowledge

1. What are the seven needs identified by Maslow? Incorporate the relevant needs in a simple diagram of the hierarchy. (see paras 2.2, 2.4)
2. List five motivator and five hygiene factors. (3.4, 3.5)
3. Explain the formula 'F = V × E'. (4.2 - 4.4)
4. Distinguish between job enrichment and job enlargement. (5.3 - 5.5)
5. Are people better motivated if they are allowed to participate in decisions? (5.12)
6. 'People will work harder and harder to earn more and more pay.' Do you agree? Why (or why not)? (7.1 - 7.6)
7. What problems of 'fairness' arise with regard to wages and salaries? (7.12)
8. Give four features of a well-designed incentive scheme. (7.24)

Now try question 13 at the end of the text

Chapter 14

JOB EVALUATION

This chapter covers the following topics.

1. Job evaluation
2. Methods of job evaluation
3. Introducing a job evaluation scheme
4. Other factors in setting remuneration levels

Signpost

- In the previous chapter we introduced pay (financial reward) and its role as a motivator. We noticed that there is a dilemma for management in the dual requirements in reward systems for pay rates that are fair in relation to others, on the one hand, and the need to be able to offer extra reward for extraordinary effort and attainment on the other.
- In this chapter we discuss how reward systems can be designed to fulfil either or both of these requirements.

1. JOB EVALUATION

1.1 Job evaluation is a systematic method of arriving at a wage or salary structure, so that the rate of pay for a job is felt to be *fair* in comparison with other jobs in the organisation.

The Institute of Administrative Management's *Office Job Evaluation* (1976) describes its purpose in the following way. 'Any job for which a wage or salary is offered has been evaluated in some way or other in order to arrive at the amount of payment to be made. To this extent it might be said that all organisations which pay employees have job evaluation. However, the term 'job evaluation' is mostly used nowadays with greater precision to describe a formal standardised method for ranking jobs and grouping them into grades. Invariably, such systems are used primarily as the basis for a payment structure....'

1.2 The British Institute of Management *(Job Evaluation)* gives the following definition. 'Job evaluation is the process of analysing and assessing the content of jobs, in order to place them in an acceptable rank order which can then be used as a basis for a remuneration system.'

1.3 The advantages of a job-evaluated salary structure are as follows.

(a) The salary structure is based on a formal study of work content, and the reasons for salary differentials between jobs has a rational basis that can be explained to anyone who objects to his salary level or grading in comparison with others.

(b) The salary structure should be well balanced, even in an organisation that employs people with a wide range of different technical skills (such as engineers, accountants and salesmen).

(c) The salary structure is based on job content, and not on the personal merit of the job-holder himself. The individual job-holder can be paid personal bonuses in reward for his efforts, and when he moves to another job in the organisation, his replacement on the job will be paid the rate for the job, and will not inherit any personal bonuses of his predecessor.

(d) Regular job evaluation should ensure that the salary structure reflects current changes in the work content of jobs, and is not outdated, so that pay differentials remain fair.

(e) A job-evaluated salary structure might protect an employer from the accusation that rates of pay discriminate between different types of worker - notably between men and women, who by law (The Equal Pay Act 1970) should be paid the same rate for 'like work', 'work rated as equivalent' or 'work of equal value'.

(f) Analysis of job content and worth are available for use in recruitment, selection, training and other personnel contexts.

1.4 Job-evaluated salary structures do have some flaws, however.

(a) They pay a fair rate for a job only in the sense that differentials are set according to *relative* worth. Job evaluation does not make any recommendations about what the general level of pay ought to be, in money terms. Indeed it cannot do so without reference to outside factors such as rates fixed by collective bargaining, statutory obligation or local custom. (Such factors are discussed in Section 4 of this chapter.)

(b) They pay a rate for the job irrespective of the personal merits of the job holder or fluctuations in his performance. If an organisation rewards individual merit with bonuses, evaluated differentials will again be distorted.

(c) Job definition for evaluation purposes supports rigid hierarchical organisation and concepts of status which suppress employee motivation and creativity. It may prevent labour flexibility and multiskilling, which are economically attractive to organisations in volatile market environments, and may undermine attempts to foster a people-centred organisation culture: job evaluation assumes that people are commodities who can be made to fit defined roles (Murlis and Fitt, *Personnel Management*, May 1991).

(d) Many job evaluation methods suggest that job evaluation is a scientific and accurate technique, whereas in fact there is a large element of subjective judgement involved in awarding points or ratings, and evaluations can be unfair.

(e) Job evaluated salary structures can get out-of-date. There ought to be periodic reviews, but in practice, an organisation might fail to review jobs often enough.

Exercise 1

Do you feel your salary is fair for the job you do, and in relation to others? Do you know how your job is evaluated? Do you know how your salary is worked out? What could the organisation do to make the system (a) clearer and (b) fairer?

2. METHODS OF JOB EVALUATION

2.1 In large organisations, it is impossible to evaluate every individual job, because the process would be too long and costly. Instead, selected key jobs are evaluated, and provide a benchmark for the evaluation of other similar jobs. Ideally, the key jobs chosen for analysis should be jobs comparable with jobs in other organisations, for which a market rate of pay is known. Some information for evaluation may already be available in the form of job descriptions.

2.2 It may be said that, even in its more quantitative or analytical forms, job evaluation is 'systematic' rather than 'scientific'. The number of different inputs and environmental variables make an element of subjectivity inevitable, despite refinements aimed at minimising it.

Non-analytical approaches to job evaluation make largely subjective judgements about the whole job, its difficulty, and its importance to the organisation relative to other jobs. (Ranking is a method of this type.)

2.3 *Analytical* methods of job evaluation identify the component factors or characteristics involved in the performance of each job, such as skill, responsibility, experience, mental and physical efforts required. Each component is separately analysed, evaluated and weighted: degrees of each factor, and the importance of the factor within the job, are quantified. (Examples of such methods include points rating and factor comparison.) These methods involve detailed analysis and a numerical basis for comparing jobs as like to like. However, there is still an element of subjectivity, in that:

(a) the factors for analysis are themselves qualitative, not easy to define and measure. Mental ability and initiative are observable in job holders, but not easily quantifiable as an element of the job itself;

(b) assessment of the importance and difficulty of a job cannot objectively be divorced from the context of the organisation and job holder. The relative importance of a job is a function of the culture and politics of the organisation, the nature of the business and not least the personal power of the individual in the job. The difficulty of the job depends on the ability of the job holder and the favourability or otherwise of the environment/technology/work methods/management;

(c) the selection of factors and the assignment of monetary values to factors remain subjective judgements.

2.4 It is undoubtedly desirable to achieve objectivity, to reduce the resentment commonly felt at the apparent arbitrariness of pay decisions. If job evaluation were truly objective, it would be possible to justify differentials on a rational basis, the organisation would have a balanced and economical pay structure based on contribution, and employers would be safe from accusations of unfair pay decisions.

2.5 Despite the element of subjectivity even in the more analytical methods of job evaluation, it may be true to say that any form of job evaluation is useful, minimising the (real or perceived) arbitrariness of pay decisions, and removing personality or discriminatory issues from pay reviews.

2.6 We shall describe five methods of job evaluation.

(a) Ranking
(b) Classification
(c) Factor comparison
(d) Points rating
(e) The HAY-MSL method

Ranking method

2.7 In a ranking system of job evaluation, each job is considered as a whole (rather than in terms of job elements) and ranked in accordance with its relative importance or contribution to the organisation. Having established a list of jobs in descending order of importance, they can be divided into groups, and jobs in each group given the same grade and salary.

2.8 The advantage of the ranking method is that it is simple and unscientific. In a small organisation, it might be applied with fairness.

2.9 However, the job evaluators need to have a good personal knowledge of every job being evaluated and in a large organisation, they are unlikely to have it. Without this knowledge, the ranking method would not produce fair evaluations. This is why more complex methods of job evaluation have been devised.

Classification method

2.10 This is similar to the ranking method, except that instead of ranking jobs in order of importance and then dividing them into grades, the classification method begins with deciding what grades there ought to be (say, grades A, B, C, D and E, with each grade carefully defined) and then deciding into which grade each individual job should be classified: is the job a grade C or a grade D job?

The advantages and disadvantages of this method are the same as those of ranking.

Factor comparison method

2.11 This is an analytical method of job evaluation. It begins with the selection of a number of qualitative factors on which each job will be evaluated. These qualitative factors might include, for example, technical knowledge, physical skill, mental skill, responsibility for other people, responsibility for assets or working conditions.

2.12 Key benchmark jobs are then taken, for which the rate of pay is considered to be fair (perhaps in comparison with similar jobs in other organisations). Each key job is analysed in turn, factor by factor, to decide how much of the total salary is being paid for each factor. So if technical skill is 50% of a benchmark job paying £10,000, the factor pay rate for technical skill (within that job) is £5,000. When this has been done for every benchmark job, all the different rates of pay for each factor are correlated, to formulate a ranking and pay scale for that factor.

2.13 Other (non-benchmark) jobs are then evaluated by analysing them factor by factor. In this way a salary or grading for the job can be built up. For example, analysis of a clerk's job factor by factor might be

Factor	*Proportion of job*		*Pay rate for factor (as established by analysis of benchmark jobs*	*Job value*
				£
Technical skills	50%	×	£12,000 pa	6,000
Mental ability	25%	×	£16,000 pa	4,000
Responsibility for others	15%	×	£10,000 pa	1,500
Other responsibilities	10%	×	£5,000 pa	500
				12,000

2.14 The Institute of Administrative Management comments about the factor comparison method that: 'the system links rates closely to existing levels for key benchmark jobs and depends heavily on careful allocation of money values to each factor of the benchmark jobs. It is not easy to explain to employees, and is best suited to situations where the range of jobs is limited and of a fairly simple nature.' It is not well-suited to the evaluation of office jobs.

Points rating method

2.15 Points rating is probably the most popular method of formal job evaluation. It begins with listing a number of factors which are thought to represent the qualities being looked for in the jobs to be evaluated. (Remember that jobs are being evaluated, not job holders themselves, and the qualities listed should relate to the jobs themselves.) In a typical evaluation scheme, there might be about 8-12 factors listed. The factors will vary according to the type of organisation, but they might include:

(a) skill - education, experience, dexterity, qualifications;

(b) initiative;

(c) physical or mental effort;

(d) dealing with others;

(e) responsibility for subordinates, or the safety and welfare of others;

(f) responsibility for equipment, for a process or product, for materials;

(g) job conditions - such as monotony of working, working in isolation, unavoidable work hazards.

2.16 A number of points is allocated to each factor, as a maximum score. In this way, each factor is given a different weighting according to how important it is thought to be. Each job is then examined, analysed factor by factor, and a points score awarded for each factor, up to the maximum allowed. The total points score for each job is found by adding up its points score for each factor. The total points scored for each job provides the basis for ranking the jobs in order of importance, for grading jobs, if required, and for fixing a salary structure.

2.17 'Points rating has the advantage of flexibility in that the factors selected are best suited for the particular types of job being evaluated, and the importance given to each factor is decided by the allocation of points. It also provides a rank order of jobs according to the numbers of points, without determining the money value of the job. This allows the pattern of grades and salary rates to be determined as separate operations.

2.18 Like all systems, it has some disadvantages; the selection of factors, the points score allocated to a job, and the points weighting given each factor remain subjective judgements.' *(The Institute of Administrative Management)*

HAY-MSL method

2.19 The HAY-MSL method of job evaluation is a points method, whereby points are awarded for significant elements of a job and the importance of individual jobs relative to others is measured by comparing their total points scores. The job elements by which jobs are compared are, in effect:

(a) *know-how:* the amount of skill, knowledge and experience needed to do the job, including the ability to handle people;

(b) *problem solving:* this is concerned with the amount of discretion and judgement the job holder must exercise, the frequency of problems that call for decisions by the job-holder and the extent to which the job-holder is expected to contribute new ideas;

(c) *accountability:* this is the assessment of whether the job-holder is responsible and accountable for small or large areas of work, and whether the activities of the job holder affect the organisation to a larger or smaller extent in terms of money (revenues and expenditures).

2.20 The HAY-MSL system lends itself better to higher levels of management than to lower-ranking jobs. For example, in the case of a financial controller, technical know-how, the ability to solve finance-related problems, and stewardship over the company's money place him high on the HAY-MSL scheme.

Job evaluation form

Key job code ______________________ **Department** ______________________

Job type ______________________ **Job holder studied** ______________________

Date ______________________ **Employee number** ______________________

Task number

Description

Factor	Rating			Comments
	Points	Weighting	Total	
Skills and knowledge Education/qualifications Experience Dexterity				
Skills sub-total				
Initiative				
Responsibility People Equipment Resources				
Responsibility sub-total				
Effort Mental Physical				
Effort sub-total				
Communication Oral Written				
Communication sub-total				
Interpersonal skills				
Conditions of work Hazards Isolation Monotony				
Conditions sub-total				
TOTAL				
RANKING				
COMMENTS				

Exercise 2

'Job evaluation methods depend to some extent on a series of subjective judgments; the progressive refinement of job evaluation techniques is an attempt to minimise the subjective element.' How far does job evaluation meet this requirement of being 'objective' rather than 'subjective'?

ICSA June 1987

Solution

(a) Job evaluation in any of its forms is 'systematic' rather than 'scientific': the number of different inputs and environmental variables makes subjectivity inevitable.

(b) Objectivity is desirable: 'No employer ....can resist an equal pay demand with any confidence if he is employing non-analytical methods of job evaluation. The equal opportunity legislation will not consider such methods as 'proper job evaluation' (Cole, *Personnel Management: Theory and Practice*)

(c) Non-analytical methods (ranking, grading and classification) are susceptible to subjectivity. Points rating, Hay-MSL etc are less subjective.

(d) Despite the subjective element, job evaluation of *any* sort helps to minimise arbitrariness and emotional trauma in pay decisions.

3. INTRODUCING A JOB EVALUATION SCHEME

3.1 Steps in introducing a job evaluation programme will include:

(a) informing and involving staff - particularly where trade union attitudes need to be considered;

(b) selecting benchmark jobs as a sample for internal and external comparison;

(c) planning the programme itself, including:

- (i) staffing - who is responsible? what training do they need?
- (ii) information - to management, staff and unions
- (iii) procedures, methods and timetable
- (iv) techniques for pay comparison and job analysis as required;

(d) communicating and negotiating the results and structure;

(e) maintaining the scheme, including machinery for regradings and appeals.

3.2 Job evaluation is a highly political exercise, and will require openness and communication - not to mention diplomacy - throughout.

(a) Staff will have to be informed of the overall purpose, objectives and potential benefits of the system. It will, in particular, have to be made clear that the employees themselves are not being judged or evaluated. Increasingly, job evaluation *committees* are used, to involve staff in setting up, conducting and maintaining the scheme, to take advantage of the job holders' knowledge, and to minimise suspicion and demoralisation.

(b) The degree of consultation and participation will depend heavily on union attitudes (where unionisation exists). Unions may consider that job evaluation should be the true basis for *job* structuring only - otherwise it undermines the traditional role of collective bargaining. Others may simply insist on full communication between management and unions throughout the programme, the active participation of union members, and the institution of revision/appeals procedures.

(c) If an evaluation committee is used, there may be a delicate balance of power between management nominees, trade union representatives, and specialists (from the personnel or, in larger organisations, salary administration department). This will have to be moulded into a team, by sorting out its collective responsibilities and its component interests.

(d) There are bound to be problems which will require appeal, and revision will in any case be necessary over time. Some of these situations may be sensitive, not only where unions are involved: managers may regard the grading of jobs in their jurisdiction to be part of their political power base, and may be sensitive to any perceived undervaluation. Appeal procedures will have to be negotiated, to involve:

(i) the immediate superior, in the first instance;

(ii) the grading committee;

(iii) if the judgement is still unacceptable, union branch officials, or a higher organisational authority (according to normal grievance procedure for non-unionised workers); and

(iv) a top management committee, for final judgement.

4. OTHER FACTORS IN SETTING REMUNERATION LEVELS

4.1 We have discussed the role of job evaluation in determining the relative value of jobs, but there are other factors not related to job content which affect the rates an organisation will actually want to pay.

(a) *Equity*. Wilfred Brown defined equity as 'the level of earnings for people in different occupations which is felt by society to be reasonably consistent with the importance of the work which is done, and which seems relatively fair to the individual.' In other words, pay must be *perceived* and felt to match the level of work, and the capacity of the individual to do it: it must be 'felt-fair'. Pay structures should allow individuals to feel that they are being rewarded in keeping with their skill, effort and contribution, and with the rewards received by others for their relative contributions.

(b) *Negotiated pay scales*. Pay scales, differentials and minimum rates may have been negotiated at plant, local or national level, according to various environmental factors:

(i) legislation and government policy (on equal pay, say, or anti-inflationary increases);

(ii) the economy (levels of inflation; unemployment, affecting labour supply and demand, and therefore market rates); and

(iii) the strength of the employers and unions/staff associations in negotiation.

(c) *Market rates*. Market rates of pay will have most influence on pay structures where there is a standard pattern of supply and demand in the open labour market. If an organisation's rates fall below the benchmark rates in the local or national labour market from which it recruits, it will have trouble attracting and holding employees.

The concept of the market rate, however, is not exact. Different employers are bound to pay a range of rates for theoretically identical jobs, especially in managerial jobs, where the scope and nature of the duties will vary according to the situation of each organisation.

(d) *Individual performance in the job*. 'A growing number of organisations, commentators and academics assert that paying for individual skills, contribution and competence is more relevant to the needs of today than traditional job-based evaluation ... Placing the heaviest emphasis on job requirements discounts the importance of other compensatable factors - particularly individual capability and performance.' (Murlis and Fitt, *Personnel Management*, May 1991).

Market rates

4.2 Thomason suggests that if an employer were free to pay what he liked, he would pay 'the lowest rate consistent with securing enough labour in quantity to satisfy his production needs and ... to ensure ... a sufficient contribution to the enterprise's tasks to allow it to survive.' This is the *market rate* for the given type of labour. It will vary with supply/demand factors, such as:

(a) the relative scarcity of particular skills; and

(b) the extent of labour mobility in response to pay levels or differentials, which may dictate the need for higher rates of pay to retain employees, or to attract them from other organisations. Pay may or may not act as an incentive to change employers, depending on the availability of work elsewhere, the employee's loyalty, willingness to face risk and change, and the attractions of his present job which may not be measurable in financial terms: work relationships, conditions and so on.

4.3 Factors which distort or dilute the effect of the forces of supply and demand on labour pricing include:

(a) the organisation's ability to pay;

(b) the bargaining strength of unions (if any);

(c) government action, including incomes policies, equal pay legislation and anti-inflationary measures. The minimum wage, in particular, prevents outright exploitation of labour, even if employers wished to pursue it;

(d) internal differentials and equity existing in the organisation. Where there are established differentials, or a job-evaluated salary structure, it will be difficult for employers to justify a conspicuously low rate of pay for one type of job, or in response to market fluctuations;

(e) the culture and value systems of the organisation, which will influence the attitude of management towards the market rate, and whether age, length of service, motivation, employee aspirations and/or other factors are taken into account in the determination of pay, rather than fluctuations in supply and demand.

4.4 One may list the general arguments for paying over market rate as follows.

(a) The offer of a notably higher remuneration package than market rate may be assumed to generate greater interest in the labour market. The organisation will therefore have a wider field of selection for the given labour category, and will be more likely to have access to the most skilled/experienced individuals. If the organisation establishes a reputation as a 'wage leader' it may generate a consistent supply of high-calibre labour.

(b) There may be benefits of high pay offers for employee loyalty, and better performance resulting from the (theoretically) higher calibre and motivation of the workforce.

(c) Even if a cheap supply of labour were available, and the employer could get away with paying a low rate, the ideology or ethical code of an organisation may make him reluctant to do so. A socially responsible employer may wish to avoid the exploitation of labour groups, such as immigrants, who may not be aware of general market rates.

(d) An employer might adopt a socially responsible position not purely for ethical reasons but to maintain a respected image and good relations with government, interest groups, employee representatives and the general public (potential customers/ consumers).

(e) Survival and immediate profit-maximisation are not necessarily the highest objective of any organisation. Employers in growth markets, or hoping to diversify into new markets, cannot afford a low-calibre, high-turnover workforce. Notably innovative organisations can be seen to be offering higher than market rate on salaries (eg Mars) or remuneration packages including profit-related bonuses (eg Sainsbury's): moreover, their financial performance bears out their view that pay is an investment. To an extent, this pay strategy stems from the culture or value system of the organisation, the importance it attaches to loyalty, innovation and initiative, and its willingness to pay more to attract and retain such higher-level attributes: quantity may not be the prime employment criterion.

4.5 On the other hand, there are substantial cost savings in paying lower rates. It cannot be assumed that high remuneration inevitably leads to higher motivation and better performance. Not everybody has an instrumental orientation to work: money may not be the prime incentive - and pay is often a source of dissatisfaction rather than satisfaction, whatever its level.

If the organisation's ability to maintain high rewards in the future is in doubt, management ought also to be aware that the disappointment and culture shock of reversing a high-remuneration policy is very great.

Exercise 3

Distinguish between the following terms.

(a) Job design
(b) Job specification
(c) Job analysis
(d) Job enrichment
(e) Job evaluation
(f) Job description

Solution

If you cannot do this (or to check, even if you can) you will need to look back to earlier chapters. Use the index to find the terms.

Chapter roundup

- 'Job evaluation is the process of analysing and assessing the content of jobs, in order to place them in an acceptable rank order which can then be used as a basis for a remuneration system'

Scheme	*Characteristics*	*Advantages*	*Disadvantages*
Ranking method	Whole job comparisons are made to place them in order of importance.	Easy to apply and understand.	No defined standards of judgement - differences between jobs are not measured.
Job classification method	Job grades are defined and jobs are slotted into the grades by comparing the whole job description with the grade definition.	Simple to operate and standards of judgement are provided in the shape of the grade definition.	Difficult to fit complex jobs into one grade without using excessively elaborate definitions.
Factor comparison method	Separate factors are identified as a proportion of the job and valued in comparison with benchmark jobs.	Uses benchmark jobs which are considered to be fairly paid.	Hard to explain. Depends on accuracy of benchmark rates. Limited applicability.
Points rating methods	Separate factors are scored to produce an overall points score for the job.	The analytical process of considering separate defined factors provides for objectivity and consistency in making judgements.	Complex to install and maintain - judgement is still required to rate jobs in respect of different factors.

- 'The constituents of a remuneration policy must therefore embrace such crucial factors as the objectives of the organisation, its finances, cash flow and profitability, the state of the labour market, expected demand and supply of various types of labour, any government regulations on pay, anticipated contraction or expansion of the organisation, as well as the personal aspirations and inclinations of the workforce.'

Test your knowledge

1. Outline the advantages and disadvantages of job evaluation. (see paras 1.3, 1.4)
2. Distinguish between analytical/quantitative and non-analytical/qualitative approaches to job evaluation. (2.2, 2.3)
3. What are the three job elements compared in the points method known as the HAY-MSL method? (2.19)
4. Outline the steps you would take if you were asked to introduce a job evaluation scheme in your office. (3.1, 3.2)
5. Job evaluation is concerned with job *content*. What other factors need to be considered in setting remuneration levels? (4.1)
6. Why might an organisation wish to offer rates of pay above the market level? (4.4)

Now try question 14 at the end of the text

Chapter 15

TERMINATION OF CONTRACT

This chapter covers the following topics.

1. Contracts of employment
2. Retirement and resignation
3. Dismissal
4. Redundancy

Signpost

- This chapter deals with one last aspect of manpower planning: employees leaving the organisation. General considerations were discussed in Chapter 8.

1. CONTRACTS OF EMPLOYMENT

1.1 A contract of employment is not necessarily a written document: all that is required to establish a contract is agreement of the essential terms by both parties. *Fixed term* contracts may be made for a clearly defined period, but most contracts of employment are *indefinite*: they run until terminated by either party, subject to notice.

1.2 A contract will contain:

(a) *express terms:* those specifically agreed upon by the parties, regarding working hours, rates of pay, the nature of the work and so on; and

(b) *implied terms:* those which are not actually stated, but still impose duties and obligations on both parties. They must be well known to all parties concerned. They include common law obligations, such as the employee's duty to perform whatever tasks he has undertaken, and the employer's to pay the agreed consideration.

(i) The *employer* also has obligations of:

(1) *trust:* to behave in a proper and responsible way;

(2) *care:* to provide safe conditions and methods, and compensation for any negligence;

(3) *provision of work:* the general obligation to ensure work, especially for those such as salesmen on commission or specialists for whom idleness might mean loss of skills; to allow the opportunity to earn overtime and to work out notice periods. (This is a fluid situation at present, and any contract should be examined for implied terms on these matters.)

(ii) The *employee* also has obligations of:

(1) *fidelity:* to give faithful service and show concern for the organisation's interests;

(2) *obedience:* to obey any lawful order within the scope of his employment, and without undue risk to himself;

(3) *care:* to carry out his work reasonably and responsibly.

1.3 Since 1963 it has been the law that, where there is not a written contract of employment, the employer must provide a statement of written particulars. This requirement is now governed by the Trade Union Reform and Employment Rights Act 1993 which states that employers must provide such written particulars to any employee who works for eight hours or more a week, within two months of that employee's start date.

1.4 Circumstances in which a contract of employment may come to an end include:

(a) *by mutual agreement:* retirement, or possibly 'constructive dismissal', where the employee is forced to resign because of irreconcilable differences with the employer;

(b) *by notice*. For example:

(i) resignation;
(ii) dismissal, or
(iii) redundancy;

(c) *by breach of contract*, entitling the employer to dismiss the employee without notice. Examples may include:

(i) wilful disobedience of a reasonable order, representing total disregard for the terms of the contract;

(ii) misconduct in employment, such as dishonesty or violence;

(iii) misconduct *outside* employment which interferes with work performance, such as persistent drunkenness;

(iv) serious negligence or incompetence;

(d) *by frustration*, through the death, illness or imprisonment of the employee or employer.

We will go on to discuss some of these circumstances.

2. RETIREMENT AND RESIGNATION

Retirement

2.1 The average age of the working population has been steadily increasing, with higher standards of living and health care. The problems of older workers and retirement are therefore commanding more attention. The time at which difficulties in obtaining or retaining jobs because of age occur will obviously vary according to the individual, his lifestyle and occupation, and the attitudes of his society and employers.

In later middle age, many workers try to move away from jobs which make demands on their agility, energy or muscular strength. However, they are still capable of less strenuous work, and particularly in the office setting may be very valuable in jobs requiring mature judgement, conscientiousness, attention to detail or experience. They may make a great contribution to the training and coaching of more junior staff.

2.2 From the organisation's point of view, however, there are various arguments for enforcing retirement.

(a) There is resistance to late retirement from younger workers, because it is felt that promotion opportunities are being blocked. The Job Release Scheme even encourages individuals to retire a year or two *earlier* than usual, so as to create job vacancies for the pool of unemployed workers.

(b) Younger employees with family responsibilities need to have their jobs secured, and in a redundancy situation, it is common for pensioners and those nearing retirement to be discharged first.

(c) The age structure of an organisation may become unbalanced for future work requirements: there may have to be an injection of 'younger blood' through the compulsory retirement of older workers.

(d) Engaging staff above middle age can be costly for the organisation: the cost of providing pensions rises according to the age at which the employee joins the superannuation scheme. Many pension funds exclude the entry of men and women above a specified age.

(e) Individual mental and/or physical shortcomings may render an older individual unfit to carry out his duties efficiently.

2.3 The personnel and/or line manager will have to consider how far any of the above factors apply in a given situation. It will depend to a large extent on the individual concerned, the type of work involved and the state of the local labour market. Retirement policies, and age limits on particular posts will have to be clearly communicated and decisions regarding particular cases discussed confidentially and tactfully with the individual concerned. Written confirmation of the decision to retire an employee should likewise be tactful, with expressions of regret and appreciation as appropriate.

2.4 Employers can give not only financial assistance to retiring employees, but also practical help and advice.

(a) The burden of work in later years can be eased by shortening hours or a transfer to lighter duties.

(b) The final stage of employee training and development may take the form of courses, commonly run by local technical colleges, intended to prepare employees for the transition to retirement and non-work.

(c) The organisation may have, or may be able to put employees in touch with, social/leisure clubs and other facilities for easing the shock of retirement.

2.5 It should be noted that the Sex Discrimination Act 1986 introduced equal retirement age for men and women. *Employers* setting a compulsory retirement age must apply it to men and women alike: women made to retire earlier than male colleagues can claim both discrimination and unfair dismissal. The *State* retirement age is still 65 for men and 60 for women at present.

'Retirement ages for women as well as men will be set at 65 if the recommendations of the Social Security Advisory Committee (SSAC) are taken up by the Government. The committee proposes that the retirement age for women is gradually equalised with that for men over a 15-year period beginning in 2000.

It suggests that the money saved by this move, around £3 billion a year, is used to improve the pensions and benefits of the low-paid, especially low-paid women....

The IPM had recommended flexible retirement around a pivotal age of 63. 'We should face the reality of the situation - very few males over the age of 63 are in employment.... If you do make people wait till 65, and many of them are unemployed between 63 and 65, you are going to have to pay them unemployment benefit.' The IPM, however, welcomes the proposal to transfer the money saved to lower-paid pensioners....

Equality at 65 would bring the UK into line with the rest of Europe.... However, the trend in the UK is towards earlier retirement, with the vast majority of women retiring at 60 or before and over 50 per cent of men retiring before 65.'

Personnel Management, September 1992

Resignation

2.6 Employees may resign for any number of reasons, personal or occupational. Some or all of these reasons may well be a reflection on the structure, management style, culture or personnel policies of the organisation itself. When an employee announces his intention to leave, verbally and/or by letter, it is important for management to find the real reasons why he is leaving, in an *exit interview*. This may lead to a review of the existing policies on pay, training, promotion, the work environment, the quality and style of supervision and so on.

2.7 The principal aspect of any policy formulated to deal with resignations must be the length to which the organisation will go to try and dissuade a person from leaving. In some cases, the organisation may decide to simply let the person go, but when an employee has been trained at considerable cost to the firm, or is particularly well qualified and experienced (no employee is irreplaceable - but some are more replaceable than others...), or has knowledge of information or methods that should not fall into the hands of competitors, the organisation may try to keep him.

Particular problems the employee has been experiencing may be solveable, though not always in the short term. It may be that the organisation will try to match or improve on a salary offer made to the individual by a new prospective employer. In that case, however, there may well be a problem of pay differentials and the individual's colleagues, doing the same work, may have to be given similar increases: can so large a cost be justified?

2.8 Various arrangements will have to be made when an employee decides to leave. There will have to be co-operation and full exchange of information between the personnel function and the leaver's superior so that procedures can be commenced upon notification of an intended departure.

(a) If attempts (if any) to make the employee stay have been unsuccessful, the exit interview will have to be arranged.

(b) The period of notice required for the employee to leave should be set out in his contract of employment, but some leeway may be negotiated on this. The time needed for recruitment and induction of a replacement may dictate that the leaving employee work out his full period of notice, and perhaps even longer if he is willing. On the other hand, if it is felt that he can be easily replaced, and that his continuing presence may be destructive of morale (or possibly of advantage to competitors, as he continues to glean information), it may be possible to persuade him to accept pay in lieu of notice and leave immediately.

(c) Details of the departure will have to be notified to the wages clerk, pension fund officer, social secretary, security officer and so on, so that the appropriate paperwork and other procedures can be completed by his date of leaving.

(d) The departmental head and/or supervisor should complete a leaving report form: an overall assessment of the employee's performance in the organisation. This can then be used to provide references to his future employer(s).

Exercise 1

(a) What value would you place on information obtained in an exit interview?

(b) Find out what are the procedures for handling the resignation of an employee in your organisation. Are exit interviews held, and if so, is anything ever done about the problems that may come to light as a result?

Solution

(a) There is clearly a strong possibility that employees who are leaving will not bother, or will still be too intimidated, to give the full reasons for their resignation. Others may simply be vindictive. When several people leave and all give similar reasons, however, there is a problem in the organisation to be addressed.

---

3. DISMISSAL

3.1 Dismissal includes not only the termination of an employee's contract by his employer, but also:

(a) the ending of a fixed term contract without renewal on the same terms; and

(b) termination by the employee himself where the employer's conduct makes him entitled to do so (constructive dismissal). Constructive dismissal is determined as being by reason of: 'Conduct which is a significant breach of the contract of employment or which shows that the employer no longer intends to be bound by one or more of the essential terms of the contract ... The conduct must be sufficiently serious to entitle him to leave at once.' *(Western Excavating v Sharp 1978)*.

3.2 If an employer terminates the contract of employment by giving notice the minimum period of notice to be given is determined by the employee's length of continuous service in the employer's service.

Employee's length of service	*Minimum notice to be given by the employer*
1 month - 2 years	1 week
2 - 12 years	1 week for each year of service
12 years and over	12 weeks

This is the statutory minimum: longer periods may be written into the contract, at the employer's discretion, and by agreement. Either party may waive his right to notice, or accept payment in lieu of notice. If an employee asks, he is entitled to a written statement of the reasons for his dismissal within 14 days.

Wrongful dismissal

3.3 A claim for wrongful dismissal is open to an employee at *common law,* if he can show he was dismissed without a reasonable cause.

An employee will have to show that his dismissal was 'without just cause or excuse', that circumstances did not justify dismissal (say, in the case of a minor breach of rules). He may then be able to claim accrued wages, payment for an entitlement to notice, or the balance of wages due under a fixed term contract. In practice, such claims are less common, now that *unfair* dismissal provisions offer wider remedies, but the common law remedy is still useful for those excluded (notably by a qualifying period of employment) from claiming unfair dismissal.

Unfair dismissal

3.4 An employee may bring a claim before the Industrial Tribunal if he considers that he has been *unfairly* dismissed, under the Employment Protection (Consolidation) Act 1978. The employee first has to prove that he has been dismissed. The onus is then on the *employer* to prove that the dismissal was *fair*. Under the 1978 Act, dismissal is fair and justified if the reason for it was:

(a) redundancy (provided that the selection for redundancy was fair);

(b) legal impediment - the employee could not continue to work in his present position without breaking a legal duty or restriction. (This is fair only if the employee was offered suitable alternative employment);

(c) non-capability (provided adequate training and warnings had been given);

(d) misconduct (provided warnings suitable to the offence have been given - so the disciplinary procedures of the organisation are vitally important); or

(e) some other 'substantial' reason: for example, the employee is married to a competitor, or refuses to accept a reorganisation made in the interests of the business and with the agreement of other employees.

3.5 Situations in which *unfair* dismissal can be claimed include:

(a) unfair selection for redundancy;

(b) dismissal because of membership (actual or proposed) and involvement in the activities of an independent trade union. (The protection previously given to employers who dismissed employees for *not* being trade union members, in certain circumstances where a closed shop operated, was removed by the Employment Act 1988);

(c) dismissal because of pregnancy, *unless* by reason of it the employee becomes incapable of doing her work adequately.

3.6 In order to qualify for compensation or other remedies for unfair dismissal the employee must:

(a) be *under the normal retiring age* applicable to his job or grade, as determined by the employer's practice;

(b) have been *continuously employed* for the required period - one year, in the case of employees whose service began on or after 1 June 1985, or two years. There is no qualifying period required in cases of dismissal because of trade union membership;

(c) have been *dismissed*;

(d) have been *unfairly* dismissed.

3.7 The Conciliation Officer or Industrial Tribunal to whom a complaint of unfair dismissal is made may order various remedies including:

(a) *re-instatement* - giving the employee his old job back;

(b) *re-engagement* - giving him a job comparable to his old one;

(c) *compensation*. This may consist of:

(i) a *basic award* calculated on the same scale as redundancy pay. If the employee is also entitled to redundancy pay, the lesser is set off against the greater amount;

(ii) a *compensatory award* (taking account of the basic award) for any additional loss (earnings, expenses, benefits) on common law principles of damages for breach of contract;

(iii) a *punitive additional award* if the employer does not comply with an order for re-instatement or re-engagement and does not show that it was impracticable to do so.

3.8 In deciding whether to exercise its powers to order re-instatement or re-engagement the tribunal must take into account whether the complainant wishes to be reinstated, whether it is practicable for the employer to comply with such an order and, if the complainant contributed to any extent to his dismissal, whether it would be just to make such an order. Such orders are very infrequent.

Dismissing incompetent workers

3.9 In an article in *The Administrator* (June 1985) John Muir wrote of the difficulties of 'firing' incompetent employees.

'It is a fairly common experience to hear managers at all levels say of a subordinate, 'I wish I could get rid of him. He's really not up to the job and costs us money. But he's been here years and I really don't see how I can do it.' Behind such a statement there are usually two themes. One is about the personal difficulty and the unpleasantness involved in going to the individual and starting the process leading to dismissal; the other is the fear of being taken to an industrial tribunal and being found to have acted unfairly.'

3.10 The solution to these difficulties lies partly in the hands of the personnel function, which may be responsible for designing the procedures for dismissal. These should include the following.

(a) Ensuring that *standards* of performance and conduct are set, clearly defined and communicated to all employees

(b) *Warning* employees where a gap is perceived between standard and performance

(c) Giving a clearly defined and reasonable *period for improvement* - with help and advice where necessary, and clear improvement targets

(d) Ensuring that disciplinary procedures and the ultimate consequences of continued failure are made clear

If such procedures are formulated, the employer will not only feel that he has given the employee every chance to redeem himself, but will also be in a strong position at a tribunal in rebutting a complaint of unfair dismissal.

3.11 Muir also notes the particular difficulty of identifying the principal reason for dismissal (as required by EP(C)A 1978). *Capability* - or, in practical terms, incapability or incompetence - is one such reason, but must be clearly identified. Incompetence means that the employee has not reached the standard required, although he has done what he considers his best, leaving aside extraneous factors like non-availability of information or materials. If the employee has deliberately not done his best, however, this is *misconduct*, which is a dismissable offence.

> 'Suppose the task is to set a machine which will then turn out work automatically. The more work that is turned out the bigger the bonus but the job should be stopped every ten minutes so as to check the settings. While the machine is stopped there is no production so no bonus accrues. In fact the machine is not stopped at all. The original setting was wrong and the entire output has to be scrapped. There is the element of misconduct in that the requirement to check the machine every ten minutes was not adhered to and there is a major element of incompetence in not setting the machine correctly in the first place. On the other hand, was the incorrect setting the result of a deliberate 'couldn't care less' attitude? If so that, too, is misconduct.' *The Administrator June 1985*

4. REDUNDANCY

4.1 Redundancy is defined by the Act as dismissal where:

(a) the employer has ceased to carry on the business;

(b) the employer has ceased to carry on the business in the place where the employee was employed;

(c) the requirements of the business for employees to carry out work of a particular kind have ceased or diminished or are expected to.

4.2 Redundant employees are entitled to compensation:

(a) for loss of security; and
(b) to encourage employees to accept redundancy without damage to industrial relations.

The employee is *not* entitled to compensation if:

(a) the employer has made a suitable offer of alternative employment and the employee has unreasonably rejected it. The 'suitability' of the offer will have to be examined in each case;

(b) the employee is of pension age or over, or has less than 2 years' continuous employment;

(c) the employee's conduct entitles the employer to dismiss him without notice.

Procedure for handling redundancies

4.3 From a purely humane point of view, it is obviously desirable to consult with employees or their representatives, and to give warning of impending redundancies.

The Employment Protection Act 1975 imposes this as a duty. Protective awards (an order of a tribunal that an employer shall continue to pay remuneration of employees) may be made against an employer who in a redundancy situation fails to consult trade unions at the earliest opportunity, or to give notice of impending redundancies to the Department of Employment.

4.4 The employer's duty is to consult with any trade union which is independent and recognised (in collective bargaining) by him as representative of employees. The consultation must begin 'at the earliest opportunity', defined as:

(a) a minimum of 90 days before the first dismissal, if 100 or more employees are to be dismissed at any one establishment;

(b) a minimum of 30 days before the first dismissal of 10 - 99 employees;

(c) at the earliest opportunity before even one (but not more than 9) employees are to be dismissed for redundancy.

These rules are applied to the total number involved and cannot be evaded by making essential dismissals in small instalments. The employer must within the same periods notify the Secretary of State in writing of proposed redundancies, with details of consultations with the trade union: a copy of this notice is given to the union representative.

4.5 In giving notice to the trade union the employer must give certain details in writing, including the reasons for the dismissals, the numbers employed and the number to be dismissed, the method of selecting employees for dismissal and the period over which the dismissals will take place. Information should be accurate, clear, realistic and positive as far as possible: ideas for retraining and redeployment, benefits and potential for voluntary redundancies or retirements should be far enough advanced that *some* good news can be mixed with the bad. The employer should allow the trade union time in which to consider what he has disclosed and to make representations or counter proposals.

Softening the blow

4.6 Measures which might be considered and discussed with employee representatives to avoid or reduce the numbers of forced redundancies that have to be made include:

(a) retirement of staff over the normal retirement age;

(b) offering early retirement to staff approaching normal retirement age;

(c) restrictions or even a complete ban on recruitment, so as to reduce the workforce over time by natural wastage;

(d) dismissal of part-time or short-term contract staff, once contracts come to sensible break-off points or conclusions;

(e) offering retraining and/or redeployment within the organisation;

(f) seeking voluntary redundancies.

4.7 Where enforced redundancies are necessary, the legal provisions discussed above, together with such collective bargaining procedures as may exist, come into force. Where possible, procedures and benefit packages should be planned for well in advance as a contingency measure, rather than as a reactive measure in the context of cost-cutting and industrial conflict. (This also benefits the employee, since benefits are likely to be higher if they are set *before* the use of economy measures which commonly necessitate workforce reduction.)

4.8 There are various approaches to selection for redundancy. If demand for a particular type of work has disappeared completely, the situation is relatively clearcut: all those previously contracted to perform that work can be dismissed. Where management have to choose between individuals doing the same work, they may take the following approaches.

(a) Enforced or early retirement.

(b) Seeking volunteers, who would be willing to take their chances elsewhere on good redundancy terms.

(c) Value to the organisation, or retention by merit - keeping those who perform well and dismissing less effective workers (although this may be harder to justify to individuals and their representatives).

(d) 'Last in, first out' (LIFO). Newcomers are dismissed before long-serving employees. This may sound fair (especially in a seniority culture such as characterises Japanese big business), but may not meet the organisation's need for 'young blood' or for particular skills or merit in the individual.

4.9 Many large organisations provide services and benefits well in excess of the statutory minimum, with regard to consultation periods, terms, notice periods, counselling and aid with job search, training in job-search skills and so on. Many, however, reduce staffing levels in order to cut costs or take advantage of technological advances, without counting the cost in human terms, in times of high unemployment. The State's minimum standards are therefore necessary - but in some cases still allow for the exploitation of short-contract and part-time workers.

Moreover, the Inland Revenue recently confirmed (*Personnel Management Plus*, September 1992) that *outplacement counselling* (counselling individuals who have been made redundant) is liable for tax, as part of the redundancy package. The IPM's spokesperson on outplacement has said that it should be considered a form of retraining and should not be taxed: 'If employers are discouraged from providing these services, then redundant workers may remain unemployed longer.'

Exercise 2

'Personnel Management will become increasingly involved in getting rid of people instead of recruiting them.' For what reasons might such a prediction be made? To what extent is it correct, in your judgment? *ICSA June 1987*

Solution

(a) John Hunt *(The shifting focus of the personnel function)*: 'In sharp contrast to the search for talent is the dramatic shift in the personnel function from people resourcing to people exiting.'

(b) Pressures on intensive use of manpower, contraction of workforce (plus expansion in unemployment in UK), due to the following.

(i) Competition
(ii) New technology
(iii) Recession and decline in world trade

(c) But this is not necessarily a trend which will continue in future. Technology does not always 'replace' human operation and lead to manpower savings.

(d) There are alternatives to redundancy: job-sharing, use of manpower agencies, part-time or temporary assignments, networking.

Unemployment in perspective

4.10 According to McIlwee *(Personnel Management in Context: the late 1980s)*, there are different types of unemployment, which will impact in different ways on individuals and job markets.

(a) *Voluntary unemployment* is where people leave an organisation because they believe their prospects are better elsewhere, or that their present employer does not put a sufficiently high value on their skills and experience.

(b) *Involuntary unemployment* is where people cannot find work for what they consider a suitable level of pay.

(c) *Unemployment caused by low levels of demand for labour* includes:

 (i) *seasonal unemployment*, the periodic job-shedding which is common in industries such as construction and tourism;

 (ii) *cyclical unemployment*; and

 (iii) *growth gap unemployment*, where there is a long recession, and little prospect of an upturn in economic activity.

(d) *Unemployment caused by changes in skill requirements* includes:

 (i) *hard-core unemployment*, that small proportion of any society that are unlikely to gain any employment because of social, physical, mental or other deficiencies;

 (ii) *frictional unemployment*, where there is a time lag between a new skill requirement in the labour market and individual adjustment to it - either by acquiring or polishing skills, or by adjusting wage expectations;

 (iii) *structural unemployment*, which is caused by long-term changes in the industrial environment. The demand for labour in a given sector (for example, a declining industry such as shipbuilding) may fall more quickly than supply readjusts itself through natural wastage, redeployment and so on.

4.11 Unemployment can represent not only an economic threat to lifestyle, but a source of insecurity, loss of self-esteem, extreme stress and hopelessness. The personnel function should therefore be concerned to *prevent* enforced redundancies where possible, and to *alleviate* the effects of unemployment where it is unavoidable, by:

(a) careful manpower planning, so that foreseen seasonal or other contractions in demand for labour can be taken into account, and the organisation is not over-supplied with labour for its needs;

(b) planning redundancy terms and measures early, to safeguard the interests of those who may be made redundant;

(c) retraining and redeployment programmes. This may be a solution where alternative jobs are available, employees have some of the skills (or at least aptitudes) required and retraining facilities are available;

(d) liaison with other employers in the same industry or area, with a view to redeployment within the linked group of organisations;

(e) provision of unemployment services, (or time, during the notice period, to seek them) such as:

 (i) counselling, to aid re-adjustment to the situation in which the newly-unemployed individual finds himself, to encourage a positive outlook;

 (ii) training in job-search skills: how to locate employment opportunities; how to carry out self-appraisal and communicate it attractively on a CV; how to use application forms, letters and phone calls to advantage; how to handle interviews;

 (ii) information on job opportunities and self-employment opportunities and funding. Individuals should be made aware of the role and accessibility of the Department of Employment's facilities and private sector services for careers counselling, recruitment and CV preparation.

4.12 Ultimately, most of these measures are palliatives rather than solutions. In the face of hard-core growth-gap or structural unemployment, the personnel function is likely to be powerless to prevent the shedding of jobs. Nevertheless, efforts to cushion the impact of unemployment, and particularly the encouragement of training and retraining, should be considered by socially responsible employers.

Chapter roundup

- *Exit* from employment takes several forms, voluntary and involuntary, but is likely to be traumatic in some degree to the leaving individual whatever the circumstances, because of the centrality of work and job security in most people's lives.
- The organisation should consider the sensitivity of the situation - not least because it may itself be traumatised by the exit of key individuals.
- Full account should be taken of the implications of and reasons for their leaving, the legal obligations with regard to employment protection and so on.

Test your knowledge

1. What procedures should be carried out when an employee resigns from the organisation? (see para 2.8)
2. In what circumstances is an employee 'dismissed' in the sense used in employment protection law? (3.1)
3. What reasons may an employer rely on in seeking to show that a dismissal was fair? (3.4)
4. What is meant by redundancy? (4.1)
5. What positive steps could a personnel department take to reduce the impact of redundancy on employees? (4.11, 4.12)

Now try question 15 at the end of the text

Chapter 16

COMMUNICATION IN ORGANISATIONS

This chapter covers the following topics.

1. The need for communication
2. The communication process
3 The management problem of communication
4. Informal communication channels
5. Communication methods
6. Face-to-face communication

Signpost

- Without constant exchanges of information organisations could not function at all. We have frequently referred to the need for effective communication and in this final chapter we explore the topic in some depth.
- Keep in mind what you learnt in Chapters 1 to 4 on organisations, their structure, and the work environment, in Chapter 5 on group working and in Chapter 13 on motivation.

1. THE NEED FOR COMMUNICATION

Information

1.1 *Information* is anything that is communicated. It is made up of *data* - facts and figures and so on which are processed so that they are meaningful to the person who receives them. Thus 0817401111 is data. 'BPP Publishing's telephone number is 081-740 1111' is information. This is a very simple example of *classification*, which involves identifying certain characteristics in items of data and grouping them in categories, according to the purpose for which the information is needed.

Communication

1.2 In any organisation, the communication of information is necessary:

(a) for management, to make the necessary decisions for *planning, co-ordination and control*; managers should be aware of what their departments are achieving, what they are not achieving and what they *should* be achieving;

(b) between departments, so that all the interdependent systems for purchasing, production, marketing and administration can be synchronised to perform the right actions at the right times to *co-operate* in accomplishing the organisation's aims;

(c) by individuals. Each employee should know what is expected from him. Otherwise he may be off-target, working without understanding, interest or motivation, and without any sense of belonging and contributing to the organisation. Effective communication gives an employee's job meaning, makes personal development possible, and acts as a motivator, as well as oiling the wheels of labour relations.

1.3 Communication in the organisation may take the form of:

(a) giving instructions;

(b) giving or receiving information;

(c) exchanging ideas;

(d) announcing plans or strategies;

(e) comparing actual results against a plan;

(f) laying down rules or procedures;

(g) job descriptions, organisation charts or manuals (communication about the structure of the organisation and individual roles).

Direction of communication

1.4 Communication within the formal organisation structure may be:

(a) *vertical:*

(i) downwards (from superior to subordinate);
(ii) upwards (from subordinate to superior);

(b) *horizontal or lateral* (between people of the same rank, in the same section or department, or in different sections or departments).

1.5 Communication is perhaps most routine between people at the same or similar level in the organisation, that is horizontal communication. It is necessary for two reasons:

(a) *formally:* to co-ordinate the work of several people, and perhaps departments, who have to co-operate to carry out a certain operation;

(b) *informally:* to furnish emotional and social support to an individual.

Horizontal communication between 'peer groups' is usually easier and more direct than vertical communication, being less inhibited by considerations of rank. It has to be tactfully handled, however, to avoid situations where one manager accuses another of putting his nose into affairs which shouldn't concern him.

1.6 Communication may be said to be 'diagonal' when it involves interdepartmental communication by people of different ranks. Specialist departments which serve the organisation in general, such as personnel, filing or data processing, have no clear line of authority linking them to managers in other departments who need their involvement. A sales supervisor, for example, may find himself 'requiring' the services of the personnel manager in a disciplinary procedure: who reports to whom? Particular effort, goodwill and tact will be required in such situations.

2. THE COMMUNICATION PROCESS

2.1 The diagram below demonstrates in outline the process that is involved in communication.

The communication process

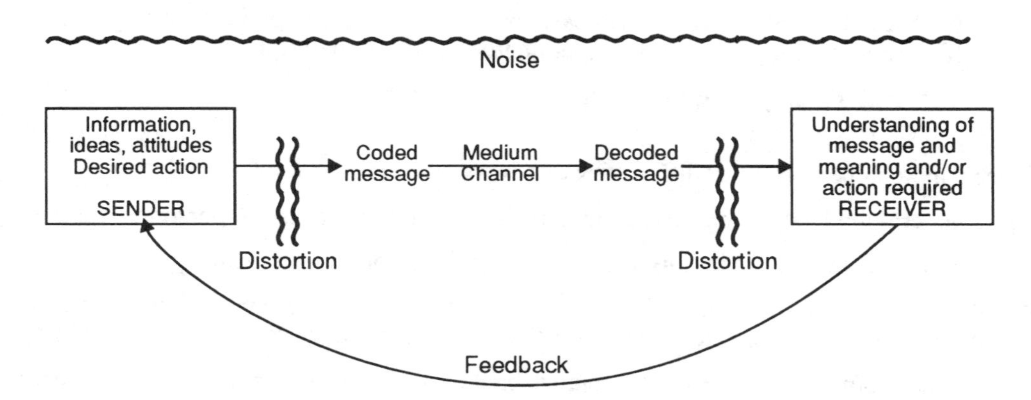

A number of points should be noted.

(a) *Coding of a message:* the code or 'language' of a message may be verbal (spoken or written) or it may be non-verbal, in pictures, diagrams, numbers or body language.

(b) *Medium for the message:* there are a number of channels for communication, such as a conversation, a letter, a notice board or via computer. The choice of medium used in communication depends on a number of factors such as urgency, permanency, complexity, sensitivity and cost. We shall consider this further below.

(c) *Feedback:* it is of vital importance that the sender of a message gets feedback on the receiver's reaction. This is partly to test his understanding of it and partly to gauge his reaction.

(d) *Distortion* refers to the way in which the meaning of a message is lost in 'handling', that is at the coding and decoding stages. Usually the problem is one of language and the medium used; most of us have found, for instance, that it is far easier to get the wrong end of the stick in a telephone call than from a letter, since non-verbal language is missing and the message is transient.

(e) *Noise* refers to distractions and interference in the environment in which communication is taking place. It may be physical noise (passing traffic), technical noise (a bad telephone line), social noise (differences in the personalities of the parties) or psychological noise (anger, frustration, tiredness).

2.2 When communicating within a business environment it is important to be aware of the actual process which is taking place. Advantage can be taken of opportunities to reduce noise and distortion by clarifying the message, choosing the appropriate medium (say by backing up a telephone conversation with a letter) and checking on feedback.

3. THE MANAGEMENT PROBLEM OF COMMUNICATION

3.1 Rosemary Stewart reported a survey of 160 managers in British companies (1967) in which it was found that 78% of their time was spent communicating (talking took up 50% and reading and writing 28% of their time).

3.2 Good communication is essential to getting any job done: co-operation is impossible without it. Difficulties occur because of general faults in the communication process:

(a) distortion or omission of information by the sender;

(b) misunderstanding due to lack of clarity or technical jargon;

(c) non-verbal signs (gesture, posture, facial expression) contradicting the verbal message, so that its meaning is in doubt;

(d) 'overload' - a person being given too much information to digest in the time available;

(e) differences in social, racial or educational background, compounded by age and personality differences, creating barriers to understanding and co-operation;

(f) people hearing only what they want to hear in a message.

3.3 There may also be particular difficulties in a work situation:

(a) a general tendency to distrust a message in its re-telling from one person to another;

(b) a subordinate mistrusting his superior and looking for 'hidden meanings' in a message;

(c) the relative status in the hierarchy of the sender and receiver of information (a senior manager's words are listened to more closely and a colleague's perhaps discounted);

(d) people from different job or specialist backgrounds (accountants, personnel managers, DP experts) having difficulty in talking on a non-specialist's wavelength;

(e) people or departments having different priorities or perspectives so that one person places more or less emphasis on a situation than another;

(f) subordinates giving superiors incorrect or incomplete information (to protect a colleague, to avoid 'bothering' the superior); also a senior manager may only be able to handle edited information because he does not have time to sift through details;

(g) managers who are prepared to make decisions on a 'hunch' without proper regard to the communications they may or may not have received;

(h) information which has no immediate use tending to be forgotten;

(i) lack of opportunity, formal or informal, for a subordinate to say what he thinks or feels;

(j) conflict in the organisation. Where there is conflict between individuals or departments, communications will be withdrawn and information withheld.

Barriers to communication

3.4 The barriers to good communication arising from differences in social, racial or educational backgrounds, compounded by age differences and personality differences, can be particularly severe.

(a) A working class person with a comprehensive school background might feel resentful towards a fellow employee with an upper class background and public school education. The ex-public school employee might consider himself superior to the other. The different backgrounds and inferior/superior attitudes will make communication difficult.

(b) A young person might be resented by an older person of the same grade or status in the organisation. The young person might look down on the older one as someone who has failed to advance his career, and is therefore second-rate. Difficulties in seeing each other's point of view might be compounded by different methods of expression (slang words and phrases and so on).

(c) Personality differences might occur where one person appears fairly happy-go-lucky, and another is more serious in his application to work and his outlook on life. Frustration may occur in communication between the two because their different values give them conflicting views about what is important and what is less so.

3.5 Differences in background might result in:

(a) failure to understand the other's point of view and sense of values and priorities;

(b) failure to listen to the information the other person is giving (the information is judged according to the person who gives it);

(c) a tendency to give the other person ready-formulated opinions (which the other does not accept) instead of factual information which will enable the other person to formulate his own opinions;

(d) lack of shared 'vocabulary' - whether linguistic or symbolic - which might lead to lack of understanding of the message.

3.6 Personal conflict or antagonisms will cause further communication problems.

(a) Emotions (anger, fear, frustration) will creep into communications and further hinder the transmission of clear information.

(b) The recipient of information will tend to:

(i) hear what he wants to hear in any message;
(ii) ignore anything he does not want to accept in a message;

and blame it on the other person if problems arise later on.

Clarity (and 'noise')

3.7 Information must be clear to the user. If the user does not understand it properly he cannot use it properly. Lack of clarity is one of the causes of a breakdown in communication, which is referred to in information system theory as *noise*.

3.8 The following three faults may cause a full or partial breakdown in communication and they are known as 'noise'.

(a) At the collecting and measuring point, more information may be collected than it is possible to use, and much of what is transmitted may be irrelevant.

(b) In communicating information, important errors or omissions may occur.

(c) At the receiving end, information may be misinterpreted, misunderstood, or simply ignored.

Noise is therefore caused by incompleteness, irrelevance, excessive volumes of information and lack of clarity.

3.9 A distinction can be made between:

(a) *semantic noise* where the meaning of the message is unclear because it is badly expressed by the sender, or not properly understood by the receiver; and

(b) *technical noise* where the message is distorted or prevented from getting through, perhaps for technical reasons such as distortion of data during transmission along a telecommunications cable.

3.10 Noise can take many forms, but some examples may help as illustrations.

(a) A manager wants to receive one item of information, and he is given a large report containing several appendices and tables of figures, somewhere in which the item he requires is held; there is so much data in the report, he cannot find what he is looking for.

(b) A typist mis-spells or omits an important word in a letter, so that its meaning is obscured or altered.

(c) A telephone call is made to a manager who is not in his office; a message is left with his secretary or assistant for passing on when the manager returns, but the secretary or assistant subsequently forgets all about it and the message is lost.

(d) A technical person writes a report to a non-technical manager, using a lot of jargon which the manager cannot understand.

(e) A report is written in bad English, and the reader, finding it difficult to follow, is adversely impressed.

(f) A flood or fire puts the telecommunications systems of the company's head office out of action.

(g) A superior tells his subordinate to do something and the subordinate, misunderstanding him, does something completely different instead.

3.11 Failure to transmit information (to get the message across) can have serious consequences on a company's operations. The problem of noise can be overcome, or at least reduced, by using the *principle of redundancy* - that is, by using more than one channel of communication, so that if a message fails to get through by one channel, it may succeed by another. For example, an agreement or decision made by telephone or at a meeting can be backed up by issuing a letter or minutes.

Exercise 1

Suggest ways in which each of the problems listed in Paragraph 3.10 could have been avoided.

Solution

In some cases there will be practical answers ((a), (b) and (f) for example), and in nearly all cases training of some description (for one or both of the parties involved) is needed.

Communication between superiors and subordinates

3.12 In a well known 'real-world' study into three companies in the USA (1962) W Read showed how the mobility aspirations of subordinates in a large organisation strongly affect the amount of communication that takes place. The more a subordinate wanted promotion, the less likely he would be to transmit 'negative' aspects of his work performance. This relationship was conditioned or modified by the degree of inter-personal trust that existed between the superior and the subordinate.

3.13 O'Reilly and Roberts (1974) developed this argument in a laboratory study from which they concluded about the upward flow of communication that: 'If the information is important but unfavourable to the sender it is likely to be blocked. This has implications for top-level decision-making and policy formulation because it may mean that organisational resources are often committed in a vacuum of relevant (particularly non-favourable) information'.

4. INFORMAL COMMUNICATION CHANNELS

4.1 The formal pattern of communication in an organisation is *always* supplemented by an informal one, which is sometimes referred to as the 'grapevine' or 'bush telegraph'. People like to gossip, and talk about rumours and events, on the telephone, over a cup of tea in the office, on the way home from work, in the corridor, at lunch, and so on.

4.2 The danger with informal communication is that it might be malicious, contain inaccurate rumour or half-truths, or simply be wild speculation. This type of gossip can be unsettling to people in an organisation, and make colleagues mistrust one another or act cautiously.

4.3 For example, suppose that you work for a company in London, and your friend from another department telephones to say that he has heard from someone who knows someone else in the personnel department that your office is going to be moved to Cardiff, and anyone refusing to go will be given the sack. This sort of news would be certain to upset you for a while, even if it turns out eventually to be wrong.

4.4 Formal communication systems need the support of a good - and accurate - informal system, and some ideas have been put forward about how this might be done.

(a) One idea is to set up 'official' corporate communications that will feed information into the informal system. House journals or briefings groups can be used to provide accurate bits of information, which individuals can pick up and gossip about.

(b) Another idea, put forward by Nancy Foy (1985), is that an organisation should encourage 'networking'. *Networking* describes 'a collection of people, usually with a shared interest, who tend to keep in touch to exchange informal information'.

The grapevine

4.5 The grapevine is one aspect of informal communication. A well-known study into how the grapevine works was carried out by K Davis (1953) using his 'echo-analysis' technique: the recipient of some information, A, was asked to name the source of his information, B. B was then asked to name his source, C etc until the information was traced back to its originator. His research findings were that:

(a) the grapevine acts fast;

(b) the working of the grapevine is selective: information is not divulged randomly;

(c) the grapevine usually operates at the place of work and not outside it;

(d) perhaps surprisingly, the grapevine is only active when the formal communication network is active: the grapevine does not fill a gap created by an ineffective formal communication system;

(e) it was also surprising to discover that higher level executives were better communicators and better informed than their subordinates. 'If a foreman at the sixth level had an accident, a larger proportion of executives at the third level knew of it than at the fourth level, or even at the sixth level where the accident happened';

(f) more staff executives were in the know about events than line managers (because the staff executives are more mobile and get involved with more different functions in their work).

4.6 Davis concluded that since the grapevine exists, and cannot be got rid of, management should learn both to accept it and to use it, by harnessing it towards achieving the objectives of the organisation.

Networking

4.7 Network is a rather vague term, probably as vague as the social group it is supposed to denote.

(a) In some circumstances, the word describes a set of external relationships (eg alliances, joint ventures).

(b) More often, the term describes informal or semi-formal alliances among managers.

4.8 A network, then, is a group of managers and will sometimes be recognised as such by the formal organisation structure. The members may be drawn from a number of corporate functions and geographical areas.

4.9 A network might be a group of middle managers whose day-to-day jobs are important in influencing the implementation of corporate plans and strategies.

4.10 A network might be institutionalised in regular meetings to discuss matters of mutual importance, to obviate the need for operating decisions to pass up and down the management hierarchy, and to ensure the relevant co-ordination of activities. However, you should note that:

(a) networks are not 'task forces' or teams brought together to solve a particular problem;

(b) networks do not simply deal with issues presented to them by senior management, but instead identify issues for examination.

4.11 The network is not a structure but instead, in the words of Chandra, a 'social architecture'. Senior management identifies key tactical and operational decision makers. Network membership may not correspond with functional position.

4.12 Finally, a network is not simply an exercise in teamwork or the raising of group morale. It deals with the fundamentals of the business, and is a means by which information and experiences can be *shared*, across functional and geographical lines, so that the organisation can learn from them.

Exercise 2

Take the opportunity now to glance back at other aspects of the 'formal' and the 'informal' in organisations. The index will show you where to look.

5. COMMUNICATION METHODS

5.1 Various media and channels of communication are available to the organisation:

(a) face-to-face communication:

(i) formal meetings;
(ii) interviews;
(iii) informal contact;
(iv) the grapevine;

(b) oral communication:

(i) the telephone;
(ii) public address systems;

(c) written communication:

(i) letters: external mail system;
(ii) memoranda; internal mail system;
(iii) reports;
(iv) forms;
(v) notice boards;
(vi) house journals, bulletins, newsletters;
(vii) organisation manual;

(d) visual communication:

(i) charts;
(ii) films and slides.

Oral and written communication are discussed below. Meetings, interviews and committees are discussed later in this chapter.

Oral communication

The telephone

5.2 This is of course the most common method of oral communication between individuals in remote locations, or even within an organisation's premises. It provides all the interactive and feedback advantages of face to face communication, while saving the travel time. It is, however, more 'distant' and impersonal than an interview for the discussion of sensitive personal matters, and it does not *by itself* provide the concreteness of written media. The latter disadvantage can be remedied by written confirmation of a telephone call.

British Telecom and Mercury offer an ever-expanding range of facilities to telephone users, for both external and internal communication. Mobile telecommunications look set to be as much of a phenomenon in the 1990s as PCs were in the 1980s.

Public address systems

5.3 A simple public address system might operate through loudspeakers placed at strategic points for example in workshops or yards, where staff cannot be located or reached by telephone and the noise would not be too disturbing. The more recent, electronic methods of paging individuals who move around and cannot be located are the 'bleeper', which alerts the user to a message waiting for him at a pre-arranged telephone number, and the radiopager, a credit card sized device to which written messages can be relayed.

Written communication

The letter

5.4 This is flexible in a wide variety of situations, and useful in providing a written record and confirmation of the matters discussed. It is widely used for external communication, via the *external mailing system*. Various facilities are provided by the Post Office for the delivery of letters, with special arrangements for guaranteed, insured or urgent deliveries in the UK or internationally. An external messenger, taxi, or courier may also be used for urgent or important letters and documents.

A direct letter may be used internally in certain situations where a confidential written record is necessary or personal handling required: to confirm an oral briefing; to announce redundancy; to urge employees against strike action; to hand in one's resignation.

The memorandum

5.5 This is the equivalent of the letter in internal communication: versatile and concrete. It is sent via the *internal mail system* of an organisation: taken by hand to recipients within the organisation or to offices for collection (say from departmental or individual pigeon holes) or forwarding. If parts of an organisation are in separate locations, provision may be made for internal messengers to collect and distribute mail by van or motorcycle. 'Internal mail envelopes', re-usable envelopes on which each successive recipient's name is simply crossed off and the next added, are a useful and cost-effective way of keeping internal mail safe and confidential.

Memoranda are useful for exchanging many sorts of message and particularly for confirming telephone conversations: sometimes, however, they are used *instead* of telephone conversations, where the call would have been quicker, cheaper and just as effective. Many memoranda are unnecessarily typed where a short handwritten note would be adequate.

Reports

5.6 The function of a formal report is to allow a number of people to review the complex facts and arguments relating to an issue on which they have to base a plan or make a decision. This is primarily an internal medium used by management, but can be used externally for the information of shareholders, the general public or government agencies (for instance the company's Annual Report, which the organisation is required by law to issue to shareholders, but is often available to employees and the general public on request).

The written report does not allow for effective discussion or immediate feedback, as does a meeting, and can be a time-consuming and expensive document to produce. However, as a medium for putting across a body of ideas to a group of people, it has the advantages that:

(a) people can study the material in their own time, rather than arranging to be present at one place and time;

(b) no time need be wasted on irrelevancies and the formulation of arguments, such as may occur in meetings;

(c) the report should be presented objectively and impartially, in a formal and impersonal style: emotional reactions or conflicts will be avoided.

5.7 Annual reports are now issued by some organisations for their employees, similar to the shareholder's report but simplified and adapted to the information needs of employees, including a review of, for example, manpower and attendance records, industrial relations, health and safety, and costs of maintaining wages, welfare and facilities.

5.8 Progress and completion reports may also be commissioned to review the progress and eventual effectiveness of long-term operations such as the conversion of files, restructuring of a department or move of office accommodation.

Forms

5.9 Routine information flow is largely achieved through the use of forms. A well-designed form can be filled quickly and easily with brief, relevant and specifically identified details of a request or instruction. They are simple to file, and information is quickly retrieved and confirmed. Examples include: expense forms, timesheets, insurance forms, and stock request forms. Staff do not usually have to exercise discretion in the selection and use of this medium: a form either is, or is not, available to meet a particular need.

Notice board

5.10 This is a channel through which various written media can be cheaply transmitted to a large number of people. It allows the organisation to present a variety of information to any or all employees: items may have a limited time span of relevance but will at least be available for verification and recollection for a while. The drawbacks to notice boards are that:

(a) they can easily fall into neglect, and become untidy or irrelevant (or be sabotaged by graffitti); and

(b) they are wholly dependent on the intended recipient's curiosity or desire to receive information.

House journal

5.11 Larger companies frequently run an internal magazine or newspaper to inform employees about:

(a) staff appointments and retirements;
(b) meetings, sports and social events;
(c) results and successes; customer feedback;
(d) new products or machinery;
(e) motivating competitions, say to elicit suggestions, or improve office maintenance or safety.

5.12 The journal usually avoids being controversial: it may not deal with sensitive issues such as industrial relations or pollution of the environment, and may stop short of criticising policy, management or products: it is, after all, designed to improve rather than threaten communication and morale, and it may be seen by outsiders (especially customers) who might get an unfavourable impression of the organisation. Journals are sometimes regarded by the workforce as predictable, uninteresting and not to be taken seriously: on the other hand, they can provide a legitimate means of expression and a sense of corporate identity.

Organisation manual or handbook

5.13 An organisation (or office) manual is useful for drawing together and keeping up to date all relevant information for the guidance of individuals and groups as to:

(a) the structure of the organisation (perhaps an organisation chart);
(b) background: the organisation's history, geography, 'who's who' of senior executives;
(c) the organisation's products, services and customers;
(d) rules and regulations;
(e) conditions of employment: pay structure, hours, holidays, notice;
(f) standards and procedures for health and safety;
(g) procedures for grievances, discipline, salary review;
(h) policy on trade union membership;
(i) facilities for employees.

5.14 The manual is thus not only a useful survival kit for new members of the organisation, but a work of reference and a symbol of corporate identity. Manuals and handbooks will have to be regularly updated and reissued, which can be expensive and time-consuming: a looseleaf format permitting additions and amendments is a common solution (though in practice the circulation of loose amendment sheets, which may or may not get inserted, may not prove efficient).

Employee attitude surveys

5.15 A communications exercise that is currently growing in popularity is the employee attitude survey, where a questionnaire is sent out to employees to find out what they think of their organisation. 'The effective survey is the organisation talking to itself. This is a source of strength on the principle that the intelligent systems which have self-knowledge are better able to survive in a changing world.' (*Financial Times*, June 1993)

5.16 The IPM have produced the following checklist for conducting an employee attitude survey.

(a) Gain the commitment of senior management, clarify objectives and define areas where the company is prepared to take action.

(b) Decide between a census of employees - probably paper-based - and a random sample - maybe face-to-face interviews.

(c) Conduct a pilot survey.

(d) Design a questionnaire - this needs experienced help.

(e) Tell employees what is being done: the more internal publicity, the greater the chances of success.

(f) Guarantee confidentiality.

(g) Consider measures to improve the response rate, for example by holding supervised completion sessions in company time.

(h) If the survey involves face-to-face interviews, conduct them with one or two employees at a time. Avoid group discussions - it is difficult to maintain confidentiality in groups and to enable everyone to have a say.

(i) Produce the final report. Distil the evidence, reach conclusions and put forward recommendations. Present the report to top management.

(j) Report back to employees with a summary of the main findings.

(k) An individual or a group should be made responsible for action following the survey findings; progress should be monitored.

Exercise 3

How would you respond to the following objections from a management traditionalist to employee surveys?

(a) Good managers already know their staff's opinions.

(b) Surveys are superficial, impersonal and bureaucratic.

(c) Surveys open cans of worms and raise false hopes.

(d) Surveys contain built in bias because of the unknown opinions of the many who do not respond.

Solution

Nigel Nicholson (*Financial Times*, January 1993) gives the following answers (briefly summarised).

(a) How do managers know if their reading of opinions is correct?

(b) Yes, they are no substitute for good personal communications, but 'Thanks for asking us' is a common response.

(c) Cans of worms probably should be opened. 'Management's commitment, clearly communicated at the outset, should be not only to take action where feasible, but also to be frank about reasons for not acting when this would be unreasonably difficult, contentious or costly.

(d) Bias is modest when return rates of over 50% are achieved, and a well-designed and carefully introduced survey can expect returns of 65% or more.

6. FACE-TO-FACE COMMUNICATION

6.1 Face-to-face communication (including meetings) plays an important part in the life of any organisation, whether it is required by government legislation or the articles of a company, or is held informally for information exchange, problem-solving and decision-making.

6.2 Face-to-face communication in general and group discussion (meetings) in particular offer several advantages for:

(a) generating new ideas;
(b) 'on the spot' feedback, constructive criticism and exchange of views;
(c) encouraging co-operation and sensitivity to personal factors;
(d) spreading information quickly through a group of people.

However, meetings can be non-productive or counter-productive if:

(a) the terms of reference (defining the purpose and power of the meeting) are not clear;
(b) the people attending are unskilled or unwilling communicators;
(c) there is insufficient guidance or leadership to control proceedings.

Meetings

6.3 Formal meetings, such as the Board meeting of a company, the Annual General Meeting (AGM) of a society, or a Local Council meeting, are governed by strict rules and conventions. These may establish procedure on such matters as:

(a) attendance rights (for members of the public or shareholders);

(b) adequate notice of forthcoming meetings;

(c) the minimum number of members required to hold the meeting (the 'quorum') and other details;

(d) the timing of meetings;

(e) the type of business to be discussed; and

(f) the binding power of decisions made upon the participants.

Many people regard such rules and conventions as a hindrance to free and effective communication, with elaborate courtesies, 'politics' and voting procedures.

6.4 However, a well-organised, well-aimed and well-led meeting can be extremely effective in many different contexts.

(a) Executive decision-making by a group of directors, managers, government officials

(b) The relaying of decisions and instructions, for example briefings

(c) The provision of advice and information for management planning and decision-making

(d) Participative problem solving, by consultation with people in different departments or fields, such as a task force or working party

(e) Brainstorming: free exchanges with a view to generating new approaches and ideas

Interviews

6.5 The interview is an excellent internal system for handling the problems or queries of individuals, allowing confidentiality and flexible response to personal factors where necessary. Interviews are, however, costly in terms of managerial time: few interviews are built into the formal communication system (although a one-to-one meeting may be requested at need).

6.6 Grievance and disciplinary interviews are important in maintaining morale and performance: the employee's complaints or dissatisfactions (about his job, conditions of work, mistreatment) can be aired in private; the organisation's displeasure with an employee (for disobedience or unacceptable behaviour, persistent lateness, dangerous recklessness) can be expressed without humiliating the individual in front of his peers.

6.7 Appraisal interviews are used to discuss the employee's performance, progress and possible need for improvement (and will therefore need to be carefully handled in an open and friendly manner if they are not to be regarded with fear and resentment). A similar type of interview is used as part of the staff selection process.

Committees

6.8 Committees are frequently used within an organisation as a means of delegating authority. The disadvantages and advantages of committees extend beyond considerations of authority and responsibility and they are described here in full.

6.9 Within an organisation, committees can consist entirely of executives. Board meetings, budget approvals, project teams and feasibility studies would be examples. Alternatively, committees can be an instrument for joint consultation between employers and employees (works councils, joint productivity groups, staff advisory councils).

Classification of committees

6.10 Committees may be classified according to the power they exercise, distinguishing between those having the power to bind the parent body and those without such power.

6.11 It is also important to consider the function and duration of committees so that the following categories may be defined as follows.

(a) *Executive committees* which have the power to govern or administer. It can be argued that the board of directors of a limited company is itself a 'committee' appointed by the shareholders, to the extent that it governs or administers.

(b) *Standing committees* which are formed for a particular purpose on a permanent basis. Their role is to deal with routine business delegated to them at weekly or monthly meetings.

(c) *Ad hoc committees* are formed to complete a particular task. An ad hoc committee can be described as a fact-finding or special committee which is short lived and, having achieved its purpose, reports back to the parent body and then ceases to exist.

(d) *Sub-committees* may be appointed by committees to relieve the parent committee of some of its routine work.

(e) *Joint committees* may be formed to co-ordinate the activities of two or more committees, for example representatives from employers and employees may meet in a Joint Consultative Committee. This kind of committee can either be permanent or appointed for a special purpose.

Advantages of committees

6.12 Committees may give rise to significant advantages.

(a) Consolidation of power and authority: whereas an individual may not have sufficient authority to make a decision himself, the pooled authority of a committee may be sufficient to enable the decision to be made. A committee may be referred to in this type of instance as a plural executive. Examples of a plural executive include a board of directors or the Cabinet of the government, which are policy-making and policy-executing committees.

(b) Blurring responsibility: when a committee makes a decision, no individual will be held responsible for the consequences of the decision. This is both an advantage and a disadvantage of committee decisions.

(c) Creating new ideas: group creativity may be achieved by a 'brainstorming committee' or 'think tank' (see below).

(d) They are an excellent means of communication. For example:

   (i) to exchange ideas on a wide number of interests before a decision affecting the organisation is taken;

   (ii) to inform managers about policies, plans and actual results.

(e) They are democratic, because they allow for greater participation in the decision-making process.

(f) Combining abilities: committees enable the differing skills of its various members to be brought together to deal with a problem. In theory, the quality of committee decisions should be of a high standard.

(g) Co-ordination: they should enable the maximum co-ordination of all parties involved in a decision to be achieved, for example:

  (i) in appraising performance throughout the organisation and modifying policy as necessary;

  (ii) in developing and improving operational procedures;

  (iii) in establishing and co-ordinating time relationships between the activities of each department, in order to achieve the optimum production cycle;

  (iv) in ensuring full co-operation between each department and special efforts, where necessary, to obviate bottlenecks arising in the flow of production;

  (v) in co-ordinating the budgets of each department and compiling a master budget.

(h) They can exercise purely executive functions, such as the management of funds of a staff superannuation scheme, as opposed to those of an advisory character which require ratification by the board of directors.

(i) Advisory capacity: a committee is frequently used to offer advice to a decision-maker.

(j) Representation: they enable all relevant interests to be involved in the decision-making process and they bring together the specialised knowledge of working people into a working combination.

(k) Through participation, they may improve the motivation of committee members (and even of their subordinates).

(l) Delay: a committee is used to gain time, eg a manager may set up a committee to investigate a problem when he wants to delay his decision, or a company may refer a labour relations problem to a committee to defer a crisis with a trade union.

(m) A committee may be set up by a board of directors to do 'spade work' and avoid detailed discussions at board meetings.

Disadvantages of committees

6.13 The disadvantages of committees may be summarised as follows:

(a) They are apt to be too large for constructive action, since the time taken by a committee to resolve a problem tends to be in direct proportion to its size. The optimum number of members will, of course, vary according to the committee's functions, so that it is difficult to dogmatise; however, in an industrial organisation, the complement should be rather lower than that found in local and public authorities, where the profit motive is absent. The ideal number may perhaps range from three to six or seven persons.

(b) Committees are time-consuming and expensive. In addition to the cost of highly paid executives' time, secretarial costs will be incurred in the preparation of agendas, recording of proceedings and the production and distribution of minutes.

(c) Delays may occur in the production cycle if matters of a routine nature are entrusted to committees; committees must not be given responsibilities which they would carry out inefficiently.

(d) Operations of the enterprise may be jeopardised by the frequent attendance of executives at meetings, and by distracting them from their real duties.

(e) Incorrect or ineffective decisions may be made, owing to the fact that members of a committee are unfamiliar with the deeper aspects of issues under discussion, which may only be fully appreciated by the directorate, or by those employees actually carrying out the work in question. Occasionally, there may be a total failure to reach any decision at all.

(f) Certain members may be apathetic, owing to pressure of work or lack of interest, resulting in superficial action.

(g) The fact that there is no individual responsibility for decisions might invite compromise instead of clear-cut decisions, besides weakening individual responsibility throughout the organisation. Moreover, members may thereby be enabled to avoid direct responsibility for poor results arising from decisions taken in committee. Weak management can hide behind committee decisions.

(h) Committees lack conscience.

(i) Proceedings may be dominated by outspoken or aggressive members, thus unduly influencing decisions and subsequent action, perhaps adversely; there may be 'tyranny' by a minority.

6.14 The writer E C Lindeman strongly disapproved of the committee function and he felt that the best kind of committee was the 'committee of one'. He felt that their function was discontinuous, contained irrelevant discussion, members tried to impress superiors, and chairmen to prohibit opinions; there is a tendency to jump to conclusions, committees are soulless, expensive and encourage irresponsibility. He also felt that committees were introduced to cover up the fact that those responsible had not decided what to do, that they enable individuals to escape responsibility and that they are consistent with the fear of assigning executive responsibility to individuals.

6.15 A committee may be misused for a number of 'wrong' purposes, such as:

(a) to replace managers. A committee cannot do all the tasks of management (eg leadership) and therefore cannot replace managers entirely;

(b) to carry out research work. A committee may be used to create new ideas, but work on those ideas cannot be done effectively by a committee itself;

(c) to make unimportant decisions. This would be expensive and time-consuming;

(d) to discuss decisions beyond the authority of its participants. This might occur, for example, when an international committee of government ministers is created, but ministers send deputies in their place to meetings, without giving the deputy sufficient authority to enable the committee to make important decisions.

Using committees successfully

6.16 It is difficult to create a truly efficient committee, but the following factors will help to foster efficiency.

(a) Well-defined areas of authority, time-scales of operations and purpose must be specified in writing.

(b) The chairman must have the qualities of leadership to co-ordinate and motivate the other committee members.

(c) The committee should not be so large as to be unmanageable.

(d) The members of the committee must have the necessary skills and experience to do the committee's work; where the committee is expected to liaise with functional departments, the members must also have sufficient status and influence with those departments.

(e) Minutes of the meetings should be taken and circulated, with any action points arising out of the meetings notified to the members responsible for doing the work.

(f) Above all, an efficient committee must provide benefits which justify its cost.

(g) Finally, if at all possible, the committee should be allowed plenty of time to reach decisions, enabling members to form sub-groups.

Brainstorming

6.17 Brainstorming sessions are problem-solving conferences of 6-12 people who produce spontaneous 'free-wheeling' ideas to solve a particular problem. Ideas are produced but not evaluated at these meetings, so that originality is not stifled in fear of criticism. Brainstorming sessions rely on the ability of conference members to feed off each other's ideas. They have been used in many organisations and might typically occur, for example, in advertising agencies to produce ideas for a forthcoming campaign.

6.18 W J Gordon of the Arthur D Little company, a consulting firm in Massachussets, developed the *Gordon technique*. The company offered, among its other business lines, the services of an invention design group which could invent a product to order. The Gordon technique relies on brainstorming sessions with the unique difference that only the conference leader knows the nature of the problem and it is his task to steer the ideas of the group towards a solution 'in the dark.' This prevents any member from getting addicted to a single idea, which is an inherent danger in a normal brainstorming session.

Conferences

6.19 A body which has a large membership spread over a wide area, such as a professional body or a trade union, may find *conferences* a useful means of improving contact between the organisation and the body of the membership. The central administration may appear too remote, and representative committees, though useful, do not always reflect the views of the general membership.

6.20 A well-organised annual conference is a means of bringing together a much larger number of members to discuss matters of current interest or concern. It can 'bridge the gap', and give members:

(a) a better understanding of what their organisation is trying to do for them; and
(b) a greater commitment to it.

Exercise 4

Michael Dixon, who writes the *Financial Times's* jobs column, occasionally publishes what he calls the Laws of Organisational Stupidity - regularly recurring problems that seem to bedevil organisations the world over.

One such law particularly relevant to communication is 'Selby's Scrambler' which states that *Myths about a chief's wishes multiply with each link in the chain of command.* The example given is of a vice chairman who casually remarked that he was going to get a car for his daughter, with the result that the organisation's purchasing department spent an entire afternoon *purchasing* a car.

Analyse Selby's Scrambler in terms of communications theory. In doing so you may appreciate how much you have learnt in this Study Text about organisations and human resources.

Chapter roundup

- Communication of information is the means by which organisations function. All parts of an organisation can communicate with each other, but because of flaws in communication channels and in the process itself information can easily become distorted or not reach those who need it at all.
- Informal channels always exist alongside formal channels.
- Methods of communication include face-to-face, oral, written and visual methods. Communication via meetings and committees is likely to involve the company secretary/administrator very heavily.

Test your knowledge

1. What are some of the problems and barriers to communication? (see paras 2.2-2.6)
2. What is informal communication, and how can the organisation utilise it? (3.1-3.6)
3. List four methods of face-to-face communication and seven methods of written communication. (4.1)
4. Why is an internal written report a useful method of communication? (4.6)
5. In what ways may meetings be useful? (5.4)

Now try question 16 at the end of the text

ILLUSTRATIVE QUESTIONS AND SUGGESTED SOLUTIONS

1 FUTURE OF BUREAUCRACY

Several years ago the death of bureaucracy was predicted. For what reasons might such a prediction have been made? How far has the prediction been fulfilled, in your view? If asked to make your own predictions about the future of bureaucracy, from the vantage point of 1988, what would you forecast, and why?

2 CONTINGENCY/SYSTEMS APPROACH

Examine the scientific status and the practical value to managers of *either* systems theory or contingency theory.

3 CULTURE AND 'EXCELLENCE'

Since the publication of Peters and Waterman's *In Search of Excellence* in the 1980s, the concept of 'culture' has become very fashionable. What exactly is meant by the term 'culture' when applied to an organisation? How can the 'culture' of an organisation be so designed as to promote 'excellence'?

4 INFORMATION TECHNOLOGY

Assess the behavioural implications of the spread of computers and information technology within large organisations.

5 WORK GROUPS

What would be the advantages and disadvantages of creating cohesive and informal work groups in each of the following situations:

(a) air crew in a commercial airline;
(b) military bomb disposal units;
(c) a collection of sales representatives scattered throughout Western Europe; and
(d) couriers working for a package holiday firm?

6 LEADERSHIP

What is 'leadership'? How far is it possible and desirable for managers to become leaders?

7 NATURE AND ROLE OF PERSONNEL FUNCTION

You have just read an article, by an accountant, which has claimed that the so called 'human resource' function is, generally speaking, 'a cost rather than a benefit' in most organisations, and that its role should be strictly confined to low-level tasks such as welfare and the administration of employee services. Produce a refutation of this argument, citing instances where the personnel function has made a positive contribution to an organisation's profitability.

8 MANPOWER PLANNING

In modern employment conditions, is manpower planning becoming easier or harder? To the extent that it is becoming harder, is it worth doing at all?

9 EVALUATION OF RECRUITMENT ADVERTISEMENT

Study the recruitment advertisement below and answer the following questions.

(a) What is good about the advertisement and why?
(b) What could be improved in the advertisement and why?
(c) What does the advertisement tell you about the organisation AOK?

Dealing with individuals demands a certain... ...um...

You've heard the old line . . . 'You don't have to be mad to work here, but it helps'. It's like that at AOK, but in the nicest possible way. We believe that our Personnel Department should operate for the benefit of our staff, and not that staff should conform to statistical profiles. It doesn't make for an easy life, but dealing with people as individuals, rather than as numbers, certainly makes it a rewarding one.

We're committed to an enlightened personnel philosophy. We firmly believe that our staff are our most important asset, and we go a long way both to attract the highest quality of people, and to retain them.

AOK is a company with a difference. We're a highly progressive, international organization, one of the world's leading manufacturers in the medical electronics field.

...Character

As an expanding company, we now need another experienced Personnel Generalist to join us at our UK headquarters in Reigate, Surrey.

Essentially we're looking for an individual, a chameleon character who will assume an influential role in recruitment, employee relations, salary administration, compensation and benefits, or whatever the situation demands. The flexibility to interchange with various functions is vital. Within your designated area, you'll experience a large degree of independence. You'll be a strong personality, probably already experienced in personnel management in a small company. Whatever your background you'll certainly be someone who likes to help people to help themselves and who is happy to get involved with people at all levels within the organization.

Obviously, in a fast growing company with a positive emphasis on effective personnel work, your prospects for promotion are excellent. Salaries are highly attractive and benefits are, of course, comprehensive.

So if you're the kind of Personnel individual who enjoys personal contact, problem solving, and will thrive on the high pace of a progressive, international organization, such as AOK, get in touch with us by writing or telephoning, quoting ref: 451/BPD, to AOK House, Reigate, Surrey.

10 RELIANCE ON SELECTION INTERVIEW

It has been said that the basic mistake made by most employers is to rely on the interview as the first and crucial stage in the recruitment and selection process. Why is this a 'mistake'? What should employers be doing?

11 JUSTIFICATION OF TRAINING COSTS

What are the direct costs of training? How can these costs be justified in terms of potential benefits?

12 PERFORMANCE APPRAISAL SYSTEMS

(a) What are the features of an effective performance appraisal system?
(b) What are the advantages and disadvantages of performance appraisal systems?

13 PURPOSES OF PAY

'Pay is a means to an end.' What end or ends? How far can pay accomplish this end or these ends on its own? To the extent that it cannot work alone, what other factors are relevant?

14 JOB EVALUATION

What are the strengths and weaknesses of job evaluation? How might the weaknesses be overcome or minimised?

15 REDUNDANCY

Why do organisations make people redundant? When they do so, what are the merits and drawbacks of using any of the following approaches to the selection of those being made redundant?

(a) Last in, First out (LIFO).
(b) Value to the organisation, ie retention by merit.
(c) The offer of voluntary redundancy.
(d) Early retirement.

Which of these methods would you regard as preferable, and why?

16 COMMUNICATIONS AND BARRIERS

Bill Brown, your friend and fellow section leader, possesses many admirable qualities and skills. He is clear-headed and logical, a good organiser and planner, and has a good understanding of the technical aspects of his work. Despite all this, his staff often seem to be puzzled because he does not always make himself clear to them and, as a result, mistakes are made and the staff become annoyed and frustrated.

How would you help Bill Brown to communicate more effectively?

1 FUTURE OF BUREAUCRACY

Tutorial note. A thought-provoking question, and one that allows you to demonstrate your organisational abilities, since there are a number of areas to be covered and a clear need for (a) an introductory paragraph defining the term 'bureaucracy' (since this is *not* part of your main answer) and (b) a conclusion giving your own judgement on the future of bureaucracy. You might also note that 'bureaucracy' is not a derogatory term solely associated with unwieldly, impersonal organisations manned by petty officials and bound by red tape. As the examiner pointed out, it is 'a description for a perfectly acceptable type of organisation appropriate to circumstances of slow change and a large customer/client base'. As a final point: please make sure you've spelt the word itself correctly!

Introduction

Bureaucracy is a formal organisation structure built on a high degree of specialisation and delegation, 'legal' authority, rules and procedures and a 'role' culture. To a layman, the term has unpleasant associations, but Max Weber was inclined to regard it as the ideal form of organisation, which is 'from a purely technical point of view, capable of attaining the highest degree of efficiency and is in this sense formally the most rational means of carrying out imperative control over human beings.'

Reasons for predicting the death of bureaucracy

Bureaucracy tends to several dysfunctions or inefficiencies, which may have led to predictions of its imminent death. Some of its features are functional and dysfunctional at the same time. Merton, for example, suggests that there is a rigidity of behaviour in a bureaucracy, due to the reduction in personalised relationships, the 'internationalisation' of rules (without reference to specific goals), the simplification of decision-making based on precedents, and the development of self-defensive esprit-de-corps. That rigidity of behaviour, Merton argued, creates reliability - but also increases difficulties with clients. In sectors led by market forces, therefore, this may be fatal.

Goulder likewise agreed that functions and dysfunctions result from rules. Rules are functional in reducing interpersonal tensions between subordinates and superiors: they are dysfunctional, however, in creating a tendency for employees to work at a defined minimum acceptable level of performance, which in turn necessitates close supervision - which *increases* tension within the work group.

The most serious 'symptoms' suggesting the imminent death of bureaucracy, however, are to do with the inability of rigid bureaucratic structures/cultures to change their behaviour in response to change in the environment and to feedback from their own control mechanisms. Warren Bennis predicted the eventual death of bureaucracy on the basis of the accelerating pace of change in the environment - social and technological - of organisations, and the inability of bureaucracies to cope with such rapid change.

Bennis' view is backed by research such as that of Burns and Stalker, who discovered that organisations exhibiting bureaucratic characteristics, which they called 'mechanistic' organisations, are unsuitable in conditions of instability and change, because they deal with change by cumbersome methods. Unfamiliar problems are referred 'higher-up', overburdening the top of the organisation. Jobs, departments and committees *are* created to deal with new problems, creating further and greater problems. Burns and Stalker contrast the mechanistic structure with the 'organic' or 'organismic' structures which are suitable to conditions of instability, where the technology of the market is changing and innovation is crucial to survival: individual tasks are related to organisational goals and continually redefined, specialised knowledge and experience are contributed to the common task, authority and communications are effected through networks.

Michael Crozier *(The Bureaucratic Phenomenon)* developed his ideas as a result of empirical studies in two large French organisations. He confirmed that 'a system of organisation whose main characteristic is its rigidity will not adjust easily to change and will tend to resist change as much as possible'. He pointed out the consequences of this rigidity for the control system of the organisation as well as for response to external pressures. The feedback of information on errors, which in other structures might be used to readjust behaviour, does not tend to initiate control action. A bureaucracy cannot learn from its mistakes! Because bureaucracies are highly centralised, constant adjustment or control action is impossible: decisions have to be taken at the top, against the grain of unwieldy upward communication mechanisms, so that corrective action is only taken 'when serious dysfunctions develop and no other alternatives remain'.

Crozier further suggests that a bureaucracy cannot adjust to inevitable environmental changes without 'deeply felt crisis', because of delays in taking action, the magnitude of change when it finally occurs and the cultural resistance caused by the personality types attracted and recruited to the system.

How far has the prediction been fulfilled?

There are indications that bureaucracy in some forms and in some environments has suffered the consequences of its dysfunctions. In the UK, the recession of the early 1980s necessitated cost-cutting which effectively wiped out middle management in many firms: thus the tall, highly 'tiered' vertical structure characteristic of bureaucracy is less common in the private sector.

Market pressures in the private sector have also created a need for client responsiveness - at least at the interface between organisation and client/customer. Sales and service units within bureaucracies have therefore had to undergo cultural and even structural change, so that thorough-going bureaucracies, without an element of what Handy calls 'differentiation' of culture within the structure, are rarer.

Management theory has in recent years emphasised the importance of flexibility and adaptability in the face of rapid environmental change. The 'fashion' is for a contingency approach to management and organisation, for differentiation of cultures and structures, for smaller task-centred units. Peters and Waterman *(In Search of Excellence)* analyse excellent or 'continuously adaptive' companies such as Mars, 3M and IBM, and find in them non-bureaucratic characteristics such as a bias for action, closeness to the customer, autonomy and entrepreneurship (or 'intrapreneurship').

However, it has to be said that bureaucracy self evidently has not died. It is much in evidence in the public sector and government, and elements of it survive in many large private sector organisations, for example in the formality of communication channels, the extent of specialisation and delegation, and the number of individuals who as administrative 'knowledge workers' are physically separated from the 'real work' of the organisation. It is not unknown in large banks, for example, for the management structure to create such emphatic formal, vertical lines of authority and communication that direct lateral communication is completely discouraged.

This survivability of bureaucracy can be explained by a number of factors, which will be discussed below.

Future of bureaucracy

There seems little likelihood of bureaucracy disappearing in the foreseeable future.

(a) In certain slow-change environments, or those in which market pressures have little short-term effect, bureaucracy can still be a perfectly acceptable structure/culture. (In some contexts, where large organisations are associated with the ready availability of labour, and alleviate potentially disastrous levels of unemployment, large bureaucratic systems may be socially beneficial.)

(b) There is still a tendency towards increasing overall size in organisations, through merger and acquisition.

(c) Levels of information used in business continue to increase and, despite the automation of many information-processing functions, administrative support for front-line activities in business will continue to be required.

(d) Bureaucratic systems tend to be self-defensive, in:

    (i) the type of people who are attracted to, and selected for, the culture. Change to the organisation itself is therefore resisted, and the self-correcting mechanism is not highly developed;

    (ii) problems and events being categorised for decision-making purposes. This, according to Merton, creates defensibility of individual action in the face of client/customer pressure;

    (iii) the development of esprit de corps. In spite of the impersonality of the relationships, Merton suggests, there is a propensity for members of the bureaucracy to defend each other against outside pressures.

(e) The dysfunctions of bureaucracy are often also functional at the same time. Rigidity of behaviour also creates reliability. Rules help to control interpersonal tension and aid the survival of the work group.

Conclusion

Bureaucracy has not become extinct. To paraphrase Mark Twain, 'reports of its death have been greatly exaggerated'. While pressures for change have altered the face of bureaucracy to some extent, there are still conditions in which it can flourish, or at least survive for the foreseeable future.

2 CONTINGENCY/SYSTEMS APPROACH

Tutorial note. We offer solutions to *both* of the options given in the question: please note when preparing your own solution that discussion of only one theory is required. Note also that you are asked to examine two specific areas: scientific status and practical value to managers.

Systems theory

The systems approach to organisation was developed at the Tavistock Institute of Human Relations in the 1950s, although General System Theory, in which it has its scientific roots, was pioneered in the 1930s.

From general systems theory come observations about: the open system and its relationship with the environment; the process of self-regulation within the system called 'homeostasis'; the degree of predictability with which states or activities follow each other in the system (from deterministic to stochastic systems); the flexibility and adaptability of open 'self organising' systems.

Trist and Bamforth took the assumption that a work organisation can be treated as an open system and developed a more complex approach which suggests that an organisation can be treated as an open socio-technical system. Any production system requires material technology (tasks, layout, equipment and tools) and social organisation (relationships between people): these two sub-systems are linked, and the system design must find a 'best fit' between the needs of both components.

(a) *Scientific status*

The theory has an imaginative appeal. An organisation has to draw in inputs (capital, labour, information, materials), transform them within the system, and return outputs to the environment (information, products, satisfied customers) in a continuing cycle. It is self-regulating, to preserve a dynamic equilibrium, flexible and adaptable. It is therefore possible to draw the analogy between living systems and organisations (the 'organic analogy') and make some assumptions about how organisations are likely to behave on that basis. The analogy provides a framework for thinking about organisations and designing their structures: it highlights important aspects of organisational behaviour (as discussed below).

However, it is only an analogy, and as such cannot be stretched too far, or provide a basis for devising testable hypotheses. It is therefore not a 'theory' at all in scientific terms. As an *approach*, it offers a useful, accessible language for discussing organisations, but - as with many behavioural frameworks - its scientific status cannot be reckoned in the same way as theories in the natural sciences, for example.

(b) *Practical value to managers*

As suggested above, the systems approach, and its root concepts from general systems theory, can provide a stimulating way of describing and analysing organisational phenomena.

General systems theory can contribute to the principles and practice of management in several ways, not least by enabling managers to learn from the experience of experts and researchers in other disciplines.

(i) It draws attention to the *dynamic* aspects of organisation, and the factors influencing the growth and development of all its sub-systems.

(ii) It creates an awareness of sub-systems, each with potentially conflicting goals which must be integrated.

(iii) It focuses attention on interrelationships between aspects of the organisation, and between the organisation and its environment, ie the needs of the system as a whole: management should not get so bogged down in detail and small political arenas that they lose sight of the overall objectives and processes.

(iv) It teaches managers to reject the deterministic idea that A will always cause B to happen. 'Linear causality' may occur, but only rarely, because of the unpredictability and uncontrollability of many inputs.

(v) The importance of the *environment* on a system is acknowledged. One product of this may be customer orientation, which is an important cultural element of successful, adaptive companies.

The practical application of socio-technical systems theory can be illustrated by the Durham coal-field research of Trist and Bamforth, which argued that work organisation is not wholly determined by technology but by organisational choices: the social system has properties independent of the technical system, and can be designed so as to meet technical demands *and* human needs. In other words, any given technical system can be operated by different social systems. The problem is to find a 'fit' that will meet technical demands and human needs. The socio-technical systems school advocated *composite autonomous group working* - 'composite' in terms of the range of skills in the group as a whole, and 'autonomous' in terms of self-determination for shifts, rotas and so on. This proved effective in the organisational redesign of the Durham pits, and the introduction of 'composite longwall' method of working.

Contingency theory

The contingency approach to organisation developed as a reaction to prescriptive ideas of the classical and human relations schools, which claimed to offer a universal 'best way' to design organisations, to motivate staff and to introduce technology. Research by Burns and Stalker, Joan Woodward, Laurence and Lorsch and others indicated that different forms of organisational structure could be equally successful, that there was no inevitable correlation between classical organisational structures and effectiveness, and that there were a number of variables to be considered in the design of organisations.

The emerging contingency 'school' rejected the universal 'one-best-way' approach, in favour of analysis of the internal factors and external environment of each organisation and the design of organisational structure as a 'best fit' between the tasks, people and environment in the particular situation. As Buchanan and Huczynski put it: 'With the coming of contingency theory, organisational design ceased to be "off-the-shelf", but became tailored to the particular and specific needs of an organisation.' John Hunt comments: 'Contingency means, "it depends".'

(a) *Scientific status*

The reaction against the universality of management/organisational principles was founded on sound research evidence.

Research by Blain in 1964 showed that the principles advanced by the classical management school - for example, small span of control, unity of command - did not necessarily correlate with organisational effectiveness.

Laurence and Lorsch compared the structural characteristics of a 'high-performing' container firm, which existed in a relatively stable environment, and a 'high-performing' plastics firm which existed in a rapidly changing environment. They concluded that in a stable environment the most efficient structure was one in which the influence and authority of senior managers was high and of middle managers low: the converse was true of the dynamic environment firm.

Joan Woodward's research with firms in Essex highlighted the importance of technology as a major factor contributing to variances in organisation structure: 'It appeared that different technologies imposed different kinds of demands on individuals and organisations and that these demands have to be met through an appropriate form of organisation.'

However, two points should be noted.

(i) John Child, among others, suggests that 'One major limitation of the contemporary contingency approach lies in the lack of conclusive evidence to demonstrate that matching organisational designs to prevailing contingencies contributes *importantly* to performance.'

(ii) 'Contingency theory' is not properly a 'theory' at all. It is more of a 'philosophy' or 'approach' - a way of thinking about organisation design - than a theory, which implies causal elements about which hypotheses can be derived and tested empirically.

(b) *Practical value to managers*

According to Tom Lupton: 'It is of great practical significance whether one kind of managerial 'style' or procedure for arriving at decisions, or one kind of organisational structure, is suitable for all organisations, or whether the managers in each organisation have to find that expedient that will best meet the particular circumstances of size, technology, competitive situation and so on.'

Awareness of the contingency approach will therefore be of value in:

(i) encouraging managers to identify and define the particular circumstances of the situation they need to manage, and to devise appropriate ways of handling them. A belief in universal principles and prescriptive theories can hinder problem-solving and decision-making by obscuring some of the available alternatives. It can also dull the ability to evaluate and choose between alternatives that are clearly open, by preventing the manager from developing relevant criteria for judgement;

(ii) encouraging responsiveness and flexibility to changes in environmental factors through organisational structure and culture. Task performance and individual/group satisfaction are more important design criteria than permanence and unity of design type. Within a single organisation, there may be bureaucratic units side by side with task-centred matrix units (for example in the research and development function) which can respond to particular pressures and environmental volatility.

3 CULTURE AND 'EXCELLENCE'

Tutorial note. Peters and Waterman's book is rather old hat now and some of the organisations they praised have since lost their way. This does not alter the fact that culture remains highly prominent in management thinking, as does excellence in its constituent parts of 'quality', customer focus, empowerment and so on.

Culture

Culture may be defined as the complex body of shared values and beliefs of an organisation.

Peters and Waterman, in their study *(In Search of Excellence)* found that the 'dominance and coherence of culture' was an essential feature of the 'excellent' companies they observed. A 'handful of guiding values' was more powerful than manuals, rule books, norms and controls formally imposed (and resisted). They commented: 'If companies do not have strong notions of themselves, as reflected in their values, stories, myths and legends, people's only security comes from where they live on the organisation chart.'

Handy sums up 'culture' as 'that's the way we do things round here'. For Schein, it is 'the pattern of basic assumptions that a given group has invented, discovered, or developed, in learning to cope with its problems of external adaption and internal integration, and that have worked well enough to be considered valid and, therefore, to be taught to new members as the correct way to perceive, think and feel in relation to these problems.'

All organisations will generate their own cultures, whether spontaneously, or under the guidance of positive managerial strategy. The culture will consist of:

(a) the basic, underlying assumptions which guide the behaviour of the individuals and groups in the organisation, for example customer orientation, or belief in quality, trust in the organisation to provide rewards, freedom to make decisions, freedom to make mistakes, the value of innovation and initiative at all levels;

(b) overt beliefs expressed by the organisation and its members, which can be used to condition (a) above. These beliefs and values may emerge as sayings, slogans, mottos and so on such as 'we're getting there', 'the customer is always right', or 'the winning team'. They may emerge in a richer mythology - in jokes and stories about past successes , heroic failures or breakthroughs, legends about the 'early days', or about 'the time the boss...'. Organisations with strong cultures often centre themselves around almost legendary figures in their history. Management can encourage this by 'selling' a sense of the corporate 'mission', or by promoting the company's 'image'; it can reward the 'right' attitudes and punish (or simply not employ) those who aren't prepared to commit themselves to the culture;

(c) visible artifacts - the style of the offices or other premises, dress 'rules', display of 'trophies', the degree of informality between superiors and subordinates.

Promoting excellence

Peters and Waterman define 'excellent' as 'continuously innovative' - the whole culture is prepared to adapt to the needs of customers, the skills of competitors and so on. Excellent companies 'experiment more, encourage more tries, and permit small failures'.

They identify various attributes of 'excellent' companies including:

(a) *A bias for action* - experiment, try, *do* rather than overanalyse, focus on problems. The 'results first' approach means not asking 'What's standing in the way?' but 'What can we do now?'

(b) *Closeness to the customer.* Customer-orientation and concern for quality at all levels can be an intense motivator, as well as a spur to innovation, because the workers' actions have measurable, even tangible, effects - a better product.

(c) *Autonomy and entrepreneurship:* giving teams and individuals control over their improvement goals and methods, and encouraging stimulating competition and adventurism.

(d) *Productivity through people* - 'turning people on' to their work and to organisational objectives, by positive reinforcement, 'reaffirming the heroic element' of the job, treating people decently; 'demanding extraordinary performance from the average man'.

(e) *Simultaneous loose-tight properties* - that is, autonomy, but control through central faith, guiding values, replacing manuals and rules.

The creation or reinforcement of all of these will depend on the culture of the organisation, and the extent to which it can be 'sold' to all the employees. Indeed, Peters and Waterman noted the success of companies with such 'strong' cultures that employees had to 'buy in or get out'. They argue that 'dominance and coherence of culture proved to be an essential quality of excellent companies.' Although they present a rather 'rosy' and unqualifiedly approving picture of their excellent companies, in pursuit of their central thesis, culture undoubtedly is becoming a real force in management thinking.

4 INFORMATION TECHNOLOGY

Impact of information technology

Seven areas can be identified where changes in technology will have an impact.

(a) Jobs. Technological innovations allow organisations to reduce the headcount by replacing several workers with one machine without a consequent loss in productivity. Inevitably such moves create in employees feelings of insecurity, fears of redundancy, changed attitudes to other workers, the new technology itself and the organisation. This can have dramatic effects on the psychological contract which exists between employer and employee.

(b) The nature of work. Frequently the new systems may tend to make obsolete or replace whole layers of skilled and semi-skilled traditional activities in the organisation. Events not long ago in the newspaper industry illustrate this point. Not only has new technology swept away traditional highly skilled tasks such as those of the compositor, but the switch to electronic based automatic equipment has meant that work traditionally within the closely guarded domain of the highly skilled print craftsman, has been transferred to the electricians, traditionally regarded as the under-dog of the newspaper industry. In less dramatic environments what remains might be challenging and demanding jobs for the highly skilled, and a number of menial jobs cannot economically be automated. This can have serious consequences for the level of job satisfaction and motivation of the employees who remain, and may create a wider schism between white and blue collar workers or confrontation between unions or groups of craft workers (as the newspaper industry again demonstrated).

(c) Opportunities for interaction. Almost inevitably the new technology will bring changes in the way in which people relate to one another at work. The new technology might disrupt relationships and reduce the opportunities for social interaction with colleagues. This will hinder the development of often essential informal groups which cross organisational boundaries and which help the organisation to achieve its goals (the Trist and Bamforth research in the coal industry). Conversely, if an individual's social needs are not met at work, or obstructed by work procedures, alienation will develop and he will try to find ways of 'beating the system' to ensure that these needs are met. Absenteeism could also increase in such conditions.

(d) Group relations. The new technology may lead to greater external control and monitoring of work performance. This will give fewer opportunities for informal group 'norms of production' to be set which may limit the volume of production. The employees have much less personal control over the output, the way in which work is done and their own working lives, which might increase feelings of powerlessness at work and again possibly foster alienation. Alternatively employees may show even greater creativity in devising new ways to beat the new sophisticated systems. Additionally 'buck passing' or a loss of accountability may develop as the coal industry studies illustrate.

(e) Leadership. It is inevitable that styles of leadership will be required to change. The supervisor's power is likely to become based much more on skill and expertise than on formal authority. With highly skilled technicians working on interdependent systems, a team management approach will probably be more effective than over-reliance on the scalar relationship.

(f) Structural changes. There are likely to be structural changes in the organisation. Departments will develop to accommodate the new skills required. Older departments which house people with redundant skills will decline in size and influence. This will create problems in inter-departmental relations as the new replace the old and the status and power of each department changes.

(g) Industrial relations. In industrial relations there could well be a substantial switch in the basis of power both between unions and management and between unions representing different workers. However, the risk of 'strategic' strike of one or two key workers will escalate.

A final aspect of the implications of technology change is that many control systems work on the basis of division of duties between clerical grades to maintain control and avoid collusion, which could conceal fraud. Such systems depend upon many clerks being employed in jobs such as wages preparation, bonus calculations and production statistics. With advanced technology automating many of these tasks, fewer people are required, and therefore the risk of collusion returns, unless rigid controls are built into the new systems. For example, the bonus calculations must be keyed in such a way, that they can be readily related to the levels of output and eligibility for bonus, to avoid bonuses being paid to operatives or staff that are not entitled to them. This may require, therefore, more detailed classification of the workers' records in the computer system as well as possible changes in access to initial data, and responsibility for input.

5 WORK GROUPS

Tutorial note. It is worth giving a brief overview of the advantages and disadvantages of cohesive and informal groups, before applying the principles to the four specific scenarios.

Advantages and disadvantages of cohesiveness

It may not be necessary for management deliberately to create or encourage the information of cohesive groups at work: even where 'teams' are not an accepted organisational mechanism, informal groups are likely to form to fulfil individuals' needs for friendship and belonging. However, as in the example of airlines cited below, some organisations have shown themselves able (and think it desirable) to impede the formation of cohesive working relationships.

It should be noted that the cohesiveness of the group is not in fact dependent on its being an 'informal' group brought together out of natural affinity. Many of the research studies into groups (for example Sherif's study of schoolboy groups) started by deliberately breaking up 'natural' attractions and relationships: in the resulting random groups, cohesiveness was still built up.

It is important to consider that groups may work for or against the organisation's interests: the cohesiveness of the group, by reinforcing its norms of attitude and behaviour, will confirm it in that co-operative or hostile position.

In an experiment reported by Deutsch, psychology students were given problems to work at in discussion groups which were divided into 'competitive' and 'co-operative' groups. The co-operative groups showed greater productivity per unit time, better quality of end result, greater co-ordination of effort and sub-division of activity, more diversity in amount of contribution per member and more attentiveness to fellow members.

Belonging to a cohesive work group may maintain employee morale, by fulfilling social needs, providing a manageably small unit for the individual to identify with and an opportunity to pool resources and share responsibility.

In an ideal working group skills, information and ideas are 'pooled' so that the group's capabilities are greater than those of the individuals. Groups have been shown to produce fewer - though better evaluated - ideas than individuals working separately, but the performance of even the group's best individual can be improved by having 'missing pieces' added by the group.

If the group has objectives which differ from those of the organisation, however, it will not be desirable from the organisation's point of view to encourage cohesion, which may provide a position of strength from which to behave in hostile or 'deviant' ways. Informal social groupings may have objectives associated with the satisfaction of social ('affiliation' or 'relatedness') needs: they may devote much energy, effort and emotion to team maintenance, rather than to the fulfilment of organisational task goals.

Brayfield and Crockett suggest that a congenial work group, creating high morale, can either raise or lower productivity, depending on the 'norm' adopted by the group. A classic piece of research into this phenomenon is the Bank Wiring Room phase of the Hawthorne research. The group developed norms related to output which represented what was felt to be 'fair' for the pay received; members of the group who 'over-produced' (as well as those who 'under-produced') were put under social pressure by the group to get back into line. The group did not follow company policy on some issues and 'fiddled' daily output reports to suppress fluctuations.

Moreover, cohesiveness itself can cause problems. Handy notes that 'ultra-cohesive groups can be dangerous because in the organisational context the group must serve the organisation, not itself.' If a group is completely absorbed with its own maintenance, members and priorities, it can become dangerously blinkered, overconfident in its decisions, blind to risk, deaf to divergent views and isolationist in its attitude to outsiders: I L Janis calls this 'group think'.

(a) *The air crew*

It appears that it is in fact the common practice of commercial airlines to allocate both cabin and flight crew on an individual basis: crews do not work together as a team for more than one trip at a time, and scheduling makes informal regular contact difficult even on the ground.

This practice presumably stems from the priority given to passenger safety, and the requirement for complete adherence to routines.

(i) Informal groups in a work context tend to find and settle into their own ways of doing things, often abbreviating or by-passing formal routines; trusting each other, they may not remain alert to faults or deviation from procedure. The airline could not risk this occurring.

(ii) The crew is small and isolated from the larger organisation. A group in such circumstances can become pre-occupied with team maintenance, especially since the lifestyle of its members is not conducive to the formation of other stable relationships. This may have the effect of drawing concern, attention and energy inwards to the group instead of outwards to the passengers - which is particularly undesirable in the event of accidents or other incidents The problem would be further exacerbated if sub-groups or cliques formed, and the crew were unable to work together in an emergency.

(iii) Crews of 'strangers' are generally more likely to stay alert and 'police' each other to an extent: this may be a consideration, since there is a large amount of valuable stock on board (for example duty free goods, gifts and toiletries, blankets, pillows) which might be subject to pilferage if the crew were in league.

On the other hand, there would clearly be disadvantages to passenger comfort and safety if personality clashes, conflict or lack of trust were to affect a crew allocated to each other at random. Rigorous selection and training would be required to ensure relative compatibility, self control, tolerance for uncertainty and ability to cope with stresses, especially on long flights. A certain cohesion will inevitably result from the need for teamwork, and from the 'culture' created by uniform, badge, shared knowledge and familiar routine. Behavioural norms of the airline will act to an extent as group norms.

(b) *The bomb disposal unit*

Bomb disposal units are at the other end of the continuum, in a sense, from the aircrews, because they deliberately set out to build up close relationships and co-operation.

The unit is likely to develop informal relationships and cohesion in any case, because the work is highly specialised and dangerous. Team members are likely to draw together against the common 'enemy' or danger, and out of mutual respect and fellow feeling. Group cohesion is a common response to external threat, and the sense of belonging may also offer compensation for deprivation of security in other areas. Under the stresses of urgency and danger in the unit's work, mutual support will be an essential maintenance activity for the survival of each individual and the group.

It is in the unit's interest to encourage cohesion as well. Mutual trust, which may be enhanced by strong informal bonds, will also be essential to the task, since the team members will need to depend on each other - literally, for their lives. In situations where the task does not conform to training or experience, it is also valuable to have a group that contributes ideas and experience freely, without political 'games' or concern for personal advancement. Cases have even been reported of extreme individual sacrifice for the good of the group, in similar situations of work risk.

The drawback to cohesion in such a group may be that strong informal bonds and concerns may intensify the awareness of and anxiety about the risks involved in the task. If the group becomes too mutually protective - putting maintenance above task - the unit may cease to function because of reluctance to endanger its members. Where injury and fatality do occur, too, the group is likely to be devastated, to the degree that its members are attached to each other. Selection, training and the culture of the unit will have to be geared to overcome these pressures, to channel fear into determination, to focus on the 'aggressor' - the common enemy - so that team maintenance is task-rather than self-oriented.

(c) *Scattered sales representatives*

The sales representatives are widely dispersed, presumably according to 'territory', and are therefore likely to have little opportunity to form informal groupings with any cohesion. (Informal groupings are by definition fluctuating and ephemeral, and sales representatives are often of a temperament to form easy, quick social rapport, but it is the *cohesion*, the sense of 'team-hood' that will be lacking.)

The organisation might wish actively to encourage informal bonding between its representatives:

(i) so that they can pool their knowledge and experience. Customer feedback on the products, given to the salesperson 'at the sharp end', should be bought to the control system of the organisation: product modification, price changes, marketing strategy adjustments and so on may need to be made. The reps might also exchange experience of selling techniques or product knowledge;

(ii) if it wishes to present a consistent corporate image to its customers, from one country to another. Reps should be trained in the company procedures and practices, and will also need to be 'sold' the company culture, if they are to present a common face to the outside world;

(iii) to compensate for the isolated nature of the reps' role. The organisation will want to generate a sense of belonging, loyalty, involvement and commitment despite the reps' physical separation. The firm may in particular be concerned that the reps might satisfy their 'affiliation' or 'relatedness' needs elsewhere - might form informal bonds with customers or even competitors in their sales area, and leave the firm altogether.

A drawback to group cohesion in this situation is not only the effort and cost involved in nurturing it, given the natural barriers, but the undesirability of creating a stability-oriented culture. Sales representatives must be able to move and respond freely to market requirements where they are, without depending on contact with and the approval of others.

Cohesion could be encouraged through: regular (not necessarily frequent) meetings, briefings or training sessions for the whole sales body or regional teams: standardisation of procedure and documentation; common culture and values, (elements of a uniform, perhaps, even if it is only a badge), shared symbols of success; free communication of sales performance information and market share to make the reps feel involved in a common effort. They will in any case be united by a common product range, and if they can be 'switched on' to that, loyalty and involvement may be enhanced.

(d) *The package tour couriers*

Again, an ambiguous situation because of the geographical dispersal and isolation of the couriers. Much will depend on whether the package tour company sends couriers out from the country of origin with each tour - to return at the end of the tour - or whether couriers are posted to foreign destinations for a given duration, to handle things at that end of the tours.

The advantages of informal grouping and cohesion will be similar to the scenario of the European sales representatives. The task of creating cohesion may be greater because there may be less common ground in the product/service: the range of situations and tasks faced by couriers in different parts of the globe, dealing with the 'general public' will be much wider. The task of creating cohesion may on the other hand be easier, because there is greater scope for using the visible symbols of a common corporate culture, for example uniform, badging, standard procedures and so on. Tour couriers need to be easily recognisable as representatives of their particular company, so they will be equally 'recognisable' as follow members to each other.

Conclusion

The work group is one way in which the organisation can satisfy an individual's need for 'affiliation' (McClelland) or 'relatedness' (Alderfer). It is true, however, that groups themselves can absorb 'E' factors (Handy's term for energy, effort, excitement etc that are generated in the process of motivation) which the organisation would prefer to be channelled in other directions. Management then faces the dilemma of whether to allow/encourage or impede the formation of cohesive work groups. The best course in most situations would be to seek to encourage such groups for the benefits they offer, within a controlling framework of organisational systems, leadership and culture that directs the energies of the group 'outward' to the task.

6 LEADERSHIP

Leadership

Leadership is the process of influencing others to work willingly towards an organisation's goals, and to the best of their capabilities. 'The essence of leadership is followership. In other words it is the willingness of people to follow that makes a person a leader' (Koontz, O'Donnell, Weihrich).

A manager is appointed to a position of authority within the organisation. He relies mainly on the (legitimate) authority of that position. Leadership of his subordinates is a function of the position he holds - but a manager will not necessarily be a 'leader'. Some leaders (for example in politics or in trade unions) might be elected; others might emerge by popular choice and through their personal drive and qualities. Unofficial spokesmen for groups of people are leaders of this style.

Leaders are *given* their roles by their putative followers; their 'authority' may technically be removed if their followers cease to acknowledge them. The *personal, physical* or *expert* power of leaders is therefore more important than position power alone.

Leaders are the creators and 'sellers' of culture in the organisation. 'The [leader] not only creates the rational and tangible aspects of organisations, such as structure and technology, but is also the creator of symbols, ideologies, language, beliefs, rituals and myths.' (Pettigrew).

Making managers into leaders

Early writers like Taylor believed the capacity to 'make others do what you want them to do' was an inherent characteristic. Early studies on leadership concentrated on personal traits of existing and past leaders. One study by Ghiselli did show a significant correlation between leadership effectiveness and the personal traits of intelligence, initiative, self assurance and individuality. It is obvious that these personal characteristics are important since it is unlikely that a person who lacks self assurance and initiative would command much respect from others. The 'helicopter ability' ('seeing the big picture') has also been identified (Hunt and others) as the only ability positively correlated with management success.

Jennings (1961), however, wrote that 'Research has produced such a variegated list of traits presumably to describe leadership, that for all practical purposes it describes nothing. Fifty years of study have failed to produce one personality trait or set of qualities that can be used to distinguish between leaders and non-leaders.' Trait theory, although superficially attractive, is now largely discredited.

There have since been many classifications of leadership 'styles'. The Hawthorne studies directed by Mayo demonstrated that the productivity shown by the girls in the Relay Assembly Room was a function of democratic (participative, people-oriented) leadership. Later studies by Lippitt and Whyte (Boys Clubs) and Likert in his Michigan studies confirmed that democratic rather than autocratic leadership appeared to encourage productivity. Most businesses (under the general influence of the Human Relations Approach) then set about training and making these new styles of leaders. This was done by intensive training in supervisory techniques and in particular human behaviour. The assumption was that if the supervisors and managers could understand the sociological theory explaining the success in the Hawthorne studies, they would automatically adopt this approach to 'good' leadership. This approach assumed leaders could be *made* through the right type of training and understanding.

Blake and Mouton (1964) suggested that an exclusive concentration on 'people' was unlikely to prove effective in productivity leadership and what was necessary was a balanced approach between 'concern for people' and 'concern for production'. Their Managerial Grid approach was a very particular approach to training and making such 'good' leaders.

It consisted of using 'peer group pressure' to correct supervisors who showed one particular orientation (for example over concern for production or over concern for people) into this new balanced concern for both people and task. The underlying assumption is that people can adopt such leadership patterns through training and development and therefore be made into good leaders by the adoption of a particular style of leadership.

The current approach to making a good leader takes more account of the factors involved in any leadership situation. This contingency approach commenced with Fiedler's study which demonstrated the importance of:

(a) the relationship between the leader and his group (eg liked, trusted);
(b) the structure of the task (eg task clearly laid down);
(c) the power of the leader in relation to the group (eg the value of rewards and punishments).

This has been followed by a further development of contingency approaches. The 'best fit' approach (C Handy) suggests that the leader to be effective must take four sets of influencing factors into consideration.

(a)	The leader:	his preferred style and personal characteristics;
(b)	The subordinates:	their preferred style in the light of the situation;
(c)	The task:	the job, its objectives and its technology;
(d)	The environment:	the organisational setting of the leader and his group.

This 'best fit' approach maintains there is no such thing as the right style of leadership but that leadership will be most effective when the requirements of the leader, group and task fit together. This latest flexible style approach still begs the question as to whether those in authority be made (by training, influence, development) to adopt such approaches or does it still require some inborn characteristics of the person?

Desirability of becoming a leader

Research has shown the apparent effect on motivation, efficiency and productivity of the chosen leadership style of managers. But does it really matter whether a manager is a leader or not?

The belief is that if a manager had indifferent or poor leadership qualities his subordinates would still do their job, but they would do it ineffectually or perhaps in a confused manner. By providing leadership, a manager should be able to use the capabilities of subordinates to better effect: leadership is by definition the 'influential increment over and above *mechanical* compliance with the routine directives of the organisation' (Katz and Kahn, *The Social Psychology of Organisations*).

Handy suggests that position power (or legitimate authority) is insufficient in itself to bring about lasting attitude change in employees, though it may be sufficient to change behaviour in a desirable direction. Personal or expert power are perceived as more 'meaningful' to the worker, and the psychological dissonance involved in opposing them more acute. The most influential managers, then, are those who wield these sorts of power - ie the leaders.

Managers therefore need - and should seek - to become leaders in situations where they require co-operation, not merely compliance, by their immediate subordinates.

7. NATURE AND ROLE OF PERSONNEL FUNCTION

Tutorial note. You are asked to produce a refutation of the accountant's arguments. If you find that you actually agree with his claims, you'll simply have to argue against your better judgement for the bulk of your answer.

The scenario in the question is a 'thumbnail sketch' of a widespread and major problem for the personnel function, the common perception of it as:

(a) 'a cost rather than a benefit';

(b) useful only for 'low-level' tasks; and

(c) alien to the world of quantifiable benefits, economic realities and 'hard' information presided over by the accountant.

It would be a mistake to try and argue with the accountant on his own terms: even if the contribution of the personnel function to the achievement of organisational objectives is understood and accepted, it is not easy to measure in quantifiable terms. Standards and budgets can be set, but mainly for the routine service activities of the function - which is probably why the accountant wishes to confine it to those areas. Value for money judgements about the personnel department are often difficult to make: while various cost indices can be used to assess the efficiency of the function, and adverse or positive trends, the guidance/advisory role of personnel is almost impossible to evaluate, and even in the service activities, there are too many non-rational and uncontrollable variables for accurate and fair performance measurement to be achieved.

The accountant should, however, be urged to open his mind to the benefits and importance of the personnel function: its success lies substantially in the absence of problems - in areas such as industrial relations or manpower planning - which means that it is all too easily taken for granted.

First of all it may be pointed out that although the prime objective of the organisation may be a financial one, various social and ethical objectives may also be included in the corporate plan, and will also require fulfilment as a measure of organisational success.

Nor is it valid to discount personnel's contribution to the economic objectives of the organisation. The IPM has described the role of personnel management as follows. 'It seeks to bring together and develop into an effective organisation the men and women who make up an enterprise, enabling each to make their [sic] own best contribution to its success both as an individual and as a member of a working group.' This is a similar orientation to what Peters and Waterman observed in 'excellent' companies, and called 'Productivity through People'. They quote IBM: 'Our early emphasis on human relations was not motivated by altruism, but by a simple belief that if we respected our people and helped them to respect themselves, the company would make the most profit.'

Organisations invest a great deal of time and resources in their employees. There are certain implications of regarding employees as organisational assets, which reflect the importance of the personnel function.

(a) People are a resource which needs to be carefully and efficiently managed with overriding concern for organisational objectives.

(b) The organisation needs to protect its investment by retaining, safeguarding and developing its human resources.

(c) Deterioration in the attitudes and motivation of employees, increases in labour turnover (followed by costs of hiring and training replacements) is a cost to the company - even though a 'liquidation' of human assets may produce short-term increases in profit.

Various techniques have been developed for 'human asset accounting', the creation of information to show the investment made in staff and the financial effects of changes in the human assets over time - whether they are increasing or decreasing in value, whether they are being used effectively and whether a satisfactory return is being made on them (in terms of 'added value', replacement costs etc). Personnel can make a measurable contribution in increasing the value of human assets through training and development, ensuring that they are effectively deployed by manpower planning and development, and retaining their services so that a satisfactory return can be made.

Major areas in which the activity of the personnel function can be identified as bringing about cost avoidance, savings or efficiency for the organisation include the following.

(a) *Manpower planning*, the efficient acquisition and deployment of manpower resources. Overemployment, or employment of unsuitable individuals, can be costly in terms of salaries, redundancy payments etc.; underemployment or under-skilling puts strain on existing staff and systems; lack of career development planning may be a source of dissatisfaction and dysfunctional behaviour.

(b) *Recruitment and selection*. The benefit of systematic recruitment and selection of suitable staff is obvious: wrong decisions in this area can be costly. There are strong arguments in favour of its being performed by personnel: consistency, elimination of duplication, expertise in selection techniques, contacts in agencies, awareness of legislation on discrimination and so on.

(c) *Training and development*. The value of human assets increases with the enhancement of their skills, and the amount of responsibility they hold. The improvement of skills and development of effective work teams has a significant effect on output and productivity, while cost-effective methods of achieving both contribute to overall efficiency. Again, the personnel function can achieve a more systematic, informed and centralised approach.

(d) *Industrial relations*. This is commonly thought to be an aspect of personnel's 'firefighting' role: the costs and dysfunctions arising from industrial disputes or problems - go-slows, strikes, absenteeism, high labour turnover - are more easily demonstrated than the positive effect on profitability of industrial harmony, which is noticed only in its absence. Nevertheless, the prevention of such problems is valuable.

(e) *Health and safety*. Again, it is not directly provable that employee appreciation and satisfaction result in increased productivity, but the costs of accidents, illness can industrial fatigue are various and high, and can be saved by the formulation of appropriate policies. This can most consistently and effectively be achieved by the personnel function, since it does require administrative effort and - ideally - specialist knowledge of legislation and regulation, industrial physiology and psychology, and ergonomics.

As to the other comment made by the accountant - that personnel's role should be confined to 'low level tasks such as welfare and the administration of employee services', it could be argued that:

(a) such tasks are not 'low level' at all, if one considers the value of human resources and recognises that action to retain their co-operation and services represents sound investment; and

(b) as discussed above, there are many other areas in which personnel can usefully contribute; these areas will multiply, as change continues, technology advances, employee aspirations grow and social responsibility becomes more appreciated.

To be fair to the accountant, some personnel departments are indeed what he describes. Charles Handy, among others, has noted that the efforts of the function to extend its 'resource power' and control in organisations often tend to cause expansion beyond an efficient size. In many cases, however, the perceived lack of effective contribution can be traced not to the personnel function itself but to line management. Brewster and Richbell comment, of personnel specialists:

'They carefully watch developments in the industrial relations, political and labour market environments, they develop sensible, well-thought-out personnel policies that would make their company one of the most progressive and highly respected of employers. And then they see their efforts continually frustrated and subverted by a management team that seems determined to ignore most of what the personnel department does.'

8 MANPOWER PLANNING

Tutorial note. Please check that you have answered the question set.

(a) Is manpower planning getting easier or harder? (It would be appropriate to consider, too, to what extent it has *ever* been easy....)

(b) If it is getting harder, why is it worth doing?

It really *isn't* worth wasting answering time on describing the manpower planning process....

Harder or easier?

'Modern employment conditions' are such that manpower planning is not currently - and is unlikely ever to become - 'easy': some of the problems with manpower planning will not 'go away', despite environmental changes. Cuming *(Personnel Management)* comments: 'Clearly, the more precise the information available, the greater the probability that manpower plans will be accurate. But, in practice, they are subject to many imponderable factors, some completely outside an organisation's control... international trade, general technological advances, population movements, the human acceptance of or resistance to change, and the quality of leadership and its impact on morale. The environment, then, is uncertain, and so are the people whose activities are being planned.'

Particularly where there are long planning horizons for manpower requirements (for example in civil engineering or aviation, where projects have long lead times, and long training periods for specialist employees), or where the market for the organisation's products/services is volatile and sensitive to unpredictable pressures (for example in fashion items), manpower planning faces difficulties.

Particular features of today's business environment, however, that may hinder or facilitate the process include:

(a) *technology*. The availability of microchip technology has in some ways made manpower planning easier, through the provision of more and better information. The speed and accuracy of data gathering and processing (sorting, analysing, selecting), and the availability of a range of formats and facilities (spreadsheets, models) have taken much sheer labour out of forecasting and planning, and facilitated more penetrating analysis of current situations and future trends.

On the other hand, the impact of technological innovation on markets (and demand for products/services and therefore demand for labour) is not easily identified in advance, and can render long-term manpower plans swiftly obsolete. The sudden 'explosion' of miniaturisation and laser technology in the music industry, for example (personal stereos and CD players) created labour demand in companies which seized the opportunity as it arose, and over-supply in companies which were still committed to large home stereo systems and record manufacture. Similarly, constant advances in hardware and software technology create new skill shortages and 'gluts' from year to year;

(b) the *demographic downturn*. The labour market is getting increasingly difficult, particularly for the recruitment and retention of young workers, and those in skilled occupations. Falling birthrates mean that although the labour force will still be growing in the coming decade, it will not be growing fast enough to meet a steeply rising demand for labour. This growth in demand is expected to be concentrated among the higher skilled occupations - but the main source of supply to such occupations (graduates) is constrained by the demographic factors, by educational policy and so on. In addition, the average *age* of the workforce will rise and employers who rely particularly on young people in certain jobs or entry grades will have to rethink their manpower plans. Because *population* growth is not 'feeding' the workforce to the same extent, readiness of individuals to work will be the key to holding off labour and specific skill shortages: employers will need to attract more women and more mature people into the workforce - overcoming prejudice and tradition along the way.

Shortage of supply for labour in some sectors will be matched by falling demand in others - for example, in teaching as school rolls fall off.

Shortage of supply not only forces employers to rethink their recruitment and retention strategies from the outset, but creates a tendency to greater career mobility, which complicates the manpower planners' assumptions about wastage rates;

(c) *economic conditions*. These tend to be unpredictable at the best of times, but the manpower planners' task will be complicated in coming years by factors such as the increased amount and variety of competition following the unification of the European market (which will also effect the size, composition and mobility of the labour force); the economic unification of East and West Germany; the opening of markets in Eastern Europe and so on;

(d) *employment trends*, such as flexibility (eg the erosion of demarcation lines and agreements) and individual (as opposed to collective) bargaining. The internal mobility of labour which flexibility offers may facilitate the manipulation of supply and demand, but also offers less predictability - as does individualism in career planning and reward negotiation;

(e) *cultural shifts* - which may be surprisingly sudden. Changes in social value systems (eg reputedly, in the late 80s, from self-interested materialism to altruistic environmentalism) effect the market environment, consumer buying patterns, employment aspirations and patterns etc.

At a less radical level, *fashion* - short term trends in taste and aspiration - can cause dramatic up- and down-turns in demand for products. This is a major thesis of Tom Peters recent book *Liberation Management*;

(f) *events* can happen with no warning, affecting demand for or supply of labour. The collapse of the Iron Curtain, for example, is likely to have a far-reaching effect on European labour and markets, on the armaments industry, tourist industry etc. On a smaller scale, an event such as the Strangeways Prison riot (April 1990), an oil rig fire or flood/storm damage may have a 'local' impact on employment.

Is it worth doing?

The factors likely to make manpower planning harder do so by increasing the unpredictability of employment conditions. This makes manpower planning more necessary, however, rather than less so, in order to minimise the disruptive effects of unpredictable events. It becomes more than ever desirable to identify trends, and plan - as far as possible - for contingencies.

Hard or easy, manpower planning is necessary, because:

(a) labour is a resource, which must be planned for and controlled as much as machines, money etc - with the added complication that people are difficult to predict and control;

(b) business organisations must react to their environment. Job contents and requirements are likely to change, and the labour resources of the organisation must be equal to that change in order for the organisation to survive;

(c) reducing demand or improving supply of labour resources is a basic logistical activity which must be performed by any organisation staffing its operations;

(d) the supply/demand equation is complicated in practice, and cannot be left to chance. Jobs often require skills and experience which are scarce in the market place: training and development may be required to make these skills available. Employment legislation, employee aspirations and trade unionism make it difficult for organisations to pursue haphazard policies of redundancies in the event of over-supply, external recruitment without promotion policies and the like.

9 EVALUATION OF RECRUITMENT ADVERTISEMENT

Tutorial note. You are required to answer three specific questions: make sure you have done so. Questions (a) and (b) are straightforward enough - although you should be clear that you are criticising the ad, and not the organisation. Question (c) is more challenging, since advertising copywriters are by definition persuasive in their style, and you need not have taken the claims made for AOK at face value. Also, the examiner in his report commented that 'some inferential comments about the company culture, style and approach were expected, not the mere repetition of the information that the company is a world leader in medical electronics and is located in Reigate, Surrey'!

(a) *Good points about the advertisement*

Given that this is an advertisement, and therefore intended to attract readers' attention and persuade them, this specimen exhibits some good techniques.

(i) It is attractively designed in terms of page layout, and in particular makes use of a strong headline presentation. The headline is not only large enough to attract attention, but the split into two halves, and the 'hesitation' in the middle ('... um ...') creates curiosity in the reader.

(ii) The tone of the headline and much of the body copy is informal, colloquial and even friendly. It starts with a joke, implying that the company has a sense of humour. The style is deliberately like speech in most areas, from the '...um ...' of the headline, to the abbreviated forms ('we're', 'you've', instead of 'we are', 'you have').

(iii) The written style is fluent and attractive. There are clearly defined paragraphs, a mixture of short/punchy and long sentences, and 'witty' phrases like 'to help people to help themselves'.

(iv) It offers quite a lot of information (at least, ostensibly) about the culture of the company - how it feels about personnel issues, where it's going etc - as well as about the job vacancy. In other words, it sets out to give a positive 'feel' about the company's ethos, and what it would be like to work for. The use of positive words and superlatives throughout ('rewarding', 'enlightened', 'highest', 'leading', 'influential', 'fast growing', 'thrive' etc) creates a very up-beat tone.

(b) *Improvements that could be made*

Advertisements cannot be seen in isolation - as in this question: they appear in the context of potentially many similar pieces, competing for attention. In addition, job advertisements carry certain 'responsibilities': they are a form of pre-selection, and as such should be not just attractive and persuasive, but accurate and complete enough to give a realistic and relevant picture of the post and the organisation, so that readers can begin to make an assessment of their suitability for the job - and its suitability for them. The ad is a resourcing mechanism, not an exercise in self-expression by the personnel department or - worse still - the copywriters.

Faults which require improvement therefore include the following.

(i) There is too much copy. Readers may not have the patience to read through so much (rather wordy) prose, particularly since:

(1) they are likely to get a sense of *déjà vu*, as the same phrases come up again ('progressive international organisation', for example), or look rather familiar in any case ('in the nicest possible way', 'our staff are our most important asset', 'a company with a difference' etc);

(2) there is very little 'hard' information contained in the ad. Not until the third paragraph of rather 'flowery' sentences do we find out what field the company is in, and not until the *fourth* do we see what job is on offer!

(ii) There are many words and expressions which sound good, and seem to *imply* good things, but are in fact empty of substance, and commit the organisation to nothing. They are usually the 'stock' expressions like 'committed to an enlightened personnel philosophy': what does that actually *mean*? 'Your prospects for promotion are excellent. Salaries... highly attractive... benefits...comprehensive' are more 'weasel' phrases, promising nothing concrete: the copywriter has had to add the words 'obviously' and 'of course' so that the reader would feel like a fool if he or she stopped to query what the words really said. (And if 'Personnel Generalist' isn't a fancy label for 'dogsbody', it certainly *sounds* like one!)

(iii) There are confusing contradictions, eg between the requirement for flexibility, 'interchange with various functions', do 'whatever the situation demands' etc and the more cautious 'within your designated area...'.

(iv) The copywriters are in places too 'clever' for their own good. The first three lines, for example, could backfire quite badly if a reader failed to catch the next line, or simply didn't appreciate the self-deprecating tone. Words like 'chameleon', too, are rather literary - and may even be discriminatory to people of different cultural and educational backgrounds.

(v) The advertisement does not give enough 'hard' information to make effective response likely - and then fails to do its job of facilitating response at all! Despite the invitation to telephone, no number is given. No named correspondent is cited, merely a reference number - despite the claimed emphasis on people as people, not numbers.

(c) *What is learnt about AOK*

The advertisement claims to say quite a lot about AOK, its culture, its people-centredness, its expansion and progressive outlook, flexibility, sense of humour etc.

Such claims should be taken with a pinch of salt, however. The advertisement is not likely to *lie* - because that is too risky under Advertising Standards - but then, it does not actually say very much in a concrete way that AOK could be 'pinned down' to.

We may, however, infer some things about the company.

(i) It has a strong cultural 'flavour', and believes in 'selling' that culture quite hard. It likes, for example, telling people what it is 'committed to', what it 'firmly believes' etc.

(ii) It is an up-beat organisation, which tends to stress its good points and opportunities: it certainly sees itself (even allowing for advertising hyperbole) as go-ahead, successful and expanding, flexible, people-oriented.

(iii) It is possibly not as deeply people oriented as it tries to project. The areas of involvement for the personnel department enumerated, for example, seem rather limited and administrative: there is no suggestion of a wider strategic role for personnel, such as would indicate that 'people issues' really do affect management outlook. Its 'enlightened personnel philosophy' does, however, extend to (unspecified) pay, benefits and prospects and to equal opportunities advertising (scrupulously using 'individual' throughout) - but not to giving the recruitment manager a name, as a personal contact. (For a supposedly flexible, adaptive, non-bureaucratic company, 451/BPD sounds a little strange!)

(iv) Given the various ways in which AOK has undermined its credibility, one might also start to wonder whether phrases like 'you don't have to be mad to work here, but it helps', 'it doesn't make for an easy life', 'you'll be a strong personality', 'or whatever the situation demands', 'high pace' and so on may not be hinting at some real problems of workload, organisation, political conflict and staff difficulties. But perhaps one should give AOK the benefit of the doubt on this one (and wait for the interview...).

10 RELIANCE ON SELECTION INTERVIEW

The objective of the recruitment and selection activity of an organisation is to get the right people for the jobs required by the tasks of the organisation, in the most efficient way possible. This will not be possible if the job itself is ill-defined (candidates are selected for the 'wrong' job) or if candidates are inefficiently assessed with reference to the job's requirements (the 'wrong' candidate is selected).

With this in mind, it is a 'mistake' to rely on the interview as the *first* stage in the process, because:

(a) the organisation must take very seriously the business of defining and describing the job itself, and the type of person required to perform it. A systematic approach to recruitment and selection must start with job analysis and description (or review of same). A person specification may also be useful preparation for certain general grades of job;

(b) interviewing is a time-consuming and costly method of sifting through applicants who may be obviously unsuitable for the post. Pre-selection and screening can be achieved through:

  (i) the placing and content of the job advertisement (so that only potentially suitable people see it, and are able to exercise some self selection from the information given);

  (ii) the consideration of application forms, employment histories (CV) etc; or

  (iii) the use of employment agencies or recruitment consultants to short-list candidates for interview.

It is a 'mistake' to rely on the interview as a crucial stage in the process, because although it does allow a face-to-face encounter with the candidate, the opportunity to elicit information about him and some experience of his skills in communication and social relations, the technique is subject to severe limitations.

(a) Interviews are highly subjective. Interviewers may each have a different opinion of a single candidate. Even where they agree, they may be prone to errors of judgement or perception:

  (i) perceptual selectivity. Interviewers may latch onto a single attribute of the interviewee and thereafter see only what they want or expect to see, by what is called 'the halo effect', or stereotyping;

  (ii) ill-defined ideas of qualitative factors such as motivation, honesty or integrity and how they can be assessed;

  (iii) a tendency to influence the behaviour of the interviewee by 'contagious bias'. Candidates tend to want to please the interviewer and falsify their responses to give the interviewer the answer that he appears to want - especially if the interviewer phrases questions 'Don't you think ...?'

(b) They are not accurate predictors of how a person will perform in the job, and within the work group. The interview is a 'false' situation, and the candidate can often sustain a 'role play' for the duration of the exercise which he could not keep up in the real context of the job. The interview may give some indication of communication and human relations skills, but there is much else the organisation ought to want to know: the candidate's expertise, initiative, soundness of judgement, capacity to handle responsibility and so on. (Even were all this information available, the candidate would not necessarily be suited to the particular systems, work group, culture and politics of the organisation.)

What should employers be doing?

As indicated in part above, the prospective employer should use interviews at a later stage of the recruitment and selection process, after suitable preparation and more cost-effective pre-screening, such as:

(a) job analysis and description or review;
(b) careful preparation of a job advertisement for a limited audience and of limited appeal;
(c) the use of agencies, where appropriate, to carry out the initial screening;
(d) scrutiny of relevant items of information on job application forms and CVs; and
(e) short-listing of only the most potentially suitable candidates for interview.

The employer should also place less reliance on the value and validity of interviews as accurate predictors of future success in the organisation. Depending on the type of job, various other selection techniques may be suitable to confirm or assess suitability. Selection tests may be used (although, again, these are often only marginally more objective and accurate than interviews): aptitude tests or proficiency tests may be appropriate where the demonstration of a particular skill or ability is essential to the job, eg typing, numeracy etc. Intelligence testing and personality testing are more problematical. For more senior posts, or where leadership/human relations/communications/group problem solving skills are important, 'assessment centre' (group selection) techniques may be appropriate, combining tests, interviews and group problem simulation and discussion sessions.

Apart from any of these alternative methods, employers should look to make the interview process itself more efficient and effective. Interviewers should be educated and trained in objective assessment and interview technique. Time should be devoted to preparation for the interview, and facilities provided for the conduct of it in appropriate surroundings.

11 JUSTIFICATION OF TRAINING COSTS

Direct costs of training include:

(a) trainees' course fees, books, study packs and other expenses;

(b) examination fees;

(c) possibly, the provision of a training department, with staff and facilities;

(d) time off with pay to attend courses;

(e) cost of organising courses and seminars and other training sessions;

(f) hidden costs, such as:

 (i) cost of lower output;
 (ii) cost (possibly) of mistakes and rectification work;
 (iii) need to provide cover - sending trainees on day release means that cover must be provided;
 (iv) possibly the cost of carrying an inexperienced trainee;
 (v) cost of lost training cost when unabsorbed trainees leave.

These costs, however, give rise to benefits.

The trade off from training can best be obtained by a correct attitude to the process.

First, consideration must be given to what kind of skills are required and when. This should involve consideration of the actual manpower plan projections.

Requirements and timescale will dictate the type of trainee selected; if the timescale is fairly long, a cheaper lower level trainee can be brought in.

Irrespective of the level of trainee that has been brought in, if the training scheme is working, then he should quickly acquire experience of the organisation and some technical knowledge of his discipline, and should demonstrate a potential for further development or even promotion. Effective training that achieves this will provide people that will grow into a job more quickly and enjoy the support of the people around them. These advantages might not be achieved by the alternative solution of costly external recruitment of people who then have to be taught the job. If, as in certain companies, the trainees are recruited at a low enough level, perhaps straight from school, then they can be carried as extras while they go through the rudiments of learning the business.

The problems of learning, however, may lead to a slower output, and possibly higher rejects and rectification costs. These can be minimised by the selection of people who have an aptitude to learn quickly. That may mean spending more initially on actual selection (a vital consideration, since it is often difficult to dismiss people these days) and perhaps spending a little more on the actual training.

The problem of costly cover can be averted either by treating the trainees as extras, or by making people train in their own time. Day release has the disadvantage that work may still require to be covered, with the result that extra headcount is needed. A better alternative might to be sponsor evening classes, block release or week-end courses. In all cases the effectiveness of the method, and indeed the merit of the trainee, can be judged by the ultimate results in examinations and at the regular appraisals. This is the approach favoured by the accountancy profession, since by releasing students for block release, they can study when they are free from the distractions of work and thus better able to absorb knowledge for passing their examinations.

Studies in both America and India have extended the understanding of training and the trade offs that can be obtained from it. Primarily, both studies emphasise the effective control of training, by highlighting what is wanted, and when. The American study found that by training people in correct health and safety procedures, savings were achieved by reduced litigation costs arising from accidents as well as the indirect costs of accidents, such as the inevitable downtime, reduced cost of any resultant penalties that arise from the delivery delays caused by downtime and the training of replacement staff.

12 PERFORMANCE APPRAISAL SYSTEMS

(a) Most large firms have a regular system of appraising staff. The objectives of staff appraisal systems are to help in developing staff members to their full potential and to enable the organisation to allocate their human resources in the most efficient way possible. To achieve these objectives an effective appraisal system is likely to incorporate certain key characteristics.

(i) Reports on employees should be made out in writing and at fixed intervals. Staff appraisal is a sensitive operation and a written record of the assessment may remove any doubts or uncertainties which arise at a later date. The report is part of a record, the personnel record, which charts an employee's progress within the organisation. The intervals at which the appraisal should be carried out depend on the nature of the employee's work. For specialist staff who move from one long-term assignment to another, appraisal may be appropriate after each assignment is completed. For staff engaged in more routine work, an interval of six months or a year may be suitable.

(ii) Written reports should be objective. An employee's superior may be inclined to assess harshly to excuse his own poor performance; alternatively, an easy-going relationship during day-to-day work may make a superior feel reluctant to be critical, especially if his subordinate's promotion prospects may be harmed. One way of improving objectivity is to make the assessment form very detailed: the more specific the assessor is required to be, the less margin there is for subjective responses.

(iii) Appraisal should be consistent throughout the organisation. This can cause problems in organisations which, like banks, have many semi-autonomous branches. Again, the use of detailed assessment forms (standard throughout the organisation) will help, but the assessment form is only the beginning of the appraisal process and care must be taken to ensure consistency in the later stages too.

(iv) Assessments should be discussed with the person assessed. If employees do not know what is being written about them they will not be able to improve in areas where shortcomings have been noted. This could cause particular frustration if the assessment system is used as part of a process of selecting staff for promotion.

(v) Persons conducting the appraisal interviews should be trained and experienced in the necessary techniques.

(vi) The employee should be encouraged to contribute to the appraisal process. Ideally, he should have sight of the written assessment in time to consider his response before being called to interview. During the interview the emphasis ought not to be on problems and obstacles, but on opportunities. The interviewee should be encouraged to talk about his career plans, his knowledge and skills and how they could be put to better use, and to make suggestions for improving the way his work is carried out.

(vii) There should be adequate follow-up after the interview has taken place. If the system is to be effective, staff must have confidence in it. This will only happen if results are seen to follow from the assessments.

(b) *Advantages of appraisal systems*

(i) They enable the organisation to gather information about the skills and potential of employees.

(ii) They provide a system on which salary reviews and promotions can be based.

(iii) They help to develop the employee's potential by directing his attention to particular strengths and weaknesses.

(iv) They allow the employee and his assessor to discuss and agree on personal objectives.

(v) They may contribute to staff motivation.

Disadvantages of appraisal systems

(i) The subjective element in such systems cannot be entirely eliminated.

(ii) They depend for their success on a mutual confidence between the assessor and the employee assessed. In practice, it is difficult to achieve that confidence.

(iii) It is difficult to go beyond appraisal of past performance, which may be an inadequate guide to future performance in a different job. If an appraisal scheme is used as a guide to promotion potential this is a serious disadvantage.

(iv) They often do not lead to improvements in performance. Criticism of areas where performance has been weak can lead to a defensive response and future performance may actually deteriorate.

13 PURPOSES OF PAY

Pay is a means to various ends, depending on your point of view. The objectives of pay from the organisation's point of view are to:

(a) attract labour of a suitable type and quality;
(b) retain the services of those staff for as long as the organisation requires them; and
(c) *motivate* employees to achieve and maintain desired levels of performance.

Items (a) and (b) are, strictly speaking, the ends of particular structures or levels of pay. Item (c) is the prime end of pay itself, according to the 'economic man' model of the worker (Schein): pay is an inducement offered in return for the contributions the organisation requires from the workforce. How far pay can accomplish this end on its own is a question of pay's value as a motivator.

From the point of view of the employee, pay may be the means to an infinite number of specific ends, ie the satisfaction of any of the various needs man has which can, directly or indirectly, be achieved through money. Insofar as man has such needs, pay will accomplish those ends - if there is enough of it. Individuals may also, however, have needs which are unrelated to money, or which money alone cannot satisfy. Here again, we are in the realm of motivation theory: to what extent is pay an inducement, a motivator?

Pay as a motivator

An employee needs income to live. The size of his income will affect his standard of living, and although he would obviously like to earn more, he is probably more concerned:

(a) that he should earn enough pay; and

(b) that his pay should be fair in comparison with the pay of others both inside and outside the organisation.

It should be apparent that pay as a motivator is commonly associated with payment by results, whereby a worker's pay is directly dependent upon his output. If pay increases don't immediately follow improved performance, or completed tasks, they lose their 'connection' with results and become accepted as the consequence of age, experience etc. This is less effective as a factor in the motivation calculus.

All such schemes are based on the principle that people are willing to work harder to obtain more money. However, the work of Elton Mayo and Tom Lupton has shown that there are several constraints which can nullify this basic principle. For example:

(a) the average worker is generally capable of influencing the timings and control systems used by management;

(b) workers remain suspicious that if they achieved high levels of output and earnings then management would alter the basis of the incentive rates to reduce further earnings;

(c) generally, the workers conform to a group output norm and the need to have the approval of their fellow workers by conforming to that norm is more important than the money urge;

(d) high taxation rates mean that workers do not believe that extra effort produces an adequate increase in pay.

Pay is a 'hygiene factor' rather than a 'motivation factor' (Herzberg). It gets taken for granted, and so is more usually a source of dissatisfaction than satisfaction.

Money *can* be a motivator, but that depends on the individual's perceived need for it. Thus a young married man with a family to support may want money very badly, so that he can afford to achieve a desired standard of living. Money is not usually an end in itself, but it provides the means of buying the things an individual wants, to satisfy his physiological, safety, social, esteem and self-actualisation needs (in the terminology of Maslow). Pay itself will not usually be an incentive if it can only be obtained at the cost of other satisfactions, eg family life.

The extent to which pay achieves the 'end' of a motivated workforce therefore depends on the individual's needs, the intrinsic satisfactions he seeks from his job and so on.

Moreover, Drucker suggests that pay is an incentive to produce better output only where a *willingness* to perform better already exists.

Pay on its own, then, cannot accomplish the organisation's aims of better motivation leading to improved performance.

Other factors in *motivation* include:

(a) the individual's needs and wants (according to 'need' theories like those of McClelland or Maslow), and the strength of them. Many of these may not be directly related to pay;

(b) the individual's expectation that by performing in the desired way he will earn a reward that will be instrumental in achieving a result that is important to him, that is, satisfying an important want (expectancy theory, Handy's motivation calculus). This expectation in turn depends on the individual's experience, the management practices and style of his organisation and so on;

(c) work group norms, and the social/non-monetary consequences of devoting extra energy, time and emotion to paid activities.

Other factors in *performance* include very many contingent variables, such as task, technology, leadership, the work group, the employee's skills and knowledge etc. The 'inclination' of the employee to work with maximum effort may not significantly enhance his capability to work more effectively.

14 JOB EVALUATION

(a) Job evaluation is a systematic method of arriving at a wage or salary structure, so that the rate of pay for a job is fair in comparison with other jobs in the organisation.

The strengths of a job evaluation system are as follows.

(i) The salary structure is based on a formal study of work content, and the reasons for salary differentials between jobs has a rational basis that can be explained to anyone who objects to his salary level or grading in comparison with others.

(ii) The salary structure should be well balanced, even in an organisation that employs people with a wide range of different technical skills (such as engineers, accountants, salesmen).

(iii) The salary structure is based on job content, and not on the personal merit of the job-holder himself. The individual job-holder can be paid personal bonuses in reward for his efforts, and when he moves to another job in the organisation, his replacement on the job will be paid the rate for the job, and will not 'inherit' any personal bonuses of his predecessor.

(iv) Regular job evaluation should ensure that the salary structure reflects current changes in the work content of jobs, and is not outdated, so that pay differentials remain fair.

(v) A job-evaluated salary structure might protect an employer from the accusation that rates of pay discriminate between different types of worker - for example between men and women, who by law (The Equal Pay Act) should be paid the same rate for the same job.

(vi) Analyses of job content and worth - job descriptions - are made available for use in recruitment and selection, training and development, and evaluation of new or revised posts.

The weaknesses of job evaluation schemes are as follows.

(i) Job-evaluated salary structures pay a fair rate for the job only in the sense that differentials are set according to *relative* worth: job evaluation does not make any recommendations about what the general level of pay should be, in money terms. Indeed it cannot do this without reference to outside factors such as rates fixed by collective bargaining, statutory obligation or local custom - which may distort objective evaluation of actual job content.

(ii) They take no account of variation in personal performance or merit displayed by particular individuals. Recognition of such variation would have to be achieved by payments additional to the job rate - which may again distort differentials.

(iii) They take no account of temporary fluctuations in the value of market labour due to supply and demand, nor to regional fluctuations in the cost of living. Job-evaluated structures can in any case get out of date as job content alters: periodic reviews are advisable, but costly.

(iv) Job evaluation techniques suggest that it is a scientific and accurate technique, while in fact there is a large subjective element in awarding points or ratings, and unfairness can still exist.

(b) *Overcoming weaknesses*

Some of the above weaknesses are such that they cannot be overcome in practice - they are inherent in the objectives of the technique. Their effects can be minimised, however, by choice of the particular method, the way in which the system is implemented and by building on the strengths of the technique. To a great extent the particular weaknesses of job evaluation, and ways in which they can be minimised, will vary according to the method used, the method of implementation, the training of the evaluator, and the attitudes of workers and management. In general terms, however, the value of the method can be enhanced as follows.

(i) Whichever method is used, evaluations must be kept up to date, with periodic reviews.

(ii) The process should (time, resources and expertise permitting) be as scientific as possible. A more qualitative approach, such as factor comparison or points rating, should be adopted, and the factors to be studied, the weightings and scores should be systematically determined. There must still be some element of subjectivity, but a visible attempt at objectivity is being made.

(iii) Job analysts and evaluators must be trained in the use of the appropriate techniques, must have access to job documentation and to job holders (for observation and interview), and their evaluations should be monitored or reviewed to ensure that they are keeping up with changes in the organisation and applying techniques correctly.

(iv) It should be recognised by management, and by job holders, that job evaluation is only part of the pay/reward setting process. Job evaluation alone allows for no incentive level of reward: bonuses and increments for merit or performance will need to be added to the salary structure if there is not to be a demotivating effect. Moreover, job evaluation may be seen to be a threatening - or simply irrelevant - basis for salary structures, if it is not recognised that there are factors unrelated to job content which affect the value of jobs, such as equity ('felt fair' rewards), market rates of pay, negotiated pay scales, age and experience and so on.

(v) Problems of acceptance by staff - since job evaluation is a highly political exercise - can be minimised by implementing the system in an open, sensitive way. The purpose, objectives and potential benefits will need to be 'sold' to staff in advance: it will in particular have to be made clear that it is the job, not the job-holder, which is being evaluated. The role of collective bargaining may have to be discussed with unions. Appeal and revision procedures will have to be built into the system, and employees must be allowed to voice any uncertainties, suspicions and objections they may have about the affect of job grading on earnings, confidentiality, promotion and so on.

15 REDUNDANCY

Tutorial note. There are three parts to this question. We have left you to answer the third part in detail (your personal preference for one method), but have discussed, as required, the merits and drawbacks of the methods, to guide your choice and its justification. The examiner warned in his report that candidates must state a preference. Pleading the impossibility of a choice on grounds of 'contingency' is not acceptable without some elaboration of particular contingencies that would dictate a given choice. 'The view that none of the four alternatives was preferable because redundancy itself should be avoided, was a position which was regarded as indefensible. There is virtually no organisation in the world which can absolve itself, permanently, from the possibility of making some of its employees redundant: certainly, no commercial organisation can afford to do so, however enlightened and humanitarian its personnel policies.'

Reasons for redundancy

Redundancy is defined by the Employment Protection Act as dismissal where:

(a) the employer has ceased to carry on the business;

(b) the employer has ceased to carry on the business in the place where the employee was employed; and

(c) the requirements of the business for employees to carry out work of a particular kind have ceased or diminished or are expected to.

These, then, are the reasons why organisations make people redundant. They cover a wide range of specific situations, such as:

(a) the introduction of new technology rendering human intervention or particular manual skills superfluous;

(b) falling demand for a product or service reducing the activity of the organisation in a particular geographical or skill area;

(c) opportunities in new product areas (requiring a different location or skill mix) moving demand for labour elsewhere in the organisation;

(d) cost-cutting necessitating the closure of a facility, or the contraction of an area of the business;

(e) poor manpower planning: recruiting excessive numbers overall or in a given area, without planned wastage or turnover occurring, and therefore having 'superfluous' labour supply;

(f) insolvency of the business.

Methods of selection

Last-in first-out (LIFO) is where newcomers are dismissed before long-serving employees. This has some merit in that:

(a) it is seen to be fair (especially in a seniority culture such as characterises Japanese big business), which may create less disruption and hostility;

(b) compensation payments are less expensive, since they are based on age and length of service. If the 'last in' individuals have been employed for less than two years, no payment need be made;

(c) the organisation retains experienced individuals, and those most established in the culture and procedures of the organisation;

(d) for the employee, it is perhaps better that those who are younger, perhaps more flexible, and less attached to the organisation, go. Personal psychological disruption is less.

However, LIFO has drawbacks where:

(a) the 'first in' people may be nearing retirement age, so early retirement may be more desirable for them *and* the organisation;

(b) the organisation needs 'fresh blood': in fast-moving fields like electronics, where innovation is required, as in slow-moving firms which need new viewpoints in order to adapt to environmental change, an increasingly aged, long-serving workforce may not be desirable;

(c) length of service does not necessarily coincide with particular skill, aptitude, enthusiasm or merit in the individual. The 'last in' could also be the best, and be lost to the organisation;

(d) it may be indirectly discriminatory to women, if recruitment of women has only recently picked up, as in many organisations.

Value to the organisation, or retention by merit, involves keeping those who perform well and dismissing less effective workers. This is obviously desirable for the organisation from a human resource management point of view: the cost effectiveness of manpower can only be enhanced by shedding less effective workers when the opportunity arises. The problems are:

(a) setting up or continuing to operate effective performance appraisal, job evaluation and assessment systems, both so that the organisation can be confident of keeping the best people in the most 'value adding' jobs, and so that selection can be seen to be fair and objective;

(b) justifying selection to employees and their trade union representatives (if any), without resentment, accusations of unfairness etc. Trade union resistance, in particular, may make negotiated redundancy benefits higher;

(c) for the *employee*, redundancy will be even more traumatic, because less impersonal. Dismissal will be perceived as personal 'failure', on top of the insecurity of leaving the job. This may be even more acute for long-serving employees.

The *offer of voluntary redundancy* is desirable because, in human relations terms, no-one is being forced against their wishes to leave the organisation. The employees make the choice to take their chances elsewhere, on favourable redundancy terms. The organisation has the benefit of less disruption and potential for industrial action, less loss of morale for remaining employees, and the PR advantages of not having to announce enforced redundancies. The drawbacks, however, are that:

(a) voluntary redundancy is unlikely to be a solution to large scale shedding of jobs unless huge costs in incentive payments are incurred, retraining offered etc;

(b) the organisation may lose key, valuable people, keeping fewer valuable individuals. In fact, this is quite likely, since it will be the better employees who have the confidence to take their chances elsewhere (though it may be argued that the organisation will be keeping the more 'loyal' ones).

The offer or enforcement of *early retirement* is similar to voluntary redundancy in many respects. Employees may feel that their career is 'winding down' in any case, and may not feel pressured to find further employment, particularly if the redundancy package is generous. Again, the organisation will feel human and public relations advantages. On the other hand, those employees *forced* into early retirement are likely to have greater problems finding other jobs if they need or want them. Again, too, the organisation may be losing more valuable individuals who are respected by juniors and have a wealth of experience.

16 COMMUNICATIONS AND BARRIERS

This is a problem involving the poor personal communications between Bill and his staff which result in mistakes in work and frustration for his subordinates.

Communication is the exchange of ideas and values between two or more people. It consists of the following.

(a) *Technical process*
Selecting the correct symbols (eg words, numbers, illustrations), selecting the correct media to transmit the message (eg verbal, written), repeating the message, allowing feedback of the message from the receiver.

(b) *Social process*
Involves social relationships of trust and friendliness, perception and values which influence the meaning of a message to both parties.

To help Bill it would be necessary to find out the exact causes of the problem. Since Bill is a friend and fellow section leader, this would involve talking to both Bill himself, to explore his own views, and (with Bill's permission) to his staff.

Questions to answer are as follows.

(a) What is the exact nature of the failures: are they caused by technical abilities or misunderstandings?

(b) What annoys and frustrates his staff?

    (i) Does he talk and explain things above their heads?
    (ii) Does he not communicate at all?
    (iii) Does he communicate too late or only when they ask?

Only by identifying the real causes will it be possible to help Bill but any advice would possibly include the points below.

Advice and help for Bill

Introduction. It is often difficult for a person who is as bright and intelligent as Bill to appreciate the limitations and inabilities of other people. Good leadership requires such an understanding.

(a) Bill must be clear in his own mind as to the content of his message and why he needs to communicate it.

(b) Bill should marshall his ideas before speaking. Consulting others may help in giving him other ideas as to how he should be communicating.

(c) He should ensure that the timing is right for his message (a relatively unimportant piece of information should not be given to a person working hard to meet a deadline); that the setting is right (a personal comment should not be delivered in front of the whole office); and that communication has taken place within the context of the organisation's customs and normal practices.

(d) Use simple and direct language and avoid technical terms and jargon whenever possible.

(e) Try to anticipate staff's technical limitations and explain using illustrations and examples whenever possible.

(f) Select the appropriate media for the message (verbal for simple or 'personal' messages; written if the message is complex).

(g) Bill must ensure that he communicates the importance of the message to the receiver so that full attention is given.

(h) Attention must be given to 'body language' (non-verbal means of communication).

(i) Always obtain feedback of the message/instruction. Ensure that what the message means to his staff is what Bill intended it to mean.

(j) Recognise that the perception and values of each individual will be different. The sense in which we see things and the importance we attach to them will depend on our education, family background and social status. Bill must develop a sensitivity to the world of his staff and understand how this influences their attitudes towards messages.

(k) Encourage Bill to use more face-to-face communications so that the other person's reaction can be seen and understood. (Non-verbal signals can be important.)

(l) Reduce the superior/subordinate relationships, so that his staff can ask questions and do not feel they are being controlled and therefore filter messages to hear only what they want to hear. Try to move from a defensive to a supportive relationship.

(m) Try to be a good listener, and show an interest in their ideas. When they are wrong, explain and discuss their point of view in a sympathetic way and avoid using:

  (i) superior position to rule out their arguments;
  (ii) superior intelligence ('I know all the answers') to reinforce his position.

(n) Bill's words must be supported by his actions.

The manager should try to ensure Bill appreciates these points and puts them into practice in the future. It should soon be apparent if this has been successful or not.

REVIEW FORM

Name:

Course:

We would be grateful to receive any comments you may have on this edition of the Study Text. You may like to use the headings below as guidelines. Tear out this page and send it to our Freepost address: **BPP Publishing Ltd, FREEPOST, London W12 8BR**.

Syllabus coverage

Presentation

'User friendliness'

Errors (please specify and refer to a page number)

Other